Current Trends in Hydraulics, Hydrodynamics and Hydroinformatics

Current Trends in Hydraulics, Hydrodynamics and Hydroinformatics

Editor: Roman Morris

Callisto Reference
www.callistoreference.com

Callisto Reference,
118-35 Queens Blvd., Suite 400,
Forest Hills, NY 11375, USA

Visit us on the World Wide Web at:
www.callistoreference.com

ISBN: 978-1-64116-827-4 (Hardback)

Cataloging-in-Publication Data

Current trends in hydraulics, hydrodynamics and hydroinformatics / edited by Roman Morris.
p. cm.
Includes bibliographical references and index.
ISBN 978-1-64116-827-4
1. Hydraulics. 2. Hydrodynamics. 3. Hydrology--Data processing. 4. Hydraulic engineering. I. Morris, Roman.
TC160 .C87 2023
627--dc23

Table of Contents

Preface

This book aims to highlight the current researches and provides a platform to further the scope of innovations in this area. This book is a product of the combined efforts of many researchers and scientists, after going through thorough studies and analysis from different parts of the world. The objective of this book is to provide the readers with the latest information of the field.

Hydroinformatics is a subfield of informatics that focuses on the use of information and communication technologies (ICTs) to address the major issues related to the effective and equitable use of water for a variety of purposes. Hydrodynamics is a branch of physics, which is engaged in the study of the motion of fluids as well as the forces acting on the solid bodies submerged in fluids. Hydraulics refers to a type of technology that involves the utilization of liquids and their mechanical properties. It is an applied science based on chemistry, engineering, and other sciences. Concepts from the fields of environmental engineering, hydraulics, and hydrology along with other disciplines are integrated into hydroinformatics. This book unravels the recent studies in the fields of hydraulics, hydrodynamics, and hydroinformatics. It will serve as a valuable source of reference for graduate and post graduate students. It strives to provide a fair idea about this discipline and to help develop a better understanding of the latest advances within this field.

I would like to express my sincere thanks to the authors for their dedicated efforts in the completion of this book. I acknowledge the efforts of the publisher for providing constant support. Lastly, I would like to thank my family for their support in all academic endeavors.

Editor

1

Influence of Erodible Beds on Shallow Water Hydrodynamics during Flood Events

David Santillán *, Luis Cueto-Felgueroso, Alvaro Sordo-Ward and Luis Garrote

Department of Civil Engineering, Hydraulics, Energy and Environment, Universidad Politécnica de Madrid, 28040 Madrid, Spain; luis.cueto@upm.es (L.C.-F.); alvaro.sordo.ward@upm.es (A.S.-W.); l.garrote@upm.es (L.G.)

* Correspondence: david.santillan@upm.es

Abstract: Flooding has become the most common environmental hazard, causing casualties and severe economic losses. Mathematical models are a useful tool for flood control, and current computational resources let us simulate flood events with two-dimensional (2D) approaches. An open question is whether bed erosion must be accounted for when it comes to simulating flood events. In this paper we answer this question through numerical simulations using the 2D depth-averaged shallow-water equations. We analyze the effect of mobile beds on the flow patterns during flood events. We focus on channel confluences where water flow and sediment mobilization have a marked 2D behavior. We validate our numerical simulations with laboratory experiments of erodible beds with satisfactory results. Moreover, our sensitivity analysis indicates that the bed roughness model has a great influence on the simulated erosion and deposition patterns. We simulate the sediment transport and its influence on the water flow in a real river confluence during flood events. Our simulations show that the erosion and deposition processes play an important role on the water depth and flow velocity patterns. Accounting for the mobile bed leads to smoother water depth and velocity fields, as abrupt fields for the non-erodible model emerge from the irregular bed topography. Our study highlights the importance of accounting for erosion in the simulation of flood events, and the impact on the water depth and velocity fields.

Keywords: hydraulic engineering; open channel flow; computational hydraulics; mobile bed; erosion

1. Introduction

The numerical simulation of flood events to model fluvial hydraulics and flood inundation areas has become a key tool to assess flood risks, effectively manage and control flood events, and evaluate practices on flood prevention, protection and mitigation—the so-called flood control. Flooding is the most common environmental risk [1], causing heavy life and economic losses. Moreover, climate change is expected to increase the intensity and frequency of extreme rainfall events that drive floods [2,3].

One of the most widely used approaches to model fluvial hydraulics at a river scale is based on the numerical solution of the one-dimensional (1D) Saint Venant equations [4]. This approach requires low computational resources, allowing us to simulate long river length. Nevertheless, the simplifying assumptions involve several drawbacks: (1) the flow is one-dimensional, (2) velocity is uniform across the cross-section, and (3) the water level is horizontal across the cross-section. Despite the simplifying assumptions of these approaches, in-channel flows are satisfactorily simulated by 1D models. However, out-of-bank flows or singular elements such as channel junctions or bridge piers incorporating two- and three-dimensional (3D) effects may not be satisfactorily described by 1D approaches [5].

Currently, the development of computational capacity has allowed us to simulate fluvial hydraulics with approaches based on the two-dimensional (2D) Shallow-Water Equations. These approaches have emerged as a robust and efficient tool for the simulation of flow on both natural and man-made environments [6]. Moreover, sophisticated 2D hydrodynamic codes have recently become available, as for instance, Delf3D-FLOW [7] or Iber [8–10]. The approach relies on the assumption of depth-averaged velocity and, in most cases, quasi-hydrostatic flow pressure, which involves neglecting vertical acceleration. Although this assumption has been identified as a source of error in flows in non-prismatic channels [11], over curved beds in 1D [12] and 2D simulations [13,14], or channel configurations with vertical water circulation [15], the approach is suitable for modeling the hydrodynamics of flood events.

While 2D Shallow-Water models may naturally incorporate well resolved bed topography on flow hydrodynamics, most numerical studies of flood events so far have neglected changes in bed topography on the water flow dynamics during flood events. Much effort has been focused on the long-term bed erosion patterns and the evolution of river morphologies over time scales of years or decades, such as the formation of bars in alluvial channels [16], the bedload transport in river confluences [17,18], the riverbank migration and island dynamics [19], or at a basin scale [20]. However, erosion during flood events may significantly modify bed topography [21,22], even leading to the formation of canyons by a single flood event [23]. The hydrodynamics, sediment transport processes and river morphology are strongly coupled in rivers and channels with erodible beds and cannot be studied in isolation [24]. Mathematical models should then interlink these three processes through fully coupled approaches.

A paradigmatic case of bed erosion on both natural and man-made environments are the river and channel confluences. Junctions of two open channels or rivers are common configurations present in many hydraulic structures such as urban channels [25], tidal channels [26], tidal river confluences [27], irrigation channels or river confluences. The morphological progress of river bathymetry is significantly influenced by turbulent flow structures associated with dunes [28], junctions of channels with different depths [29] or discordant channels [30–32], indicating that these structures may play a key role in the upstream erosion patterns under subcritical flows. Flow dynamics at channel junctions have been characterized with laboratory or field experiments, aimed at the characterization of the flow-separation zone [33] and the effect of curved profiles [34], the contaminant transport with lab experiments [35] or field studies [36,37], the development computational approaches [38,39], or providing benchmark experimental data sets [40]. Erosion patterns at junctions have also been experimentally characterized in channels with [41,42] or without [43–45] sediment supply.

Flow dynamics at junctions are characterized by six regions [43], as shown in Figure 1; of particular interest among them are the separation zone in the main channel intermediately downstream of the junction [33], and the shear plane between the two flows in the main channel [46,47]. In the separation zone the water depth depresses and a recirculating flow appears. The shear plane is characterized by high turbulence levels that generate quasi-two-dimensional vortices with near-vertical axes [48,49].

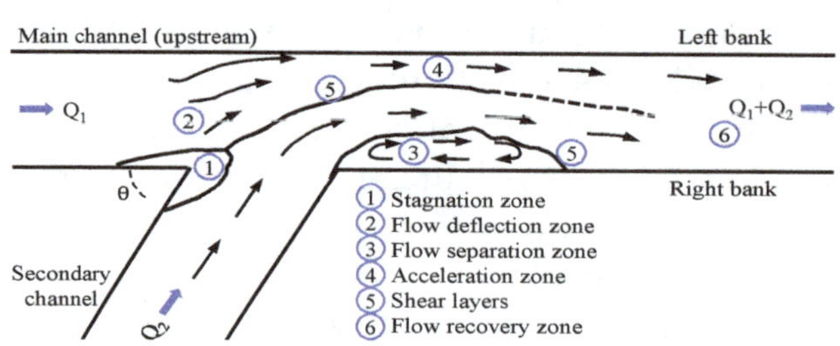

Figure 1. Hydrodynamics at open-channel junctions. Here we plot the six characteristic regions.

Steady-state bed bathymetry at junctions presents two salient morphological features [50]: a scour-hole on the downstream corner of the junction, and a bar on the left bank closed to the downstream side of the confluence, denoted as a bank-attached lateral bar. Scour hole depth varies non-linearly with the confluence angle [45,51,52]. The bar appears due to the lower water depth and recirculating flow in the separation zone, as well as the flow detaches from the left sidewall downstream of the junction. Moreover a helicoidal flow emerges in the main channel downstream of the junction [50], which is responsible for the movement of water in the vertical direction and the concentration of the finest sediments in the bar [45].

In this study we assess the effect of mobile beds on the flow patterns during flood events. Our case study is a river confluence in a semiarid basin located in southeast Spain. Our main objective is to determine whether erosion should be accounted for or not to simulate fluvial hydraulics during flood events. To validate our numerical model, we compare our simulated results with laboratory experiments of sediment transport at channel confluences. In particular, we focus on two laboratory setups: (1) a set of experiments without mobile beds to validate the hydraulic model, and (2) an experiment of dynamics of bed morphology to validate the erosion model. We test those predictions against the experiments and identify the sources of model discrepancy. Lastly, we take advantage of the mathematical model to evaluate the effect of mesh size and bed properties on the erosion patterns.

2. Materials and Methods

We use the freely available Iber code [8,53]. Iber is a 2D hydraulic model that solves the hydrodynamic equations coupled to the sediment transport equations, among others. In the following sections, we describe the hydrodynamic equations and the erosion model.

2.1. Two-Dimensional Depth-Integrated Shallow-Water Equations (SWE)

Iber solves the 2D SWE using a finite volume scheme. This scheme is able to handle unstructured meshes, irregular topographies, friction losses, and wet-dry fronts [53]. The 2D SWE are derived from the Navier–Stokes equations by assuming quasi-hydrostatic flow and incompressibility of water. The 2D mass conservation equation in a Cartesian coordinate system reads:

$$\frac{\partial h}{\partial t} + \frac{\partial h U_x}{\partial x} + \frac{\partial h U_y}{\partial y} = S_q \tag{1}$$

where h is the water depth, t is time, (U_x, U_y) is the depth-averaged velocity in the x and y-directions, and S_q are the sources or sinks.

The linear momentum conservation equation in a Cartesian coordinate system is given by:

$$\frac{\partial h U_x}{\partial t} + \frac{\partial}{\partial x}\left(h U_x^2 + g\frac{h^2}{2}\right) + \frac{\partial}{\partial y}\left(h U_x U_y\right) = -gh\frac{\partial Z_b}{\partial x} - \frac{\tau_{b,x}}{\rho} + \frac{\partial}{\partial x}\left(\nu_t h \frac{\partial U_x}{\partial x}\right) + \frac{\partial}{\partial y}\left(\nu_t h \frac{\partial U_x}{\partial y}\right), \tag{2}$$

and

$$\frac{\partial h U_y}{\partial t} + \frac{\partial}{\partial y}\left(h U_y^2 + g\frac{h^2}{2}\right) + \frac{\partial}{\partial x}\left(h U_x U_y\right) = -gh\frac{\partial Z_b}{\partial y} - \frac{\tau_{b,y}}{\rho} + \frac{\partial}{\partial x}\left(\nu_t h \frac{\partial U_y}{\partial x}\right) + \frac{\partial}{\partial y}\left(\nu_t h \frac{\partial U_y}{\partial y}\right), \tag{3}$$

where g is the gravity, Z_b is the channel-bottom elevation, τ_b is the bed friction, ρ is the density of the water, and ν_t is the eddy viscosity. We neglect wind surface tractions and Coriolis force. τ_b is computed through the Manning formulation as:

$$\tau_{b,j} = \rho g h \frac{n^2 U_j \left|U\right|}{h^{1/3}}, \tag{4}$$

where $j = x, y$, and n is the Manning's coefficient.

The last two terms in Equations (3) and (4) are the turbulent stresses. They are determined from the so-called k-ϵ turbulence model of Rastogi and Rodi [54]. The model consists of two differential transport equations for the turbulence kinetic energy, k, and the rate of its dissipation, ϵ. The turbulence kinetic energy is given by:

$$\frac{\partial k}{\partial t} + \frac{\partial U_x k}{\partial x} + \frac{\partial U_y k}{\partial y} = \frac{\partial}{\partial x_j}\left(\left(\nu + \frac{\nu_t}{\sigma_k}\right)\frac{\partial k}{\partial x_j}\right) + 2\nu_t S_{ij} S_{ij} + c_k \frac{u_f^3}{h} - \epsilon, \tag{5}$$

and the rate of its dissipation by:

$$\frac{\partial \epsilon}{\partial t}+\frac{\partial U_x \epsilon}{\partial x}+\frac{\partial U_y \epsilon}{\partial y}=\frac{\partial}{\partial x_j}\left(\left(\nu+\frac{\nu_t}{\sigma_\epsilon}\right)\frac{\partial \epsilon}{\partial x_j}\right)+C_{\epsilon 1}\frac{\epsilon}{k}2\nu_t S_{ij}S_{ij}+C_\epsilon\frac{u_f^4}{h^2}-C_{\epsilon 2}\frac{\epsilon^2}{k}, \tag{6}$$

where S_{ij} is the velocity gradient, given by:

$$S_{ij}=\frac{1}{2}\left(\frac{\partial U_i}{\partial x_j}+\frac{\partial U_j}{\partial x_i}\right), \tag{7}$$

and $\nu_T = C_\mu \frac{k^2}{\epsilon}$, $C_k = \sqrt{C_f}$, $C_\epsilon = 3.6 C_k^{3/2} C_{\epsilon 2}\sqrt{C_\mu}$, and $C_f = \frac{\tau_b}{\rho |U|^2}$. The constants of the model are: $C_\mu = 0.09$, $C_{\epsilon 1} = 1.44$, $C_{\epsilon 2} = 1.92$, $\sigma_k = 1.0$, and $\sigma_\epsilon = 1.31$.

2.2. Erosion Model

Iber erosion and sediment transport module solves the non-cohesive sediment transport equations with uniform granulometries, in a non-stationary regime. The hydrodynamics, sediment transport processes, and river morphology interplay at mobile-bed confluences [55,56] and cannot be studied in isolation [24].

Iber software couples these three processes. The sediment transport process is simulated through the 2D Exner equation [57–59]. The equation provides the bed elevation evolution in response to the erosion and sedimentation and is given in the form of bedload and suspended load. It reads:

$$(1-\Phi)\frac{\partial Z_b}{\partial t}+\frac{\partial q_{b,j}}{\partial x_j}+\omega_s(E_s-C_b)=0, \tag{8}$$

where Φ is the porosity, $q_{b,j}$ is the bedload discharge along the j direction, ω_s is the fall velocity of the suspended sediments, E_s is the dimensionless factor of the inlet of suspended sediment, and C_b is the concentration of suspended sediment on the bed.

Several formulations are available in Iber to compute the bedload solid flow rate due to the bedload transport. We have conducted our simulations with the van Rijn equation [60], given by:

$$q_b^* = \frac{0.053}{D^{*0.3}}\left(\frac{\tau^*}{\tau^*_{crit}}-1\right)^{2.1}, \tag{9}$$

where q_b^* is the dimensionless bedload solid volumetric flow rate, D^* is the dimensionless diameter of solid particles, τ^* is the dimensionless shear stress acting on the particles, and τ^*_{crit} is the dimensionless critical shear stress.

The dimensionless bedload solid volumetric flow rate, q_b^*, reads:

$$q_b^* = \frac{q_b}{\sqrt{(s-1)gD^3}}, \tag{10}$$

being q_b^* the bedload solid volumetric flow rate, s is the ratio between the densities of the particles and water, $s = \rho_s/\rho$, and D is the characteristic diameter of sediments, usually taken as the median diameter $D = D_{50}$.

The dimensionless diameter of solid particles, D^*, reads:

$$D^* = D\left(\frac{g(s-1)}{\nu^2}\right)^{1/3}, \tag{11}$$

where ν is the kinematic viscosity of water.

The dimensionless shear stress, τ^*, has the form:

$$\tau^* = \frac{\tau_b}{\rho(s-1)gD}, \tag{12}$$

where τ_b is the bed friction shear stress, given by Equation (4).

The solid bedload flow rate equation is combined with the Shields-stress criterion for incipient motion [61] and a bed slope correction for bedload transport. Moreover, Iber simulates the suspended sediment transport with the depth-averaged convection-diffusion equation for the sediment concentration. The model accounts for the turbulent diffusion and the equilibrium suspended concentration. The latter, together with the sedimentation velocity, are computed with the van Rijn formulation [62]. In the following, we do not consider the suspended sediment transport since it is negligible compared with the bedload sediment transport.

2.3. *Solution of the System of PDEs*

We solve the system of Partial Differential Equations (PDEs) with a second-order numerical scheme based on the Monotonic Upwind Scheme for Conservation Laws (MUSCL) scheme [63]. High-resolution finite volume methods [64–68] are suitable to minimize numerical dissipation and capture complex flow patterns [69].

3. Results and Discussion. Model Validation with Laboratory-Scale Channel Junctions

3.1. *Validation of the Hydraulic Model*

We test our hydraulic model by comparing the numerical outputs with laboratory experiments reported on the literature [70]. Although the Iber code has been previously validated with experiments of open-channel contractions with several geometries and flow regimes [71], here we compare the outputs of the model with a set of lab experiments of open-channel junctions [70]. The experiment setup is a rectangular channel confluence of 0.30 m wide, as shown in Figure 2a. The channel was made of acrylic plates and metal sheets. The slope in both channels was 0.14%. Water was introduced in the system through both inlets located upstream in the main and secondary channels. Water level was measured at five points. Here, we simulate the experiments under subcritical flow. The confluence angle θ between the main and secondary channels θ is 30° and 60°. The simulated experiments are listed on Table 1, as well as the flow rates and water levels measured at five points.

Laboratory experiments have characterized the flow behavior at the junction of rectangular channels [33,34,38–40]. Briefly, flow dynamic is characterized by six regions [43], of which the separation zone in the main channel intermediately downstream of the junction [33], and the shear plane between the two flows in the main channel are of particular interest [46,47].

Our model is able to reproduce the main flow features at channel junctions, as shown in Figure 2b,c. The momentum of the water in the secondary channel makes the flow detach from the sidewall as water gets in the main channel, and water depth decrease in the vicinity of the wall. Our model replicates this effect, as shown in Figure 2b where a depression in water depth appears at the downstream intersection of both channels and extends along the downstream direction. Another side effect of the intersection is the increase in water depth upstream the junction in both channels. Our model reproduces this effect, Figure 2b.

The interaction between both channel streams generates a shear plane that divides the main and lateral flow. We identify this plane with our numerical results by plotting the x-velocity gradient, given by $\partial U_x / \partial y$, in Figure 2c. The gradient is high along the intersection of both water streams in the main channel. As expected, the gradient is especially high downstream of the junction, in the separation zone. However, since our model is 2D depth-averaged, it is unable to reproduce the helicoidal flow that appears downstream of the junction in the main channel [50]. The helicoidal flow is responsible for the movement of water in the vertical direction and may influence the erosion patterns.

We compare the output of our numerical model with the results of the laboratory experiment listed ih,n Table 1 in Figure 2d,e for the channel junction with $\theta = 30°$. The agreement between the five experiments and the simulations is remarkable. Such agreement is also satisfactory for the channel junction with $\theta = 60°$, depicted in Figure 2f,g. Nevertheless, the numerical results at point P5 slightly deviate from the lab experiments for this channel configuration. However, the agreement between experiments and numerical simulations at point P5 is very satisfactory for the channel junction with $\theta = 60°$.

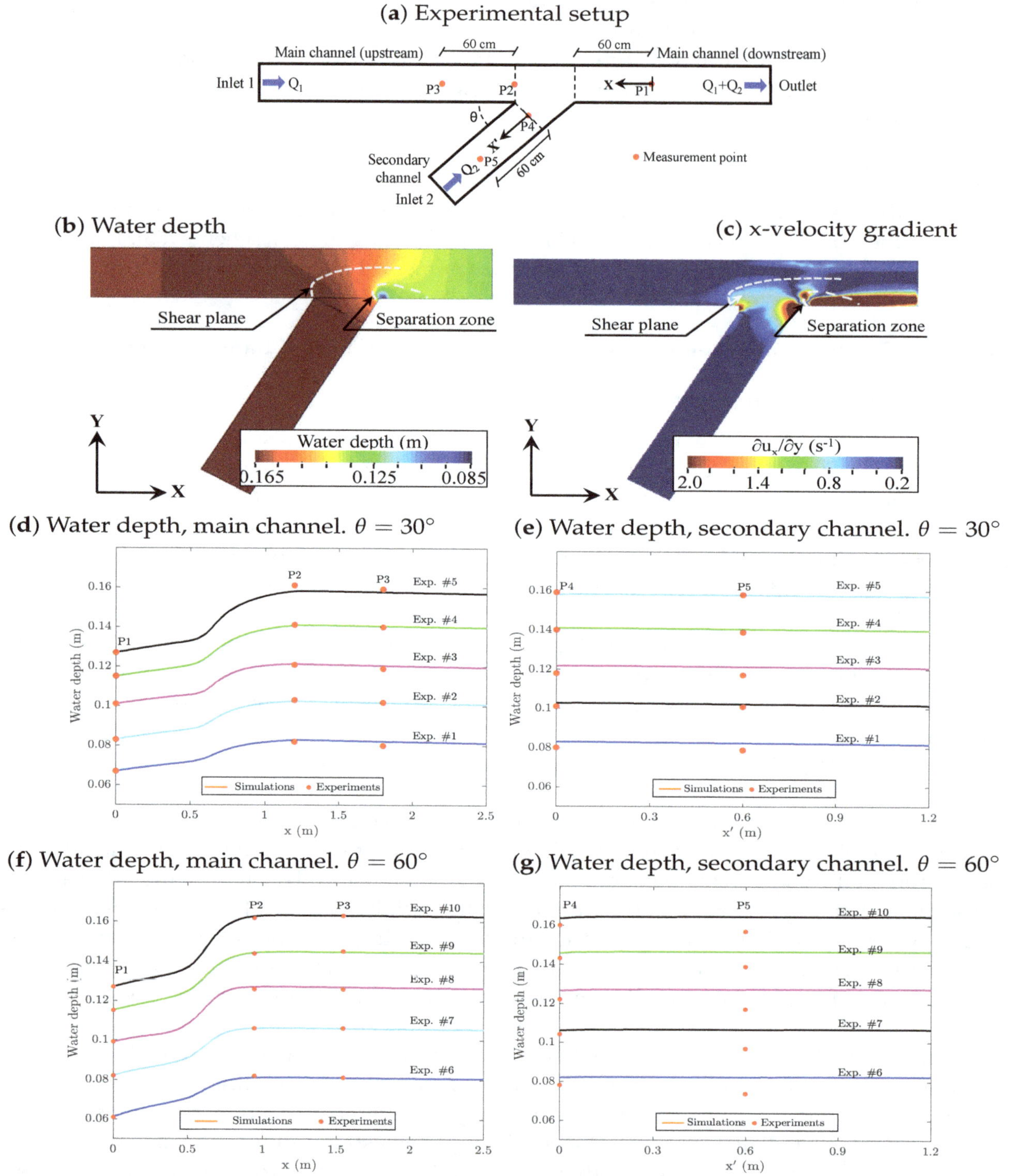

Figure 2. Validation of the numerical hydraulic model with laboratory experiments. (**a**) Setup of the laboratory experiments. (**b**) Contour plot of the water depth computed with the numerical model. Experiment # 10. (**c**) Contour plot of the x-velocity gradient given by $\partial U_x/\partial y$ computed with the numerical model. Experiment # 10. (**d**) Water depth for the model with $\theta = 30°$ along the main channel, and (**e**) along the secondary channel. (**f**) Water depth for the model with $\theta = 60°$ along the main channel, and (**g**) along the secondary channel.

Table 1. Experimental results taken from [70].

# Experiment	θ	Q_1 (L/s)	Q_2 (L/s)	y_1 (cm)	y_2 (cm)	y_3 (cm)	y_4 (cm)	y_5 (cm)
1	30°	5.1	5.4	8.0	8.2	7.9	8.0	6.7
2	30°	5.0	10.1	10.2	10.3	10.1	10.1	8.3
3	30°	5.0	14.9	11.9	12.1	11.7	11.8	10.1
4	30°	14.9	9.8	14.0	14.1	13.9	14.0	11.5
5	30°	14.9	14.9	15.9	16.1	15.8	15.9	12.7
6	60°	5.1	4.9	8.1	8.1	7.4	7.8	6.1
7	60°	5.1	9.9	10.6	10.6	9.7	10.4	8.2
8	60°	5.1	14.8	12.6	12.6	11.7	12.2	9.9
9	60°	15.1	9.9	14.5	14.4	13.9	14.3	11.5
10	60°	15.1	14.8	16.3	16.2	15.7	16.0	12.7

3.2. Validation of the Erosion Model

Our next validation experiment was conducted by Yuan et al. [45]. The authors studied the dynamics of bed morphology in channel confluences through laboratory experiments under subcritical flow conditions. The setup consisted of a perpendicular confluence of rectangular concordant-bed open-channels, the dimensions of which are shown in Figure 3a. The inlet flow rates of the simulated experiment are $Q_m = 9$ l/s and $Q_t = 6$ l/s. The bed sediment was poorly sorted with $D_{50} = 0.9$ mm, and $D_{90} = 2.5$ mm, and the bed was initially covered with a 5.5 cm-thick layer of sediment. In our numerical simulations, we got the best agreement between experiments and numerical simulations with the following properties of the sediments: density of grains $\rho = 2500\ \mathrm{kg/m^3}$, friction angle $\phi = 25°$, and Manning's coefficient $n = 0.018\ \mathrm{s{\cdot}m^{-1/3}}$. The steady-state bed elevation contour is shown in Figure 3b, we remind the initial bed elevation was 5.5 cm.

Our 2D numerical model is suitable to replicate this experiment. The eddy development process in the shear layer takes place in a quasi-horizontal plane [46] in concordant-bed confluences, and vortices are then quasi-two-dimensional with near-vertical axes [48,49]. The two-dimensional nature of the shear layer may disappear in discordant-bed confluences [30,31].

We depict the steady-state bed elevation contour plot computed with our mathematical model in Figure 3c. The profile presents two morphological features [50]: a scour-hole at the downstream corner of the confluence, and a sediment bar on the left bank immediately downstream of the confluence. The mathematical model is able to reproduce both features, as shown in Figure 3c.

The scour hole depth is similar in both models, experimental and numerical, with a maximum depth of about 2 cm. The experiment is highly sensitive to the depth, since it increases with the confluence angle [45,51] in a non-linear fashion [52]. However, the extension of the hole slightly differs in both models. The experiment shows the hole extends up to the central axis of the main channel, whereas the numerical model provides a hole that extends up to the left bank. Under certain conditions, experiments also show scour holes that expand up to the bank [50].

The sediment bar on the left bank downstream confluence, also denoted as a bank-attached lateral bar, is also reproduced in both the experiment and the numerical model. The sediment bar appears due to the lower water depth and recirculating flow in the separation zone, as well as from the flow detached from the left sidewall. The experiment results in a higher and more concentrated bar than the dimensions computed with the mathematical model. Since our mathematical model is depth-averaged, it is unable to reproduce the helicoidal flow that emerges downstream of the junction in the main channel [50]. The helicoidal flow is responsible for the movement of water in the vertical direction and may concentrate the sediments resulting in a higher and narrower bar. Moreover, the van Rijn model for bedload transport [60] characterizes the sediments with the median diameter (D_{50}) and disregards the variance of the diameter distribution. Bank-attached lateral bars are made of the finest sediments [45], which our model is unable to consider because the variance of the sediment diameter is ignored.

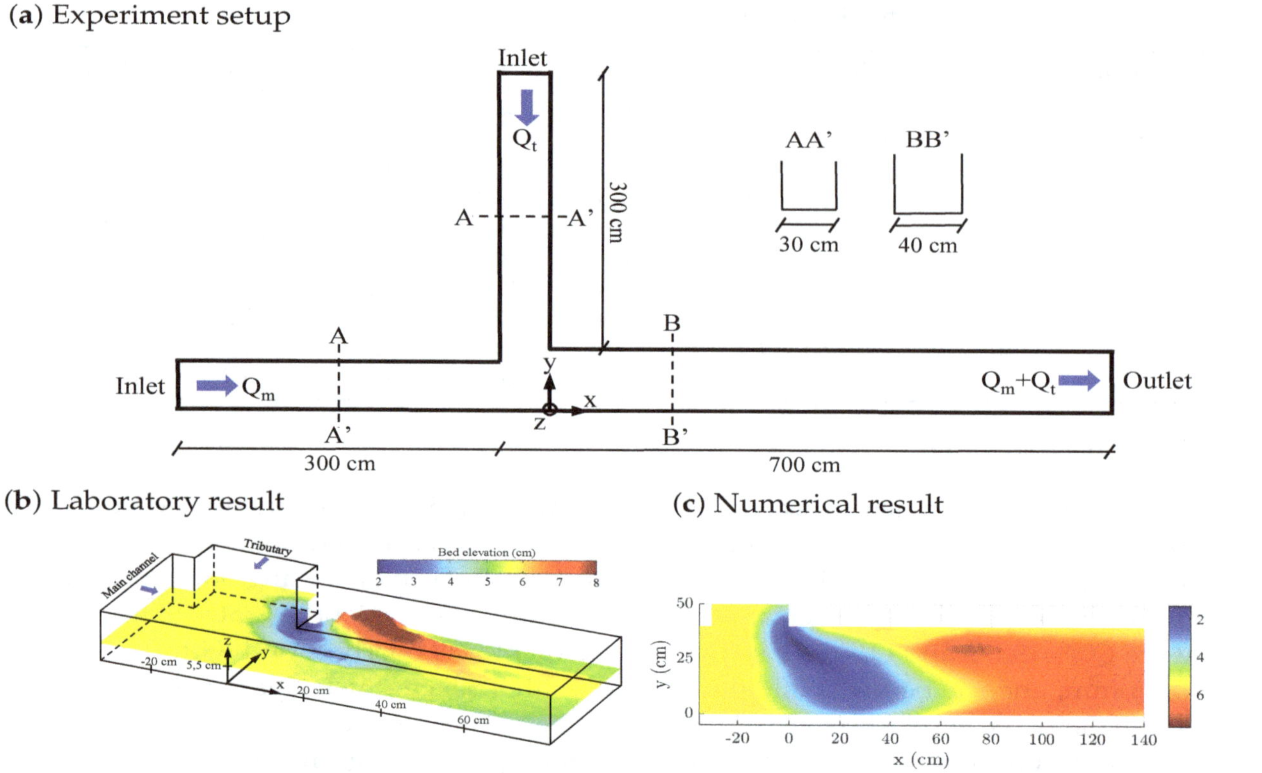

Figure 3. Sketch of bed erosion experiment setup and results. Panel (**a**) depicts the setup. It consists of a perpendicular confluence of rectangular concordant-bed open-channels. The dimensions of the channels are included in the figure. Panel (**b**) depicts a contour of the steady-state bed elevation provided by the laboratory experiment. The initial bed elevation was 5.5 cm. The plot is reproduced from [45]. Panel (**c**) includes the steady-state bed elevation computed with the numerical model. Units are in cm.

The evolution of the bed elevation, water head, water velocity, and stream lines provided by the numerical model are depicted in Figure 4. Initially, the bed elevation is 5.5 cm, as shown in Figure 4a, which provides an almost constant water head everywhere except in the separation zone, i.e., the downstream corner of the confluence (Figure 4d). Under this situation, the velocity contour plot shows a clear shear layer in the separation zone, as shown in Figure 4g, and the highest velocities are given downstream of the confluence in the main channel. The streamline plot shows how water recirculates inside the circulation zone, as shown in Figure 4i. Moreover, this zone extends up to 1.6 m downstream of the junction.

After 3600 s, the bed topography has changed significantly. The scour hole depth is up to 3 cm and the extension is almost the definitive, and slight sedimentation appears downstream of the hole (Figure 4b). The water depth increases in the area of the scour hole due to the erosion, but increases downstream of the hole because of sedimentation (Figure 4e). The water velocity is also affected by the bed elevation evolution. The velocity decreases in the scour hole area thanks to the erosion, but also water slows down downstream of the hole due to the contraction of the recirculation zone that increases the effective channel width (Figure 4h). The contraction of the recirculation zone is well represented in the streamline plot (Figure 4j).

The steady-state is given at time 10,800 s. The bed elevation has evolved up to create a scour hole of about 4 cm deep that extends up to the left bank and a bar in the left bank downstream of the hole (Figure 4c). The water depth increases in the scour hole due to erosion, as shown in Figure 4f, and water slows down due to the larger depth (Figure 4i). The recirculation zone contracts slightly compared with the previously analyzed time step (Figure 4l).

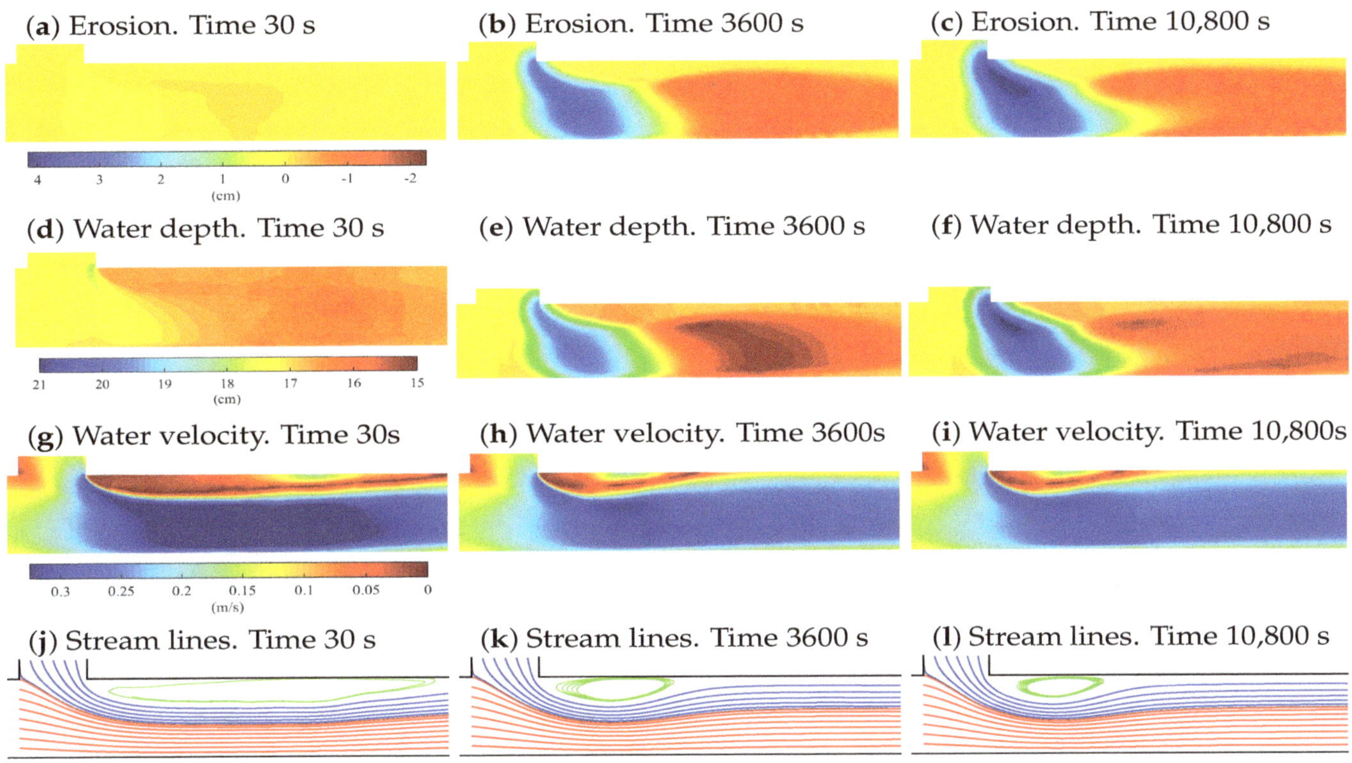

Figure 4. Evolution of bed elevation, water head, water velocity, and stream lines computed with the mathematical model. The first row plots the contour plot of the erosion at time 30 s—panel (**a**), 3600 s—panel (**b**), and 10,800 s—panel (**c**). Erosion is positive and sedimentation negative. The second row shows the water depth at time 30 s—panel (**d**), 3600 s—panel (**e**), and 10,800 s—panel (**f**). The third row includes the water velocity at time 30 s—panel (**g**), 3600 s—panel (**h**), and 10,800 s—panel (**i**). The last row includes the stream lines at time 30 s—panel (**j**), 3600 s—panel (**k**), and 10,800 s—panel (**l**).

3.3. *Sensitivity Analysis*

We take advantage of our mathematical model to study the erosion patterns as a function of the mesh computational size and the bed properties—friction angle of sediments and roughness. In the following subsection we present and discuss the results of a mesh refinement study and the effect of the friction angle of the sediments and roughness on the erosion patterns. The properties of the sediments and the bed are: $D_{50} = 0.9$ mm, $n = 0.018\ \mathrm{s{\cdot}m^{-1/3}}$, $\phi = 25°$, $\rho = 2500\ \mathrm{kg/m^3}$, unless otherwise indicated.

3.3.1. Mesh Refinement Study

We test the grid convergence of our model results with a mesh refinement study. We simulate the erosion experiment depicted in Figure 3 using four mesh sizes, $\Delta\delta = (0.5, 1, 2, 4)$ cm, in the erodible areas of the channel. The erosion patterns and two bed profiles at the x-coordinates 0 and 20 cm (see reference system in Figure 3a) are included in Figure 5.

The overall erosion patterns are grid independent, but quantitatively we observe smoother topographies (less erosion) in coarser grids. Note that the finest grid already has a grid size that is comparable with the characteristic size of the sediment particles (around 1 mm), so refining below the sub-millimeter scale may raise questions about the validity of the continuum hypothesis. For fixed hydro-mechanical properties the steady state bed elevation converges for four increasing mesh sizes, from $\Delta\delta = 4$ cm to $\Delta\delta = 0.5$ cm (Figure 5a–d). The deepest scour hole and highest downstream bar are computed with the finest mesh: as the mesh is refined, the scour hole depth increases.

The difference between the computed erosion depths decreases as the grid is refined in both profiles, as shown in Figure 5e,f, as it remains relatively small for the finest grids, $\Delta\delta = 1$ cm and $\Delta\delta = 0.5$ cm. Nevertheless, the scour hole depth is quite short for the coarsest mesh, 1 cm deep, in contrast with the estimated depth with the finest mesh, 4 cm deep. Based on these convergence results, we adopt a mesh size of 0.5 cm.

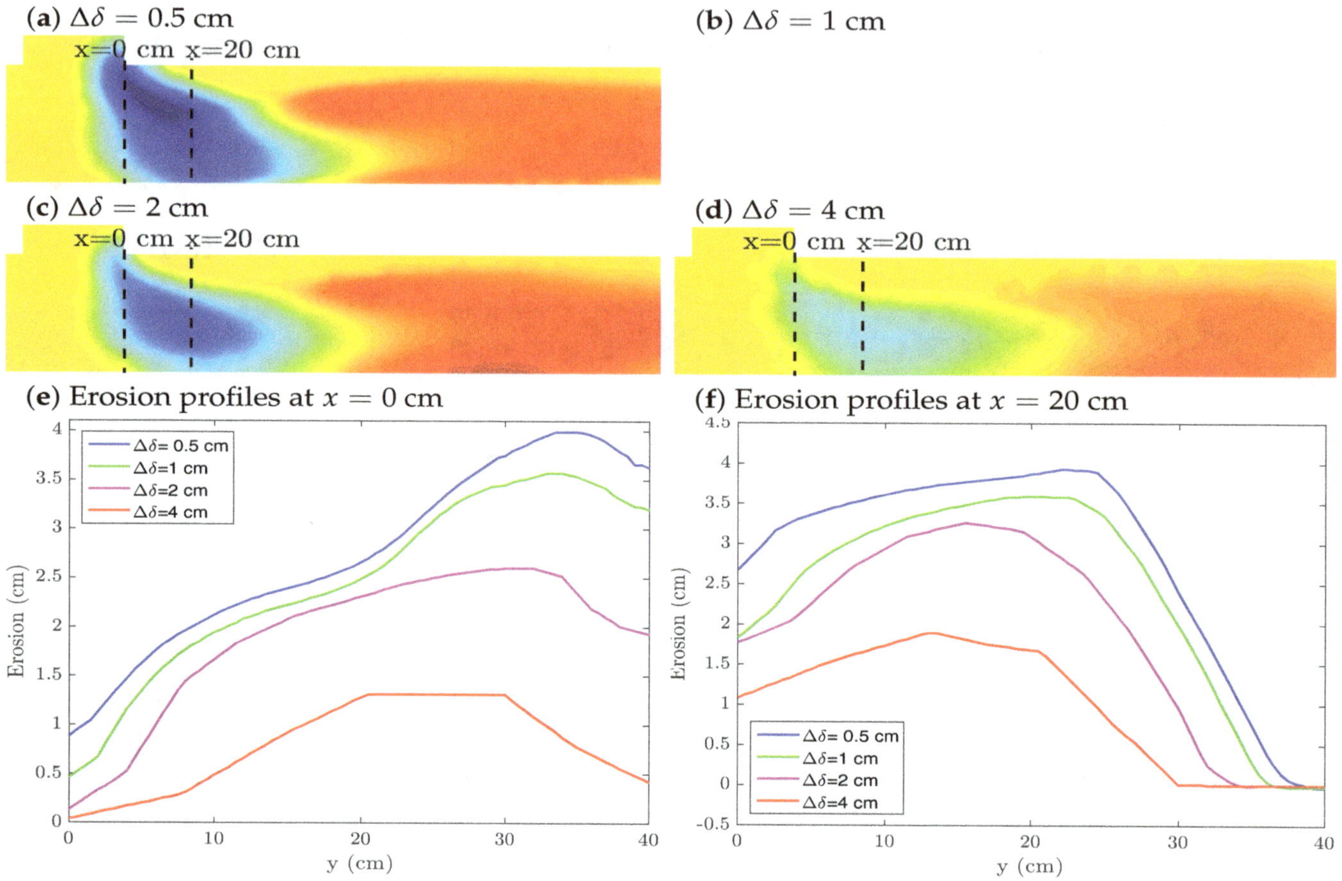

Figure 5. Mesh refinement study. We plot the erosion patterns for four increasingly refined meshes: in panel (**a**) the mesh size, $\Delta\delta$, is 0.5 cm, in panel (**b**) 1 cm, in panel (**c**) 2 cm, and in panel (**d**) 4 cm. We plot two erosion profiles of the main channel at x-coordinate 0 cm in panel (**e**) and $x = 20$ cm in panel (**f**). The reference system is depicted in Figure 3a. Erosion is positive and sedimentation negative.

3.3.2. Bed Properties: Friction Angle of Sediments

The model assumes that the sediments can be described as a cohesion-less granular material. We study the effect of the internal friction angle, ϕ, of sediments on the erosion pattern through four mathematical simulations of the experiment depicted in Figure 3. We adopt four angles, from $\phi = 20^\circ$ up to $\phi = 35^\circ$ and run the mathematical simulations. The erosion patterns and two bed profiles at the x-coordinates 0 and 20 cm (see reference system in Figure 3a) are included in Figure 6.

The friction angle plays an important role in the erosion pattern: the higher friction angle, the steeper scour hole and bar. The sediment with the lowest angle, $\phi = 20^\circ$, provides a flatter pattern, as shown in Figure 6a, than the sediment with the highest angle, $\phi = 35^\circ$ and Figure 6d, for both the scour hole and the bar.

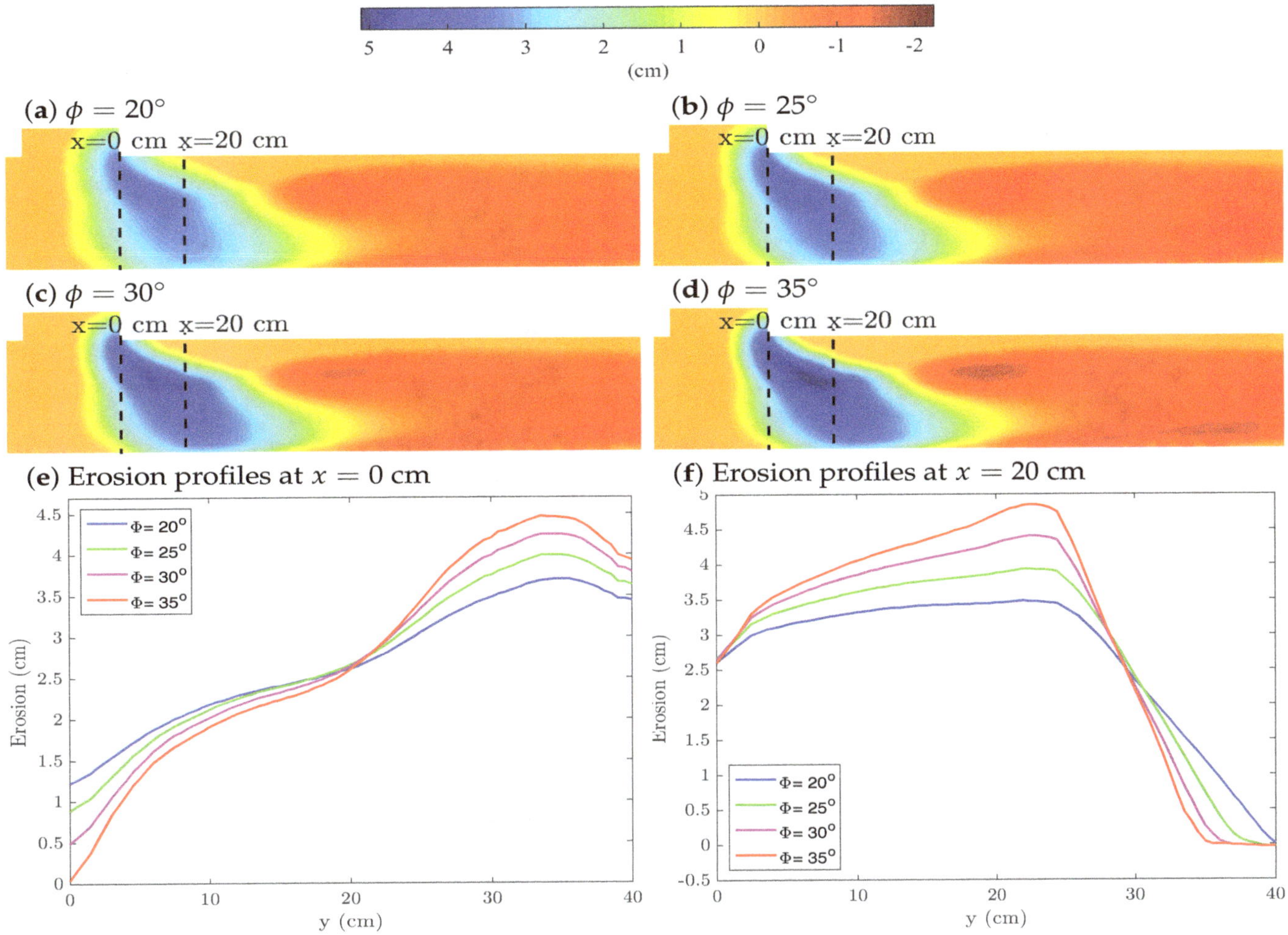

Figure 6. Effect of the friction angle of sediment on the erosion pattern. We plot the erosion contour plot for four friction angles, ϕ: in panel (**a**) ϕ is 20°, in panel (**b**) $\phi = 25°$, in panel (**c**) $\phi = 30°$, and in panel (**d**) $\phi = 35°$. We plot two erosion profiles of the main channel at x-coordinate 0 cm in panel (**e**) and $x = 20$ cm in panel (**f**). The reference system is depicted in Figure 3a. Erosion is positive and sedimentation negative.

The deepest point of the scour hole is given near the downstream corner of the junction close to the left bank, and ranges from almost 5 cm for the sediment with the highest friction angle to 3.5 cm for the sand with the lowest ϕ (Figure 6e,f). Sediments with high friction angles allow for steeper steady state holes and, consequently, deeper ones. Nevertheless, the erosion in the right bank of the main channel, y-coordinate equal to 0 cm, is deeper as the friction angle decreases. Indeed, the low friction angle of the sand hampers a deep scour hole in the downstream corner of the junction and sediment is conveyed by sliding from the right bank to the left (Figure 6e). A similar behavior is given in the profile at x-coordinate equal to 20 cm, but in the left bank where $y = 40$ cm (Figure 6f). Since the sediment with the lowest angle does not allow for a steep hole, the erosion conveys up to the left bank.

3.3.3. Bed Properties: Effective Roughness

We simulate the roughness of the bed channel with the Manning's coefficient, n. We simulate four erosion experiments with four coefficients with values ranging from 0.015 to 0.025 s·m$^{-1/3}$. The erosion patterns and two bed profiles at the x-coordinates 0 and 20 cm (see reference system in Figure 3a) are depicted in Figure 7.

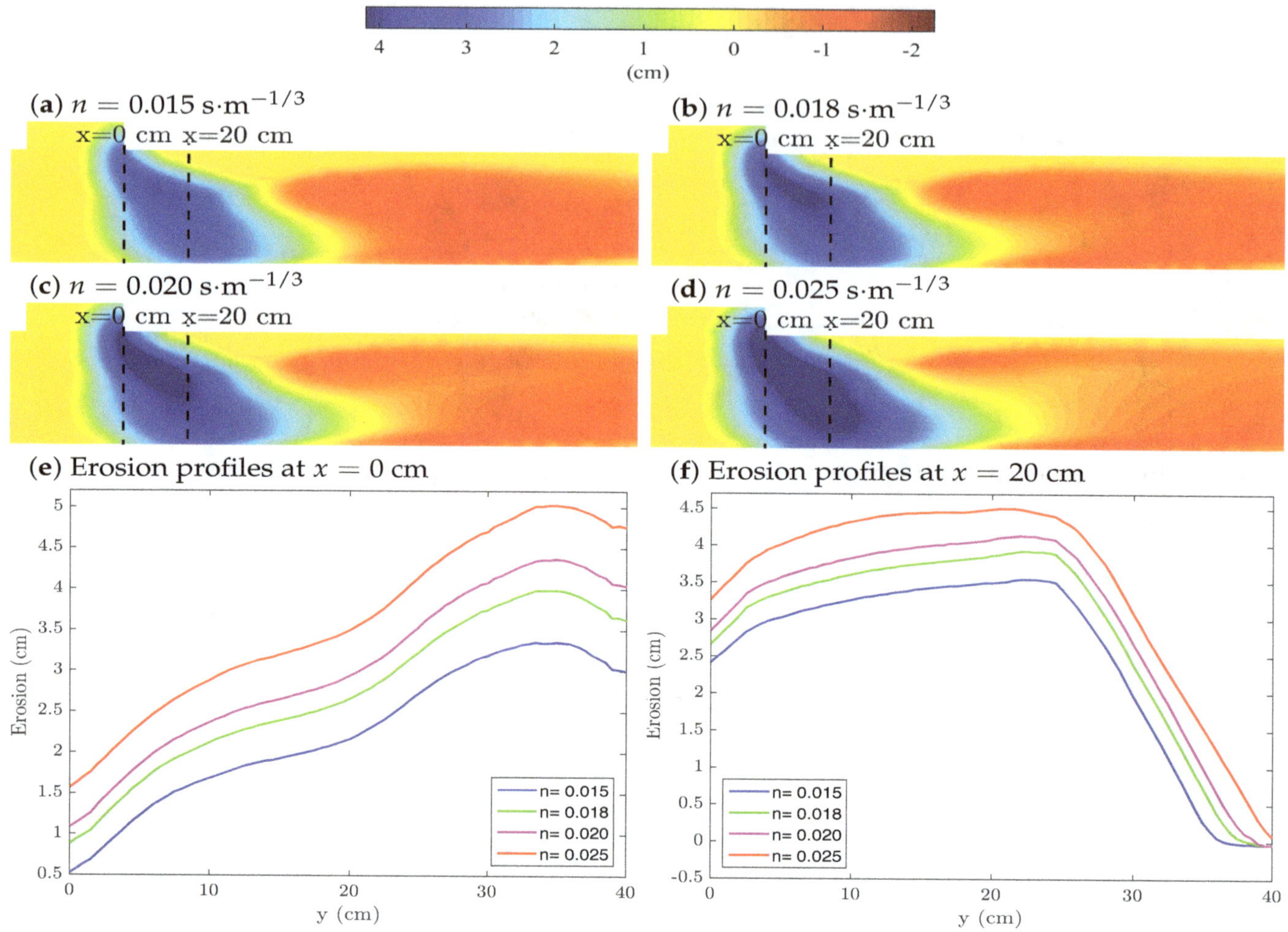

Figure 7. Effect of roughness on the erosion pattern. We plot the erosion contour plot for four Manning's coefficients, n: in panel (**a**) n is 0.015 s·m$^{-1/3}$, in panel (**b**) $n = 0.018$ s·m$^{-1/3}$, in panel (**c**) $n = 0.020$ s·m$^{-1/3}$, and in panel (**d**) $n = 0.025$ s·m$^{-1/3}$. We plot two erosion profiles of the main channel at x-coordinate 0 cm in panel (**e**) and $x = 20$ cm in panel (**f**). The reference system is depicted in Figure 3a. Erosion is positive and sedimentation negative.

Roughness has a significant influence on the erosion patterns. The larger the Manning's coefficient, the deeper and larger scour hole, and the lower and smaller downstream bar (Figure 7a–d). The maximum depth of the scour hole varies from 3.5 cm for the experiment with the lowest friction to 5 cm for the simulation with the highest n. The shear stress due to friction between the flow and the bed is proportional to the roughness and increase as n does. Consequently, the scour hole is larger in both sections of the main channel as the n-coefficient increases (Figure 7e,f).

The bar in the main channel downstream of the junction is affected only slightly by roughness. As the n-coefficient increases, the area of the bar decreases and concentrates in both banks of the channel (Figure 7a compared to Figure 7d).

The effective roughness can be estimated through formulae as function of the particle size. Two examples are the Muller formula [72], $n = D_{90}^{1/6}/26$, or the Strickler equation [73], $n = D_{50}^{1/6}/21.1$. Both formulae provides n-values of 0.0142 and 0.0147, respectively, for the sediment size used in the experiments. We got the best agreement between experiments and numerical simulations for $n = 0.018$ s·m$^{-1/3}$, which is higher than the previous values. The discrepancy may arise from the grain-size distribution of the sediments, since the formulae account for the whole distribution with only one variable, D_{90} or D_{50}.

4. Results and Discussion. Model Application to Field-Scale Channel Junction

To test the implications of sediment erosion and transport for river hydraulics during flash floods, we simulate the erosion on a real river confluence during a flood event. The simulations allow us to study the interplay between bed erosion and water flow patterns and to discern whether erosion has to be taken into account in eroded beds during flood events. Our case study is a channelized junction in southeast Spain. The confluence is located within a semiarid basin with major soil erosion problems situated between Sierra Nevada and Sierra de Los Filabres mountain ranges in Almería, Spain [74,75].

4.1. Case Study

Our case study is the confluence of the Nacimiento and Andarax rivers. Both streams are located in Almería province, Spain, between Sierra Nevada and Sierra de Los Filabres mountain ranges. The Nacimiento river is a left tributary of the Andarax river, and both meet at Alhabia, Almería province, Spain. The junction is channelized and the bed of the river is sand, whereas banks are protected with vertical concrete walls. The topography of the simulated stretch is depicted in Figure 8a. Since the river typically has a very small natural discharge, we know the actual bathymetry, which is typically unknown in most practical scenarios. The width of the Andarax river before the confluence is 25 m, and after it is 40 m, and the width of the Nacimiento river is 30 m.

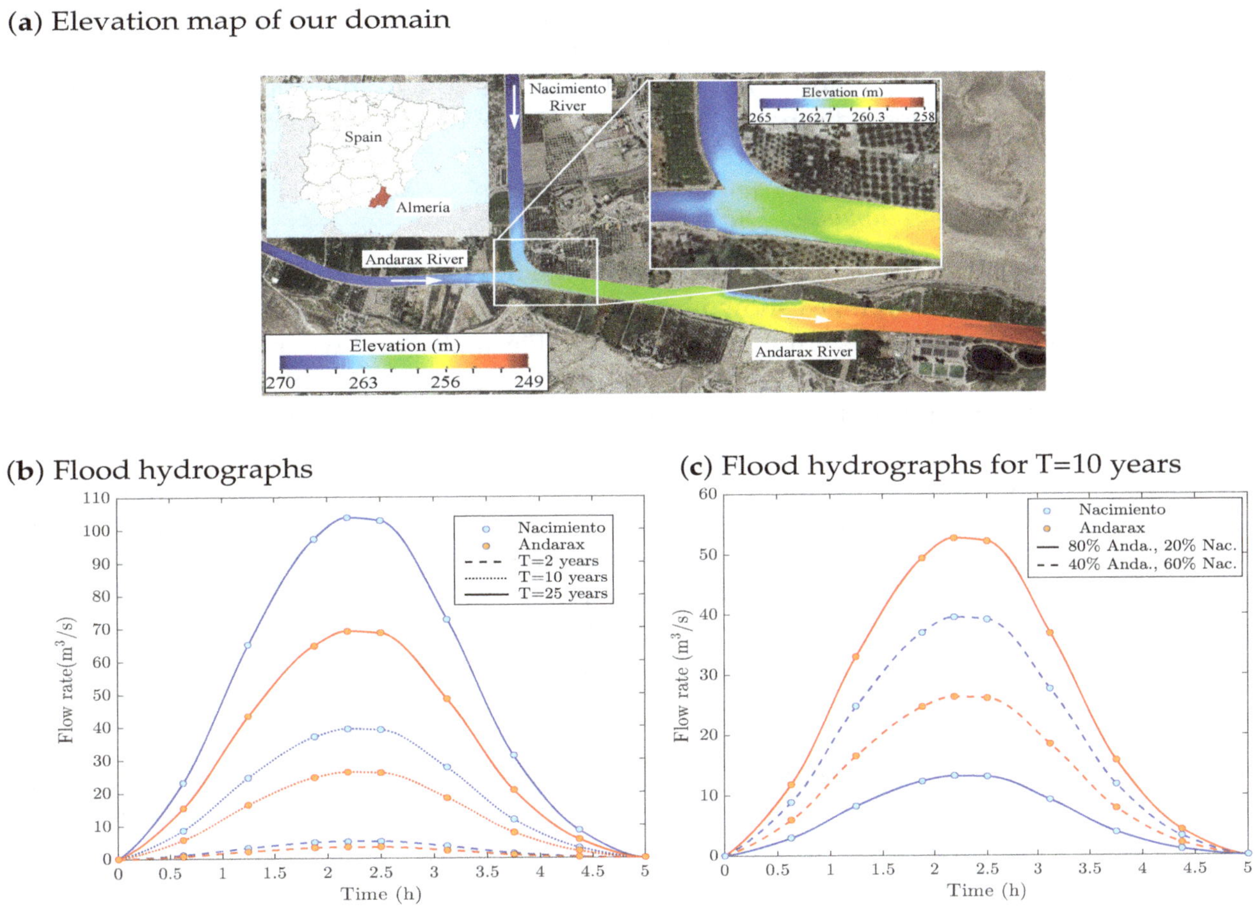

Figure 8. Model geometry and input flow-rates. In panel (**a**) we plot the location of the study area and the elevation map of our domain. Panel (**b**) includes the flood hydrographs for three return periods: T = 2, 10, and 25 years. Panel (**c**) depicts two flood hydrographs for the return period T = 10 years. The plot includes two discharge ratios, i.e., two flow distribution between the Nacimiento and Almarax rivers.

The Nacimiento river is 40 km in length at the junction. The catchment area is 600 km^2, the terrain elevation at the junction is about 263 m and the maximum elevation is about 1100 m. The Almarax river is 33 km in length at the junction. The basin area is 288 km^2, and the maximum elevation is approximately 1160 m. The median diameter of the sediments at the junction is $D_{50} = 1$ mm, the porosity is 0.4, the density is 2650 kg/m^3, and the friction angle is 30°. The roughness of the rivers, measured with the Manning's coefficient, is $n = 0.025$ s·m$^{-1/3}$. We include an elevation map of our simulation domain in Figure 8a.

We estimate the annual maximum flood quantile for the desired return period through the map of maximum discharges in Spain [76,77]. The map divides Spain into a set of homogeneous regions and provides the most likely regression model to estimate the annual maximum flood quantiles for the desired return period. The map was built using flood series recorded at gauged basins [76]. We compute the flood quantiles downstream of the junction for the return periods, T, of 2, 10, and 25 years. These quantiles are 8.97, 65.84, and 173.06 m^3/s, respectively. We consider that the flood event lasts 5 h with a parabolic shape, and assign the total flow rate to each basin proportional to its area. The simulated flood events in both basins for the selected return periods are plotted in Figure 8b. Since the overall water flow regime in our domain is subcritical, as boundary conditions we impose the total discharge at the upstream boundary, and critical flow (free discharge) at the downstream end. The outlet of the model is far enough from the junction to guarantee that water reaches uniform flow downstream the confluence. The imposed flow rates are plotted in Figure 8b.

We take advantage of our model and study the effect of the discharge ratio on the erosion patterns. In this sense, we study two situations for the total flood quantile of 10-year return period, i.e., 65.84 m^3/s: (1) the flow distributes 80% in the Andarax river and 20% in the Nacimiento river, which gives a discharge ratio of 0.2; and (2) the flow distributes 20% in the Andarax river and 80% in the Nacimiento river, which gives a discharge ratio of 0.8. The evolution of the flood events, assuming 5-h duration and parabolic shape, are depicted in Figure 8c.

4.2. Effect of Bed Erosion on Hydrodynamics During Flood

We study the interplay between the bed erosion on the flow pattern and vice-versa. We simulate the 25-year return period flood event, as shown in Figure 8b, in our computational domain (Figure 8a). The erosion at the end of the flood event is represented in Figure 9a, and we include the water depth and velocity contour plots at time step 2.5 h in Figure 9b,c, respectively. We also depict the water depth and velocity contour plots at time step 2.5 h in Figure 9d,e, respectively, for a simulation with non-erodible bed.

The initial bed topography is irregular, as represented in Figure 8a. The cross sections of all the channels are proximately U-shape in the central part and flat in the banks. The irregular initial bed shape results in an irregular erosion pattern, as shown in Figure 9a. The junction between both rivers is curved to avoid the scour hole that appears in rectangular confluences.

The curved shape produces erosion in the left or inlet bank and sedimentation in the right or outer bank. Erosion experiments and numerical models of sediment transport in curved channels have demonstrated that erosion occurs in the opposite way: erosion in the outer bank and sedimentation in the inlet bank [78,79]. However, the influence of the lateral flow is not considered in these studies, as well as the bed topography was initially flat. The flow from the Andarax river pushes the Nacimiento flow and detaches it from the outer wall. This effect, together with the bed topography produces erosion on the inner bank. Moreover, the widening of the Andarax river at the confluence promotes the sedimentation at the junction, as two bars appear: the first one in the upstream corner of the junction and the second one in the central area of the Andarax river just downstream of the upstream corner.

The flood event slightly erodes the right bank downstream of the confluence, whereas several sediment ridges emerged in the mid-channel area as well as a bar in the left bank (Figure 9a). Sedimentation patterns downstream of the junction have been observed in laboratory experiments [45], and they are caused by the decrease in velocity since the shear layer vanishes [46].

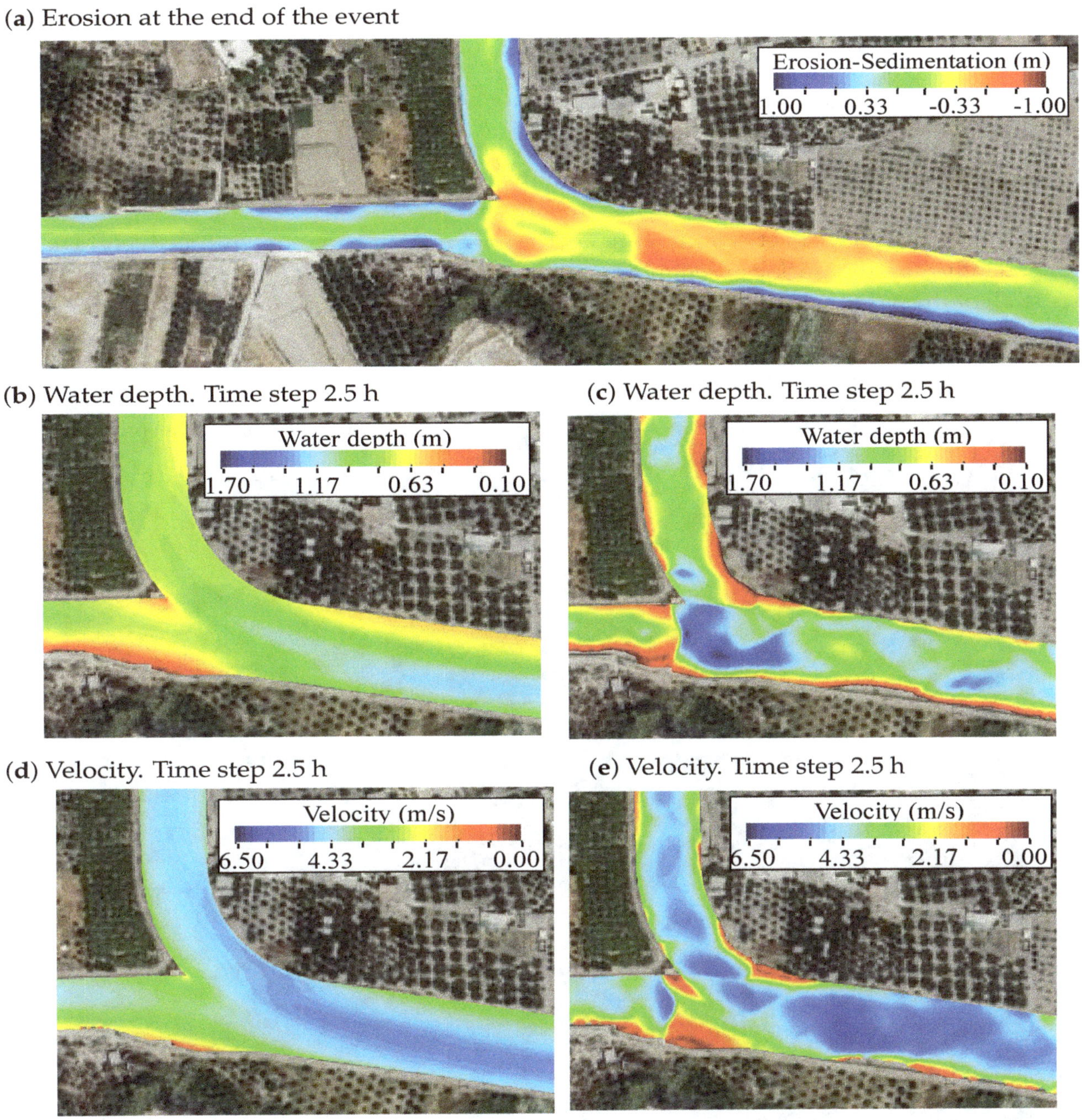

Figure 9. Effect of erosion on the flow patterns. In panel (**a**) we depict the contour plot of erosion after the 25-year flood event. Erosion is positive and sedimentation negative. Panel (**b**) depicts the water depth at time step 2.5 h for the 25-year flood event computed with bed-mobile model, whereas panel (**c**) plots the result provided by a model without bed erosion. Panel (**d**) depicts the water velocity at time step 2.5 h for the 25-year flood event computed with a bed mobile model, and panel (**e**) plots the result provided by a model without bed erosion.

Erosion has a great influence on the water depth and flow velocity patterns. The water depth pattern computed with a bed-mobile model at a given time step is smooth, as shown in Figure 9b, whereas the output provided by a model without bed erosion is very irregular, with areas with high depths close to other ones with low depths, as shown in Figure 9c. The abrupt pattern for the non-erodible model emerges from the irregular bed topography. Since the bed topography in the bed mobile model can evolve according to the flow conditions, the water depth pattern is smoother.

Water velocity patterns are also greatly influenced by erosion. The water velocity pattern computed with the bed-mobile model at a given time step is much smoother than the one provided by the fixed bed approach, as shown in Figure 9d,e. As in the water depth pattern, the initial abrupt bed topography causes the water velocity pattern to also be abrupt.

4.3. *Effect of Flooding on Erosion Patterns*

We take advantage of our model to study the effect of the flood event on the erosion pattern. We analyze the effect of the flood return period, and the flood discharge ratio—the distribution of the total flow between the two rivers—on the erosion patterns.

We simulate three flood events with return period of 5, 10, and 25 years, respectively. The flood return period influences the erosion or sedimentation intensity, but not the erosion pattern, which remains constant in all of them (Figure 10). As the return period increases, so does the intensity of the erosion and sedimentation.

(**a**) Erosion. 5-year flood event

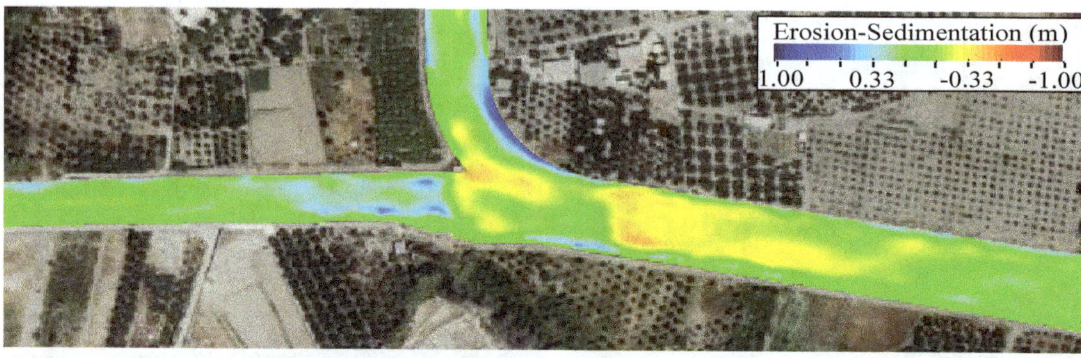

(**b**) Erosion. 10-year flood event

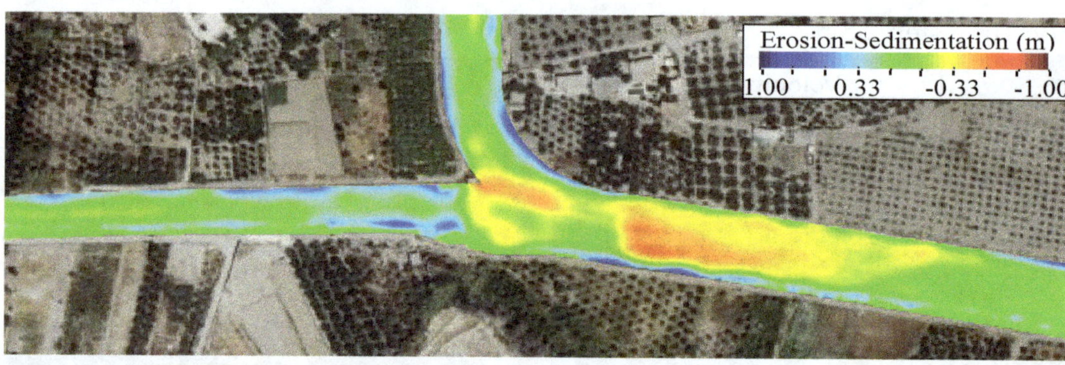

(**c**) Erosion. 25-year flood event

Figure 10. Effect of flood return period on the erosion pattern. We plot the erosion patterns for three flood events with return periods: (**a**) T = 5 years, (**b**) T = 10 years, (**c**) T = 25 years. Erosion is positive and sedimentation negative.

We also study the effect of the flood discharge ratio, or total flow distribution between both rivers, on the erosion patterns. We simulate three new scenarios for the 10-year flood event: (1) the total flow distributes 80% in the Andarax river and 20% in the Nacimiento river, which results in the pattern included in Figure 11a, (2) the flow distributes 40% in the Andarax river and 60% in the Nacimiento river, which has the outcome of the pattern depicted in Figure 11b, and (3) the flow distributes 20% in the Andarax river and 80% in the Nacimiento river with the pattern depicted in Figure 11c.

(**a**) Discharge flow ratio: 80% Andarax river and 20% Nacimiento river

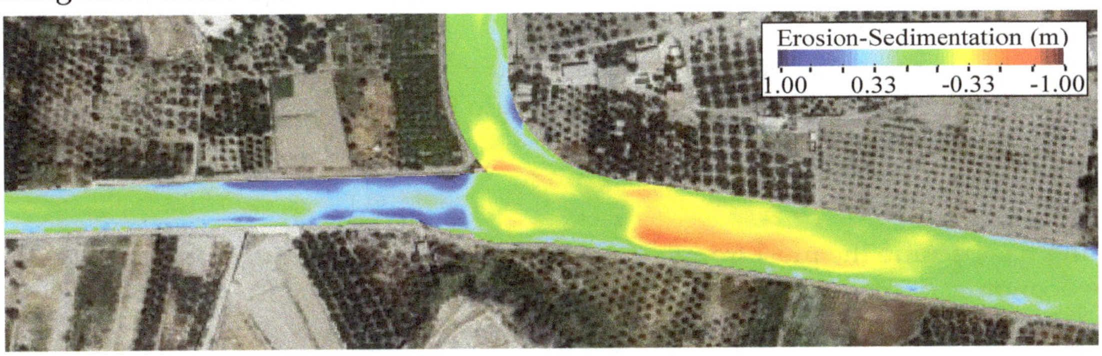

(**b**) Discharge flow ratio: 40% Andarax river and 60% Nacimiento river

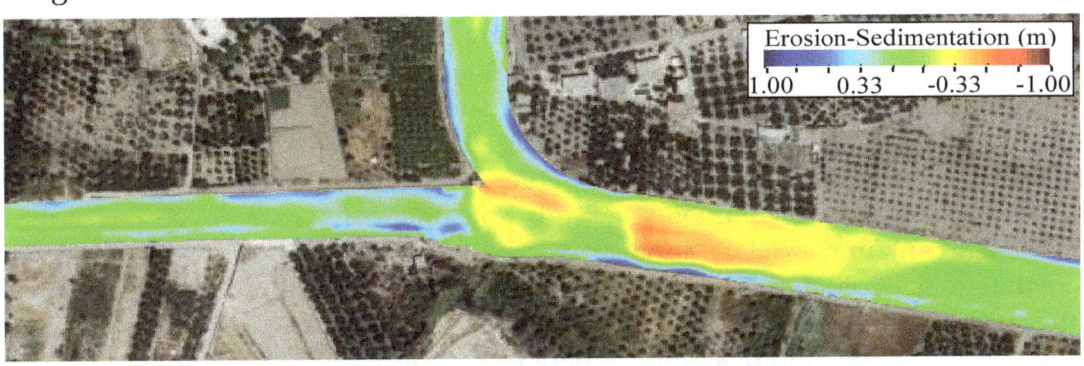

(**c**) Discharge flow ratio: 20% Andarax river and 80% Nacimiento river

Figure 11. Effect of the discharge ratio on the erosion patterns. Here we plot the erosion patterns under the 10-year flood event. In panel (**a**) the total flow distributes 80% in the Andarax river and 20% in the Nacimiento river, in panel (**b**) the distribution is 40% in the Andarax river and 60% in the Nacimiento river, and in panel (**c**) it is 20% in the Andarax river and 80% in the Nacimiento river.

The comparison between the erosion patterns for the three flood discharge ratios allow us to reach the following conclusions. As the flow rate in the Andarax river increases, the erosion in both banks of the river upstream the junction also does. In the same way, the erosion in the banks of the Nacimiento river upstream the junction rises up as the flow rate in the rives does.

The sedimentation patterns downstream of the junction are nearly the same for the three flood discharge ratios, since the flow rate in this stretch is identical in all the simulations. Small differences emerge in the tail of the bar, which may arise from the total volume of sediment eroded that is different in every situation.

The erosion in the right bank downstream of the junction increases as the flow in the Nacimiento river does. The Andarax river flow displaces the Nacimiento river flow at the junction, pushing the main stream away from the right bank downstream of the confluence. The larger the Nacimiento river flow rate, the lower the Andarax river flow and the more water hits the right bank that promotes erosion.

Two bars emerge from the upstream corner of the junction. The first one is parallel to the Andarax river and extends in some cases upstream along the Nacimiento river. This bar shrinks as the Nacimiento river flow increases. The second bar is perpendicular to the Andarax river and in some cases changes its direction and becomes parallel to the flow. Opposite to the other bar, this one contracts as the Nacimiento river flow increases.

5. Conclusions

Flooding has become the most frequent environmental hazard, causing significant losses of life and severe economic damage. Mathematical models are useful tools for flood risk forecasting and control, and the current computational resources allow us to simulate flood events with two-dimensional (2D) approaches. An open question is whether erosion should be accounted for or not when it comes to high-fidelity simulation of flood events.

In this paper we have addressed the question by assessing the effect of mobile beds on the flow patterns during flood events through numerical simulations. We have simulated the erosion in a river confluence located in a semiarid basin of southeast Spain. We have also validated our simulations by comparing our numerical results with laboratory experiments of channel confluences reported in the literature. Our 2D simulation is able to reproduce the main features of the erosion patterns in a rectangular channel confluence: a scour-hole on the downstream corner of the confluence, and a bar on the left bank immediately downstream of the junction. However, laboratory experiments have shown that a helicoidal flow emerges downstream of the junction in the main channel and concentrates the sediments. Since our mathematical model is 2D depth-averaged, it is unable to reproduce the helicoidal flow and discordance between experiments and simulations arises.

We have conducted a mesh refinement study that has shown the well-posedness of the problem. Moreover, our sensitivity analysis has concluded that the bed roughness has a great influence on the erosion patterns: the higher the effective bed roughness, the deeper and larger the scour hole, and the lower and smaller the downstream bar. The internal friction angle of the granular river bed sediments plays an important role in the erosion pattern: the larger the friction angle, the steeper the scour hole and bar, which in turn controls the overall hydrodynamics.

We have simulated the erosion process in a channelized river confluence during flood events. The case study is located in a semiarid basin in southern Spain. Erosion has a great influence on the water depth and flow velocity patterns. The water depth pattern computed with a bed-mobile model at a given time step is smooth, whereas the depth provided by a model without bed erosion is very abrupt. Water velocity patterns are also greatly affected by erosion: the pattern provided by a bed-erodible model is smoother than the one computed with the fixed bed approach.

We have analyzed the effect of the flood return period on the erosion pattern. The return period influences the erosion or sedimentation intensity, but not the erosion pattern, which remains constant for all the simulated periods. We have studied the distribution of a given flow rate between both rivers, i.e., the discharge ratio. The increase in flow rate in one river also increases the erosion upstream from the junction in that river. Moreover, the formation of bars in the junction is quite influenced by the discharge ratio. Nevertheless, the erosion pattern downstream of the junction is quite independent of the discharge ratio.

Our study elucidates the interplay between shallow water hydrodynamics and sediment transport during flood events. Erosion plays a key role in these processes, leading to smoother water profiles and overall smaller flow velocities. Current computational resources can deal with 2D numerical simulations of river flow that incorporate sediment transport. Hydrodynamic models in conjunction with sediment transport arise as a useful tool for engineers, practitioners, and stakeholders to evaluate flood control problems.

Author Contributions: Literature review, L.C.-F.; research design, D.S. and L.C.-F.; validation cases, L.C.-F., and A.S.-W.; numerical experiments, D.S.; results discussion, D.S., A.S.-W., L.C.-F., and L.G.; manuscript drafting, D.S.; manuscript proofing, L.C.-F., A.S.-W., and L.G. All authors have read and agreed to the published version of the manuscript.

Acknowledgments: D.S. gratefully acknowledges funding from the BBVA Foundation through the 2019 Leonardo Grant for Researchers and Cultural Creators. The BBVA Foundation accepts no responsibility for the opinions, statements and contents included in the project and/or the results thereof, which are entirely the responsibility of the authors.

References

1. Smith, K. *Environmental Hazards: Assessing Risk and Reducing Disaster*; Routledge: London, UK; New York, NY, USA, 2013.
2. Kundzewicz, Z.W.; Kanae, S.; Seneviratne, S.I.; Handmer, J.; Nicholls, N.; Peduzzi, P.; Mechler, R.; Bouwer, L.M.; Arnell, N.; Mach, K.; et al. Flood risk and climate change: Global and regional perspectives. *Hydrol. Sci. J.-J. Sci. Hydrol.* **2014**, *59*, 1–28. [CrossRef]
3. Mediero, L.; Santillán, D.; Garrote, L.; Granados, A. Detection and attribution of trends in magnitude, frequency and timing of floods in Spain. *J. Hydrol.* **2014**, *517*, 1072–1088. [CrossRef]
4. Tayefi, V.; Lane, S.; Hardy, R.; Yu, D. A comparison of one-and two-dimensional approaches to modelling flood inundation over complex upland floodplains. *Hydrol. Process.* **2007**, *21*, 3190–3202. [CrossRef]
5. Sellin, R.; Ervine, D.; Willetts, B. Behaviour of meandering two-stage channels. *Proc. Inst. Civ. Eng.-Water Marit. Energy* **1993**, *101*, 99–111. [CrossRef]
6. Toro, E.F. *Shock–Capturing Methods for Free–Surface Shallow Flows*; John Wiley & Sons, Ltd.: Chichester, UK, 2001.
7. Deltares Systems. *Simulation of Multidimensional Hydrodynamic Flows and Transport Phenomena, Including Sediments*; User Manual: Delft, The Netherlands, 2011.
8. Bladé, E.; Cea, L.; Corestein, G.; Escolano, E.; Puertas, J.; Vázquez-Cendón, E.; Dolz, J.; Coll, A. Iber: Herramienta de simulación numérica del flujo en ríos. *Rev. Int. Métodos Numér. Cálc. Diseño Ing.* **2014**, *30*, 1–10. [CrossRef]
9. Pérez-Sánchez, M.; Sánchez-Romero, F.; Redón-Santafé, M.; Torregrosa Soler, J.; Ferrer Gisbert, C.; Ferrán Gozálvez, J.; Ferrer Gisbert, A.; López-Jiménez, P. Numerical study for the development of flood maps considering the break hypothesis in irrigation reservoirs. *Ing. Agua* **2019**, *23*, 1–18. [CrossRef]
10. Sanz-Ramos, M.; Bladé, E.; Escolano, E. Optimization of the Floodplain Encroachment calculation with hydraulic criteria. *Ing. Agua* **2020**, *24*, 203–218. [CrossRef]
11. Zerihun, Y.T. A Numerical Study on Curvilinear Free Surface Flows in Venturi Flumes. *Fluids* **2016**, *1*, 21. [CrossRef]
12. Zerihun, Y.T.; Fenton, J.D. One-dimensional simulation model for steady transcritical free surface flows at short length transitions. *Adv. Water Resour.* **2006**, *29*, 1598–1607. [CrossRef]
13. Berger, R.C.; Carey, G.F. Free-surface flow over curved surfaces Part II: Computational model. *Int. J. Numer. Methods Fluids* **1998**, *28*, 201–213. [CrossRef]
14. Shimozono, T.; Sato, S. Coastal vulnerability analysis during tsunami-induced levee overflow and breaching by a high-resolution flood model. *Coast. Eng.* **2016**, *107*, 116–126. [CrossRef]
15. Sánchez-Cordero, E.; Gómez, M.; Bladé, E. 3D numerical analysis of flow characteristics in an open—Channel bend. *Ing. Agua* **2020**, *24*, 157–168. [CrossRef]
16. Duró, G.; Crosato, A.; Tassi, P. Numerical study on river bar response to spatial variations of channel width. *Adv. Water Res.* **2016**, *93*, 21–38. [CrossRef]
17. Martín-Vide, J.; Plana-Casado, A.; Sambola, A.; Capapé, S. Bedload transport in a river confluence. *Geomorphology* **2015**, *250*, 15–28. [CrossRef]

18. Herrero, H.S.; Lozada, J.M.D.; García, C.M.; Szupiany, R.N.; Best, J.; Pagot, M. The influence of tributary flow density differences on the hydrodynamic behavior of a confluent meander bend and implications for flow mixing. *Geomorphology* **2018**, *304*, 99–112. [CrossRef]
19. Baki, A.B.M.; Gan, T.Y. Riverbank migration and island dynamics of the braided Jamuna River of the Ganges—Brahmaputra basin using multi-temporal Landsat images. *Quat. Int.* **2012**, *263*, 148–161. [CrossRef]
20. Hicks, D.M.; Gomez, B.; Trustrum, N.A. Erosion thresholds and suspended sediment yields, Waipaoa River basin, New Zealand. *Water Resour. Res.* **2000**, *36*, 1129–1142. [CrossRef]
21. R. Grove, J.; Croke, J.; Thompson, C. Quantifying different riverbank erosion processes during an extreme flood event. *Earth Surf. Process. Landf.* **2013**, *38*, 1393–1406.
22. Croke, J.; Fryirs, K.; Thompson, C. Channel–floodplain connectivity during an extreme flood event: implications for sediment erosion, deposition, and delivery. *Earth Surf. Process. Landf.* **2013**, *38*, 1444–1456. [CrossRef]
23. Lamb, M.P.; Fonstad, M.A. Rapid formation of a modern bedrock canyon by a single flood event. *Nat. Geosci.* **2010**, *3*, 477–481. [CrossRef]
24. Bilal, A.; Xie, Q.; Zhai, Y. Flow, Sediment, and Morpho-Dynamics of River Confluence in Tidal and Non-Tidal Environments. *J. Mar. Sci. Eng.* **2020**, *8*, 591. [CrossRef]
25. Arega, F.; Lee, J.H.; Tang, H.W. Hydraulic jet control for river junction design of Yuen Long Bypass Floodway, Hong Kong. *J. Hydraul. Eng.* **2008**, *134*, 23–33. [CrossRef]
26. Ferrarin, C.; Madricardo, F.; Rizzetto, F.; Kiver, W.M.; Bellafiore, D.; Umgiesser, G.; Kruss, A.; Zaggia, L.; Foglini, F.; Ceregato, A.; et al. Geomorphology of scour holes at tidal channel confluences. *J. Geophys. Res. Earth Surf.* **2018**, *123*, 1386–1406. [CrossRef]
27. Xie, Q.; Yang, J.; Lundström, S.; Dai, W. Understanding morphodynamic changes of a tidal river confluence through field measurements and numerical modeling. *Water* **2018**, *10*, 1424. [CrossRef]
28. Bennett, S.; Best, J. Mean flow and turbulence structure over fixed, two-dimensional dunes: Implications for sediment transport and bedform stability. *Sedimentology* **1995**, *42*, 491–513. [CrossRef]
29. Best, J.L.; Roy, A.G. Mixing-layer distortion at the confluence of channels of different depth. *Nature* **1991**, *350*, 411–413. [CrossRef]
30. Biron, P.; Best, J.L.; Roy, A.G. Effects of bed discordance on flow dynamics at open channel confluences. *J. Hydraul. Eng.* **1996**, *122*, 676–682. [CrossRef]
31. Biron, P.; Roy, A.; Best, J. Turbulent flow structure at concordant and discordant open-channel confluences. *Exp. Fluids* **1996**, *21*, 437–446. [CrossRef]
32. Boyer, C.; Roy, A.G.; Best, J.L. Dynamics of a river channel confluence with discordant beds: Flow turbulence, bed load sediment transport, and bed morphology. *J. Geophys. Res. Solid Earth* **2006**, *111*. [CrossRef]
33. Best, J.L.; Reid, I. Separation zone at open-channel junctions. *J. Hydraul. Eng.* **1984**, *110*, 1588–1594. [CrossRef]
34. Webber, N.B.; Greated, C.A. An investigation of flow behaviour at the junction of rectangular channels. *Proc. Inst. Civil. Eng.* **1966**, *34*, 321–334. [CrossRef]
35. Tang, H.; Zhang, H.; Yuan, S. Hydrodynamics and contaminant transport on a degraded bed at a 90-degree channel confluence. *Environ. Fluid Mech.* **2018**, *18*, 443–463. [CrossRef]
36. Yuan, S.; Tang, H.; Xiao, Y.; Xia, Y.; Melching, C.; Li, Z. Phosphorus contamination of the surface sediment at a river confluence. *J. Hydrol.* **2019**, *573*, 568–580. [CrossRef]
37. Cheng, Z.; Constantinescu, G. Stratification effects on hydrodynamics and mixing at a river confluence with discordant bed. *Environ. Fluid Mech.* **2019**, *20*, 843–872. [CrossRef]
38. Gurram, S.K.; Karki, K.S.; Hager, W.H. Subcritical junction flow. *J. Hydraul. Eng.* **1997**, *123*, 447–455. [CrossRef]
39. Hsu, C.C.; Lee, W.J.; Chang, C.H. Subcritical open-channel junction flow. *J. Hydraul. Eng.* **1998**, *124*, 847–855. [CrossRef]
40. Weber, L.J.; Schumate, E.D.; Mawer, N. Experiments on flow at a 90 open-channel junction. *J. Hydraul. Eng.* **2001**, *127*, 340–350. [CrossRef]
41. Guillén-Ludeña, S.; Franca, M.; Cardoso, A.; Schleiss, A. Evolution of the hydromorphodynamics of mountain river confluences for varying discharge ratios and junction angles. *Geomorphology* **2016**, *255*, 1–15. [CrossRef]

42. Leite Ribeiro, M.; Blanckaert, K.; Roy, A.; Schleiss, A.J. Flow and sediment dynamics in channel confluences. *J. Geophys. Res. Earth Surf.* **2012**, *117*. [CrossRef]
43. Best, J.L. Sediment transport and bed morphology at river channel confluences. *Sedimentology* **1988**, *35*, 481–498. [CrossRef]
44. Nazari-Giglou, A.; Jabbari-Sahebari, A.; Shakibaeinia, A.; Borghei, S.M. An experimental study of sediment transport in channel confluences. *Int. J. Sediment Res.* **2016**, *31*, 87–96. [CrossRef]
45. Yuan, S.; Tang, H.; Xiao, Y.; Qiu, X.; Xia, Y. Water flow and sediment transport at open-channel confluences: An experimental study. *J. Hydraul. Res.* **2018**, *56*, 333–350. [CrossRef]
46. Rhoads, B.L.; Sukhodolov, A.N. Spatial and temporal structure of shear layer turbulence at a stream confluence. *Water Resour. Res.* **2004**, *40*. [CrossRef]
47. Rhoads, B.L.; Sukhodolov, A.N. Lateral momentum flux and the spatial evolution of flow within a confluence mixing interface. *Water Resour. Res.* **2008**, *44*. [CrossRef]
48. Rhoads, B.L.; Sukhodolov, A.N. Field investigation of three-dimensional flow structure at stream confluences: 1. Thermal mixing and time-averaged velocities. *Water Resour. Res.* **2001**, *37*, 2393–2410. [CrossRef]
49. Sukhodolov, A.N.; Rhoads, B.L. Field investigation of three-dimensional flow structure at stream confluences: 2. Turbulence. *Water Resour. Res.* **2001**, *37*, 2411–2424. [CrossRef]
50. Mosley, M.P. An experimental study of channel confluences. *J. Geol.* **1976**, *84*, 535–562. [CrossRef]
51. Wallis, E.; Nally, R.M.; Lake, P.S. A Bayesian analysis of physical habitat changes at tributary confluences in cobble-bed upland streams of the Acheron River basin, Australia. *Water Resour. Res.* **2008**, *44*. [CrossRef]
52. Best, J.L.; Rhoads, B.L. *River Confluences, Tributaries and the Fluvial Network*; John Wiley & Sons, Ltd.: Chichester, UK, 2008; Chapter Sediment transport, bed morphology and the sedimentology of river channel confluences; pp. 45–72.
53. Cea, L.; Puertas, J.; Vázquez-Cendón, M.E. Depth Averaged Modelling of Turbulent Shallow Water Flow with Wet-Dry Fronts. *Arch. Comput. Methods Eng.* **2007**, *14*, 303–341. [CrossRef]
54. Rastogi, A.K.; Rodi, W. Predictions of heat and mass transfer in open channels. *J. Hydraul. Div.* **1978**, *104*, 397–420.
55. Ashworth, P.; Ferguson, R. Interrelationships of Channel Processes, Changes and Sediments in a Proglacial Braided River. *Geogr. Ann. Ser. A Phys. Geogr.* **1986**, *68*, 361–371. [CrossRef]
56. Leeder, M.R. On the Interactions between Turbulent Flow, Sediment Transport and Bedform Mechanics in Channelized Flows. In *Modern and Ancient Fluvial Systems*; John Wiley & Sons, Ltd.: Hoboken, NJ, USA, 2009; pp. 3–18. [CrossRef]
57. Anderson, R.S. Modeling the tor-dotted crests, bedrock edges, and parabolic profiles of high alpine surfaces of the Wind River Range, Wyoming. *Geomorphology* **2002**, *46*, 35–58. [CrossRef]
58. Mudd, S.M.; Furbish, D.J. Influence of chemical denudation on hillslope morphology. *J. Geophys. Res. Earth Surf.* **2004**, *109*. [CrossRef]
59. Paola, C.; Voller, V.R. A generalized Exner equation for sediment mass balance. *J. Geophys. Res. Earth Surf.* **2005**, *110*. [CrossRef]
60. Van Rijn, L.C. Sediment Transport, Part I: Bed Load Transport. *J. Hydraul. Eng.* **1984**, *110*, 1431–1456. [CrossRef]
61. Shields, A. Anwendung Der Aehnlichkeitsmechanik Und Der Turbulenzforschung Auf Die Geschiebebewegung. Ph.D. Thesis, Technical University Berlin, Berlin, Germany, 1936.
62. Van Rijn, L.C. Sediment Transport, Part II: Suspended Load Transport. *J. Hydraul. Eng.* **1984**, *110*, 1613–1641. [CrossRef]
63. Van Leer, B. Towards the Ultimate Conservative Difference Scheme, V. A Second Order Sequel to Godunov's Method. *J. Comput. Phys.* **1979**, *32*, 101–136. [CrossRef]
64. Alcrudo, A.; García-Navarro, P. A High Resolution Godunov–Type Scheme in Finite Volumes for the 2D Shallow Water Equations. *Int. J. Numer. Methods Eng.* **1993**, *16*, 489–505. [CrossRef]
65. García-Navarro, P.; Hubbard, M.E.; Priestley, A. Genuinely Multidimensional Upwinding for the 2D Shallow Water Equations. *J. Comput. Phys.* **1995**, *121*, 79–93. [CrossRef]
66. Toro, E.F.; García-Navarro, P. Godunov–Type Methods for Free–Surface Shallow Flows: A Review. *J. Hydraul. Res.* **2007**, *45*, 736–751. [CrossRef]
67. Cueto-Felgueroso, L.; Colominas, I. High-order finite volume methods and multiresolution reproducing kernels. *Arch. Comput. Methods Eng.* **2008**, *15*, 185–228. [CrossRef]

68. Cueto-Felgueroso, L.; Colominas, I.; Fe, J.; Navarrina, F.; Casteleiro, M. High-order finite volume schemes on unstructured grids using moving least-squares reconstruction. Application to shallow water dynamics. *Int. J. Numer. Methods Eng.* **2006**, *65*, 295–331. [CrossRef]
69. Navas-Montilla, A.; Murillo, J.; García-Navarro, P. High order simulation models for the resolution of wave propagation phenomena in turbulent free surface flows. *Ing. Agua* **2019**, *23*, 275–287. [CrossRef]
70. Pinto Coelho, M.M.L. Experimental determination of free surface levels at open-channel junctions. *J. Hydraul. Res.* **2015**, *53*, 394–399. [CrossRef]
71. Cueto-Felgueroso, L.; Santillán, D.; García-Palacios, J.H.; Garrote, L. Comparison between 2D Shallow-Water Simulations and Energy-Momentum Computations for Transcritical Flow Past Channel Contractions. *Water* **2019**, *11*, 1476. [CrossRef]
72. Muller, R. *Theoretische Grundlagen der Flussund Wildbachverbauungen*; Leemann: Zurich, Switzerland, 1943.
73. Strickler, A. *Beitrage zur Frage der Geschwindigkeitsformel und der Rauhigkeitszahlen fur Strome, Kanale und Geschlossene Leitungen*; Amt. f. Wasserwirtschaft: Berna, Switzerland, 1923
74. Bellin, N.; Vanacker, V.; Van Wesemael, B.; Solé-Benet, A.; Bakker, M. Natural and anthropogenic controls on soil erosion in the Internal Betic Cordillera (southeast Spain). *Catena* **2011**, *87*, 190–200. [CrossRef]
75. Martinez-Mena, M.; Castillo, V.; Albaladejo, J. Hydrological and erosional response to natural rainfall in a semi-arid area of south-east Spain. *Hydrol. Process.* **2001**, *15*, 557–571. [CrossRef]
76. Álvarez, A.J.; Montañés, C.G.; Orduña, L.M.; Caballer, L.I.; Revilla, J.G. El mapa de caudales máximos de las cuencas intercomunitarias. *Rev. Obras Públicas* **2012**, *3533*, 7–32.
77. Santillán, D.; Mediero, L.; Garrote, L. Modelling uncertainty of flood quantile estimations at ungauged sites by Bayesian networks. *J. Hydroinform.* **2014**, *16*, 822–838. [CrossRef]
78. Odgaard, A.J.; Bergs, M.A. Flow processes in a curved alluvial channel. *Water Resour. Res.* **1988**, *24*, 45–56. [CrossRef]
79. Wu, W.; Rodi, W.; Wenka, T. 3D numerical modeling of flow and sediment transport in open channels. *J. Hydraul. Eng.* **2000**, *126*, 4–15. [CrossRef]

2

Numerical Analysis on Hydraulic Characteristics of U-Shaped Channel of Various Trapezoidal Cross-Sections

Ruichang Hu and Jianmin Zhang *

State Key Laboratory of Hydraulics and Mountain River Engineering, Sichuan University, Chengdu 610065, China; hrc@stu.scu.edu.cn

* Correspondence: zhangjianmin@scu.edu.cn

Abstract: Curved channel with trapezoidal cross-section is approximate to the common form in nature fluvial networks and its hydraulic characteristics are considerably complex and variable. Combined with volume of fluid (VOF) method, renormalization group (RNG) k-ε turbulence model was employed to numerically investigate the flow properties in the U-shaped channel with various trapezoidal cross-sections. Analyses were performed from the aspects of the water surface transverse slope in bend apex (WTS-BA), longitudinal velocity, secondary flow, shear stress and turbulent kinetic energy (TKE) under several scenarios, specifically, four types of radius-to-width ratio and seven types of slope coefficient with a constant aspect ratio. The calculated results suggested that the maximums of shear stress and TKE in the bend were observed in the convex bank and the maximal intensities of secondary flow were observed within the range of 60 to 75 degrees for various varieties. As the radius-to-width ratio increased, the maximums of shear stress, TKE and WTS-BA decreased; but increased with increasing slope coefficients. The intensity of secondary flow decreased as slope coefficients increased and the angle of maximum intensity of secondary flow moved to the upstream for the increasing radius-to-width ratios. In addition, a new equation concerning the vertical distribution of longitudinal velocity in trapezoidal cross-sectioned channel was presented.

Keywords: trapezoidal cross-section; U-shaped channel; radius-to-width ratio; slope coefficient; hydraulic characteristics

1. Introduction

The water surface transverse slope (WTS), violent turbulence, forceful secondary flow and shear stress aggravate the erosion and siltation of the channel [1]. Consequently, detailed study regarding the hydraulic characteristics of the channel is an essential aspect for solving problems related to river restoration and navigability.

The secondary flow was observed by the movement of impurities in straight and curved open channel [2] and then the investigations referring to the secondary flow and shear stress in the channel were extensively conducted [3–6]. The structure of secondary flow was changed for the varying shape of cross-section in straight channel, which was investigated using three slope coefficients of 0.58, 1 and 1.73 by Tominaga et al. [7]. And the spanwise distribution of boundary shear stress was extraordinarily affected by the secondary flow and the shape of the cross-section [8] in the straight trapezoidal channel. The value of shear stress increased in the bottom of the wall and a dramatical variation from the apex to the middle of the sidewall was observed with increasing slope coefficients [8,9] in straight trapezoidal channel.

The curved trapezoidal channel is approximate to the common form in nature. The phenomenon of multi-vortex was also observed in curved trapezoidal channel [2], however, the influence of slope coefficient had not been explained completely. Farhadi et al. [10] investigated the transformation law of multi-vortex in the cross-section for the varying Froude number with the slope coefficient of 2.375 and Termini and Piranino [11] suggested that counter rotating circulation was conspicuous only in the lower aspect ratio, while the effect of various slope coefficients and radius-to-width ratio on secondary flow had not been studied. Shukla et al. [12] predicted the free surface flow velocities and shear stress reasonably well, using the Reynolds-averaged Navier-Stokes and continuity equations and the calibration of empirical coefficients and constants in standard k-ε turbulence model was recommended to improve accuracy. The modified k-ε model adopted in a 180 degree curved trapezoidal channel by Mosalman et al. [13] demonstrated good agreement with the experimental data. Also, k-ε turbulence model has been used in other different numerical approaches, such as the grid-less models [14,15]. The investigations on the water-air flow properties were conducted by Peng et al. [16–20]. The secondary flow and vortex structures can also be investigated via the direct numerical simulations (DNS) [21–23].

The slope coefficient and radius-to-width ratio are the essential parameter for the investigation on hydraulic characteristics of the channel [8]. The research on small slope coefficient, which is common in city, is not perfect and the objective of the present work is to bridge the gap in previous literature. Simultaneously, the effect of flow properties on the radius-to-width ratios was performed. The present study verified the feasibility of RNG k-ε turbulence model by the existing experiment [24]. Next, the analysis of the effect of varying radius-to-width ratios and slope coefficients on the WTS-BA, longitudinal velocity, secondary flow, shear stress and TKE was presented.

2. Numerical Model

A large number of U-shaped channel of trapezoidal cross-sections were modeled using Gambit software [25]. Fluent [26] was used to perform the numerical simulation and the renormalization group (RNG) k-ε turbulence model originated by Yakhot and Orszag [27] was employed. The U-shaped channel with symmetrical arrangement of trapezoidal cross-section is consisted of a half circle bend of 180° with different centerline radiuses and two straight cross-sections up and down the bend of 4m in length.

The water depth of the inlet and outlet of the numerical model were 0.203 m and 0.19 m, respectively. 3D view of the numerical model and vertical view were shown in Figure 1a,b, respectively. The slope coefficient of the sidewall in the curved channel remains uniform and the shape of the cross-section or centerline radius were varied for the study. The cross-section with different slope coefficients were presented in Figure 1c. Simultaneously, the area of the cross-section which was 0.08 m^2 and the flow rate, 0.038 m^3/s, remained constant. And constant equivalent width (B_0) is 0.4 m, which was defined in Equation (1).

$$B_0 = Q/Hv \tag{1}$$

where H is water depth, v is the velocity and Q is the flow rate. In this study, H and v of the inlet of the U-shaped channel were adopted to calculate the equivalent width.

The slope coefficient ($\boldsymbol{m}$) is the ratio of the projection of the wall on the horizontal plane to the vertical plane, as shown in Figure 1d. In order to distinguish the slope coefficient ($\boldsymbol{m}$) from the unit meter (m), the slope coefficient ($\boldsymbol{m}$) was shown in bold. The radius-to-width ratio (λ) can be defined as follows:

$$\lambda = R/B_0 \tag{2}$$

where R is centerline radius of the bed in bend, which is defined in Equation (3); B_0 is equivalent width.

$$R = 0.5(R_1 + R_2) \tag{3}$$

where R_1 and R_2 are the inner radius and outer radius of the bed in bend, respectively.

The relationships between bed width, width of the surface and equivalent width are as follows:

$$B_0 = B + mH \tag{4}$$

$$B_{up} = B + 2mH \tag{5}$$

where B is the bed width, B_{up} is the width of the surface and m is slope coefficient.

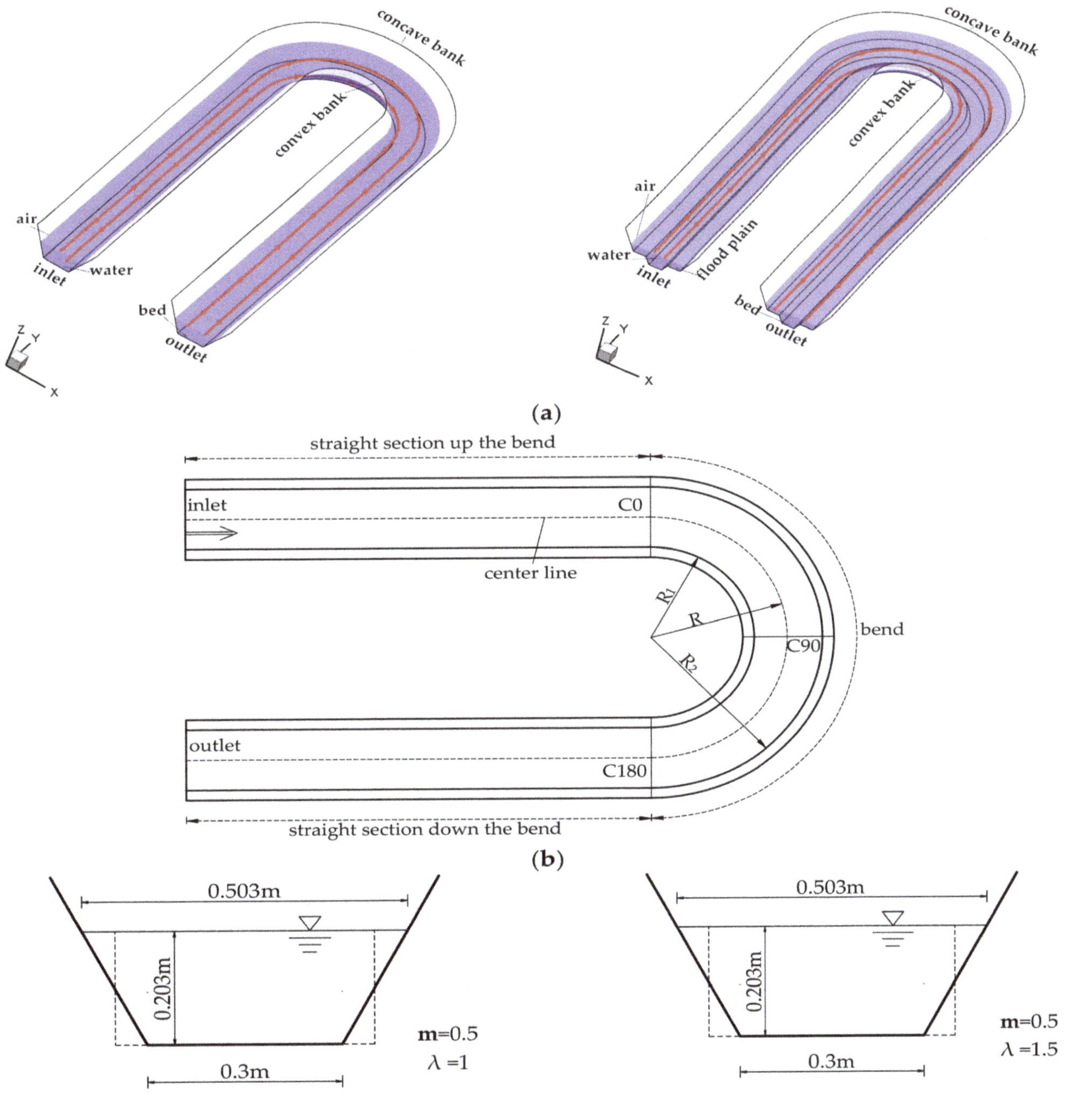

Figure 1. *Cont.*

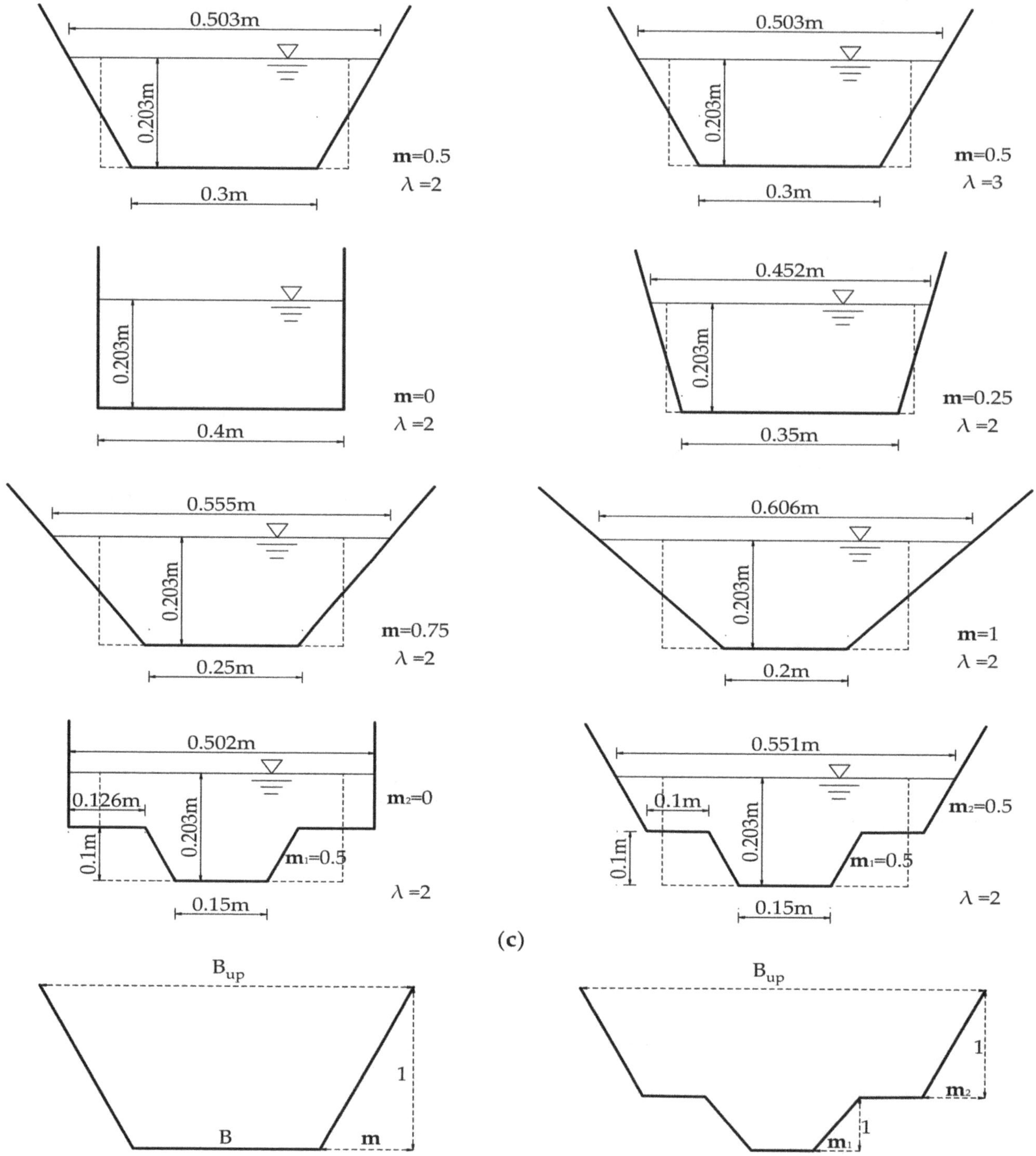

Figure 1. U-shaped channel and investigated cross-sections. (**a**) 3D view of the U-shaped channel; (**b**) vertical view of the channel; (**c**) geometry of the cross-section; (**d**) Definition of the slope coefficient.

The aspect ratio, the ratio of the equivalent width (B_0) to the inlet water depth (H) of U-shaped channel, remained 1.97 in this study. The numerical simulations were performed for four radius-to-width ratios (λ) of 1.0, 1.5, 2.0 and 3.0 and seven slope coefficients ($\boldsymbol{m}$) of 0, 0.25, 0.5, 0.75, 1.0, 1.10 and 1.24. The test runs program were summarized in Table 1. The m_1 and m_2 of case9 were 0.5 and 0, respectively and the m_1 and m_2 of case10 were 0.5 and 0.5, respectively. For convenience, the slope coefficients ($\boldsymbol{m}$) of case9 and case10 were calculated by the definition (shown in Figure 1d) and the values were 1.10 and 1.24, respectively.

Table 1. Summary of the test runs for the simulation, where m is slope coefficient; λ is radius-to-width ratio; θ is the central angle of the bend; B and B_0 are the width of the bed and equivalent width; R, R_1 and R_2 are centerline radius, inner radius and outer radius of the bend, respectively; Q is the flow rate; Fr is Froude number; Re is Reynolds number; B_0/H is the aspect ratio; and cells are the volume elements.

Series	m	λ	θ (°)	B (m)	R (m)	R_1 (m)	R_2 (m)	Q (L s^{-1})	B_0 (m)	Fr	Re	B_0/H	Cells
case1	0.50	1.0	180	0.30	0.40	0.25	0.55	38	0.40	0.33	4.98×10^4	1.73	180,000
case2	0.50	1.5	180	0.30	0.60	0.45	0.75	38	0.40	0.33	4.98×10^4	1.73	180,000
case3	0.50	2.0	180	0.30	0.80	0.65	0.95	38	0.40	0.33	4.98×10^4	1.73	180,000
case4	0.50	3.0	180	0.30	1.20	1.05	1.35	38	0.40	0.33	4.98×10^4	1.73	180,000
case5	0.00	2.0	180	0.30	0.80	0.60	1.00	38	0.40	0.33	4.66×10^4	1.73	180,000
case6	0.25	2.0	180	0.35	0.80	0.625	0.975	38	0.40	0.33	4.88×10^4	1.73	180,000
case7	0.75	2.0	180	0.25	0.80	0.675	0.925	38	0.40	0.33	4.95×10^4	1.73	180,000
case8	1.00	2.0	180	0.20	0.80	0.70	0.90	38	0.40	0.33	4.85×10^4	1.73	180,000
case9	1.10	2.0	180	0.15	0.80	0.725	0.875	38	0.40	0.33	4.03×10^4	1.73	262,500
case10	1.24	2.0	180	0.15	0.80	0.725	0.875	38	0.40	0.33	4.15×10^4	1.73	262,500

2.1. Numerical Methods

The volume of fluid (VOF) method, treating complicated free boundary configurations more flexible and efficient via volume fraction in stationary Euler mesh, was applied to track the air-water interface in the present work [28]. The differential equation in each control volume of the discrete region was integrated by the VOF method firstly and next the integral equation was linearized further to obtain the algebraic equation set of the corresponding parameters. Finally, all unknown quantities were solved. The solver employed pressure-based calculation, absolute velocity formulation and transient time and the SIMPLE algorithm was adopted for the pressure-velocity coupling. The computational domain was discretized using structured grid (shown in Figure 2) and the boundary layer cells were meshed in the vicinity of the wall and bed to ensure that they fell within the y^+ values required in the turbulence model. The number of the single trapezoidal U-shaped channel were $20 \times 30 \times 300$, whereas for the compound channel, were $25 \times 35 \times 300$. The number of volume cells for ten tests investigated in the paper were listed in Table 1.

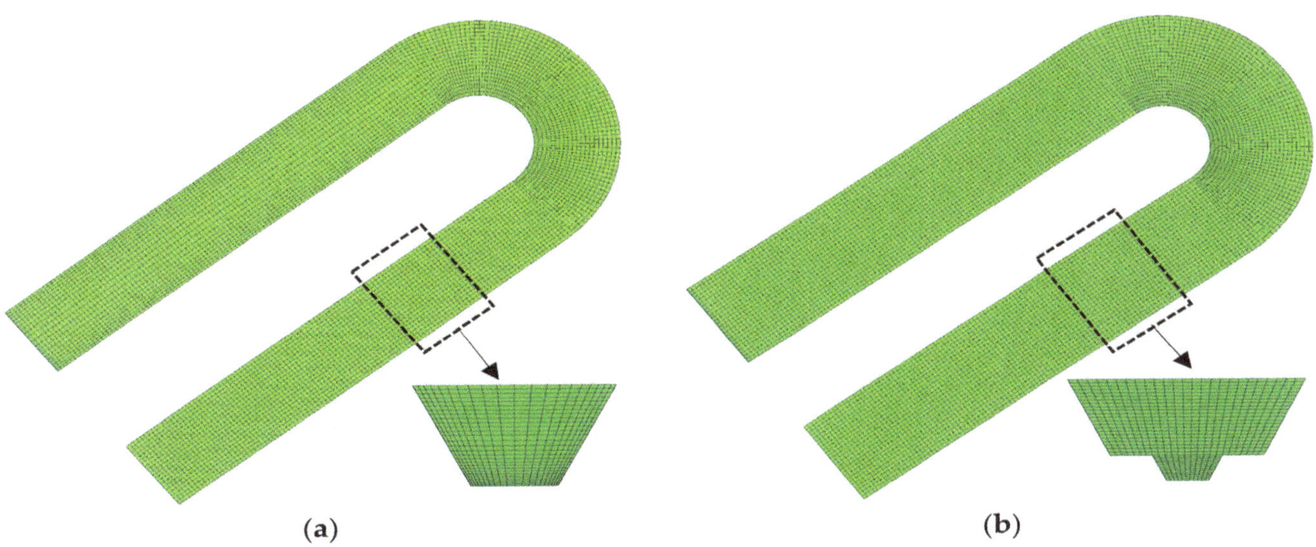

Figure 2. Meshing pattern of the U-shaped channel. (**a**) case3; (**b**) case10.

2.2. Governing Equations

The continuity equation, momentum equation, TKE budget [23] and the turbulent dissipation rate (ε) [29], adopted in the model, are shown below:

Continuity equation:

$$\frac{\partial \rho}{\partial t} + \frac{\partial \rho \widetilde{u_i}}{\partial x_i} = 0 \tag{6}$$

Momentum equation:

$$\frac{\partial(\rho \widetilde{u_i})}{\partial t} + \frac{\partial(\rho \widetilde{u_i}\widetilde{u_j})}{\partial x_j} = -\frac{\partial p}{\partial x_i} + \frac{\partial}{\partial x_j}\left[\mu\left(\frac{\partial \widetilde{u_i}}{\partial x_j} + \frac{\partial \widetilde{u_j}}{\partial x_i}\right)\right] + \frac{\partial}{\partial x_j}\left(-\rho\overline{u_i' u_j'}\right) \tag{7}$$

TKE budget:

$$\frac{\partial \overline{\rho} K}{\partial t} = \underset{\text{convection term}}{-C} + \underset{\text{production term}}{P} + \underset{\text{turbulent transport term}}{T} + \underset{\text{pressure dilatation term}}{\Pi} + \underset{\text{compressible mass flux contribution term}}{M} + \underset{\text{viscous diffusion term}}{D} - \underset{\text{TKE dissipation rate term}}{\overline{\varepsilon}} \tag{8}$$

$$C = \frac{\partial \overline{\rho}\widetilde{u_j}\overline{K}}{\partial x_j} \tag{9}$$

$$p = -\overline{\rho u_i' u_j'}\frac{\partial \widetilde{u_i}}{\partial x_j} \tag{10}$$

$$T = -\frac{\partial}{\partial x_j}\left[\overline{\rho}\frac{\overline{u_i' u_i' u_j'}}{2}\right] + \overline{p' u_i'}\sigma_{ij} \tag{11}$$

$$\Pi = -\overline{p'\frac{\partial u_i'}{\partial x_j}} \tag{12}$$

$$\mathrm{M} = -\overline{\rho u_i'}\left(\frac{\partial \overline{\sigma_{ij}}}{\partial x_j} - \frac{\partial \overline{p}}{\partial x_i}\right) \tag{13}$$

$$\mathrm{D} = \frac{\partial}{\partial x_j}\left(\overline{\sigma_{ij}' u_i'}\right) \tag{14}$$

k equation:

$$\frac{\partial\left(\rho\overline{k}\right)}{\partial t} + \frac{\partial\left(\rho\overline{k}\overline{u_i}\right)}{\partial x_i} = \frac{\partial}{\partial x_j}\left[\left(\mu + \frac{\mu_t}{\sigma_k}\right)\frac{\partial \overline{k}}{\partial x_j}\right] + G_k - \rho\varepsilon \tag{15}$$

ε equation:

$$\frac{\partial(\rho\overline{\varepsilon})}{\partial t} + \frac{\partial(\rho\overline{\varepsilon}\overline{u_i})}{\partial x_i} = \frac{\partial}{\partial x_j}\left[\left(\mu + \frac{\mu_t}{\sigma_\varepsilon}\right)\frac{\partial \varepsilon}{\partial x_j}\right] + \frac{C_{1\varepsilon}^*\overline{\varepsilon}}{\overline{K}}G_k - C_{2\varepsilon}\rho\frac{\overline{\varepsilon}^2}{\overline{K}} \tag{16}$$

$$\mu_t = \rho C_\mu \frac{\overline{K}^2}{\overline{\varepsilon}} \tag{17}$$

$$G_k = \mu_t\left(\frac{\partial \overline{u_i}}{\partial x_j} + \frac{\partial \overline{u_j}}{\partial x_i}\right)\frac{\partial \overline{u_i}}{\partial x_j} \tag{18}$$

$$C_{1\varepsilon}^* = C_{1\varepsilon} - \frac{\eta(1 - \eta/\eta_0)}{1 + \beta\eta^3} \tag{19}$$

$$\eta = \sqrt{\frac{G_k}{\rho C_{\mu\varepsilon}}} \tag{20}$$

where K is the TKE, μ is the dynamic viscosity, μ_t is the turbulence viscosity, ρ is the corresponding density, u_i is the velocity component in the ith direction, t is the time and C_μ, C_μ, $C_{1\varepsilon}$, $C_{2\varepsilon}$, β and η_0 are empirical constants. In the RNG k-ε model, the empirical constants are given as $C_\mu = 0.0845$, $C_\mu = C_\varepsilon = 0.7179$, $C_{1\varepsilon} = 1.42$, $C_{2\varepsilon} = 1.68$, $\beta = 0.012$, and $\eta_0 = 4.38$ [29].

The compressible mass flux contribution term is zero for the incompressible fluid. Reynolds stress independent components can be solved via using Equations (9)–(14) above and $\varepsilon = \sigma'_{ik}\frac{\partial u'_i}{\partial x_k}$, however, supererogatory components, that is, the trible correlation matrix for fluctuating velocity $(\overline{u'_i u'_i u'_j})$ and correlation matrix for fluctuating velocity and pressure $(\overline{p' u'_i})$, are added in the turbulence kinetic energy budget, which made the budget is not closed. So K and ε equations are recommended to close the budget.

2.3. Boundary Conditions and Assumptions

In this study, without considering the dissolution of each other, the open channel flow was employed by VOF sub-models with air-water two phases. Pressure-inlet was adopted for the water surface and the pressure of the air-water interface was standard atmospheric pressure. Both sidewalls and bed of the channel, whose roughness height were 0.25 mm, were stationary wall and the fluent standard wall function approach was adopted. The velocity-inlet and pressure-outlet were adopted for the inlet and outlet of the U-shaped channel, respectively and the mean velocity of the inlet cross-section was 0.462 m/s.

2.4. Model Verification

It is necessary to validate the numerical model before conducting further research. Therefore, the mesh convergence study were performed via using grid convergence index (GCI) [30] with grid numbers of approximately 0.12 million, 0.20 million and 0.30 million. The effect of the grid sizes on the uncertainty of the computational velocity distribution in vertical direction was shown in Figure 3, which showed that the discretization uncertainties were little in most positions. The maximum of the GCI for the velocity profiles in concave bank, center line and convex bank were 6.3%, 2.4% and 8.2% and corresponding to 0.05 m/s, 0.06 m/s and 0.07 m/s, respectively. The grid number was set as 0.20 million to optimize the computational time.

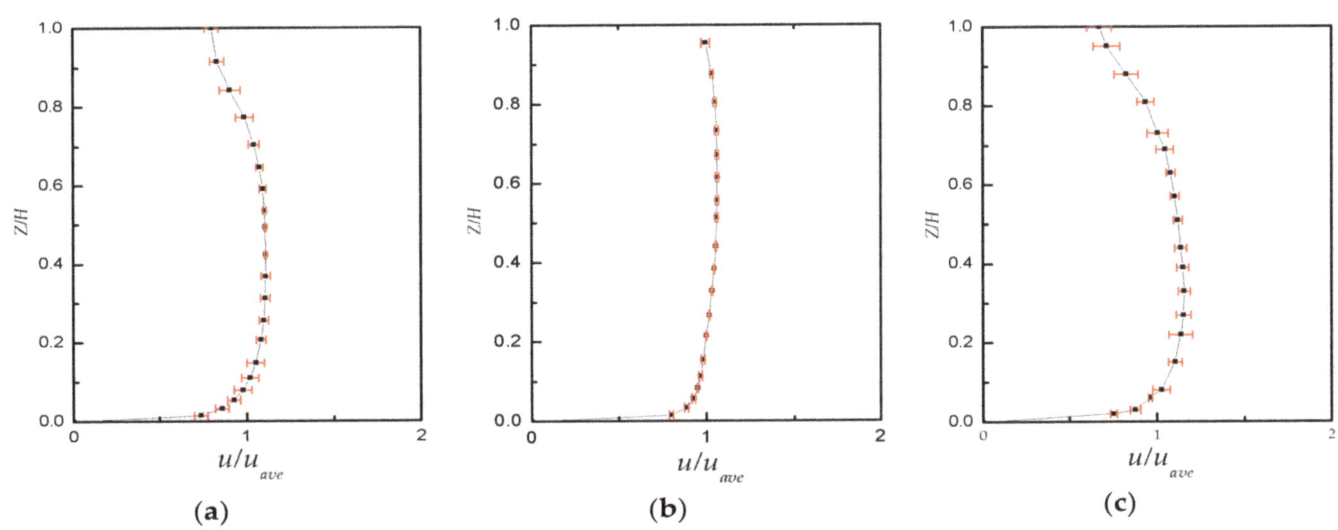

Figure 3. Discretization error bars computed using the GCI index for different grid densities, (**a**) the concave bank, (**b**)center line and (**c**)convex bank of the bend apex.

In addition, the results calculated by RNG k-ε turbulence model were compared with the experimental data obtained from Ma [24]. The geometry of the experimental curved channel was exactly the same as case5 in the present article, except the lengths of two straight cross-sections of the experiment were 6 m. The sidewalls and bed composed of cement had a low roughness and the velocity was obtained by ADV in experiment.

In the present work, the length of two straight cross-sections all were 4 m or 6 m in the U-shaped channel, which were compared with the experiment, respectively. The RNG k-ε model was adopted for simulation with the same experimental conditions and the data of simulation and experiment were dimensionless and compared. The longitudinal velocity in different position of cross-section of C0, C90 and C180, the structure of secondary flow of the cross-section in the bend apex and shear stress in the bend were analyzed to verify the numerical model. In Figure 4, the vertical coordinate is the relative depth, where Z is the instantaneous height and H is the water depth in corresponding position; the horizontal coordinate is the ratio of the longitudinal velocity (u) to the corresponding average velocity (u_{ave}).

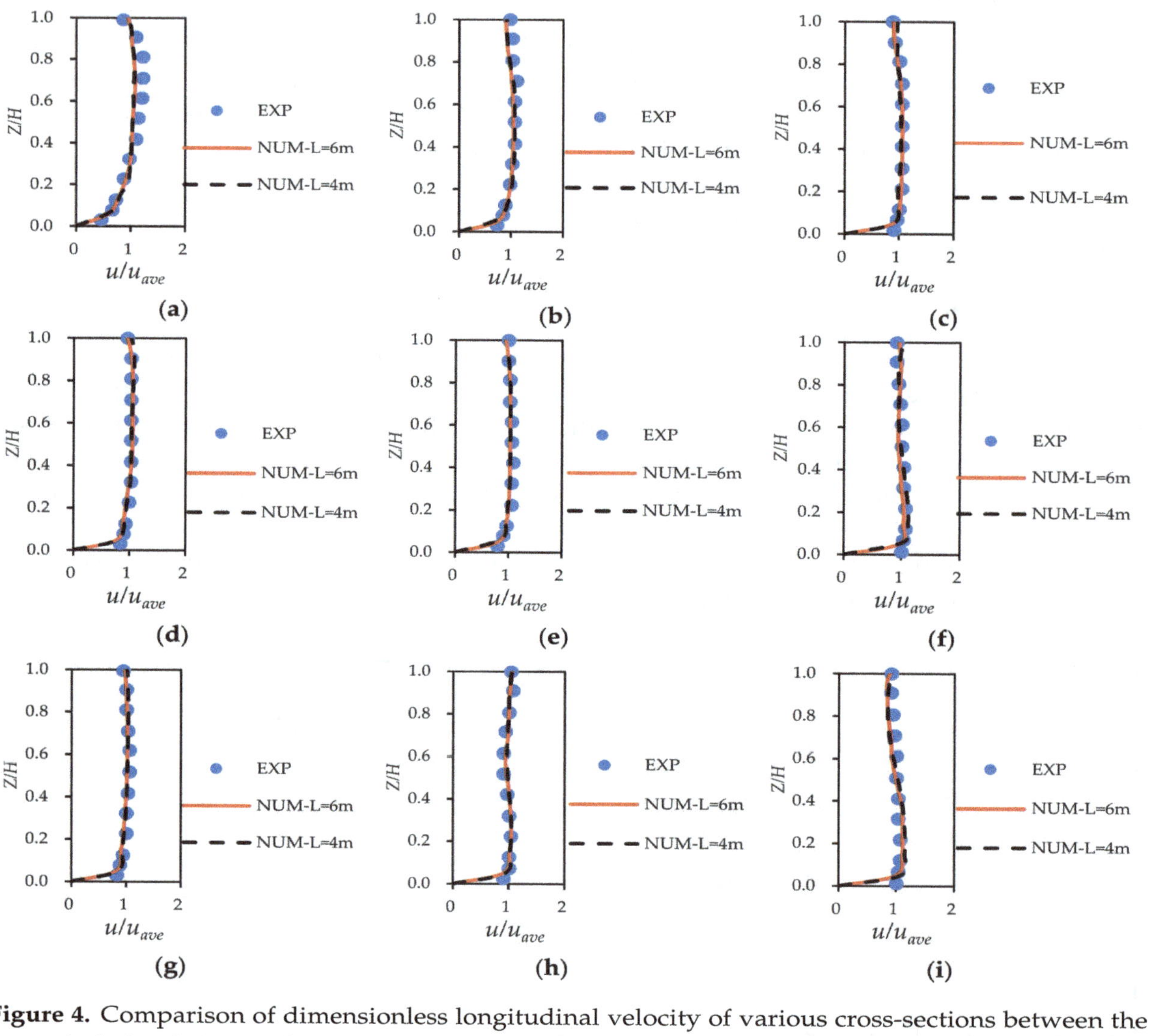

Figure 4. Comparison of dimensionless longitudinal velocity of various cross-sections between the numerical simulation and experiment, (**a–c**) the concave bank, center line and convex bank of the inlet of the bend, respectively; (**d–f**) the concave bank, center line and convex bank of the bend apex, respectively; and (**g–i**) the concave bank, center line and convex bank of the outlet of the bend, respectively.

It can be seen from Figure 4 that, in most locations, the longitudinal velocity distribution of the simulation exhibited a good agreement with the experimental in the vertical direction. The maximums of absolute error (the absolute value of the difference between experiment and numerical simulation) and relative error (the ratio of the absolute error to experiment) in different location were presented in Table 2. A well consistency was shown for the structure of secondary flow in Figure 5. Although the simulated shear stress in the bed was slightly larger than the experimental data, shown in Figure 6b, the commutated results were fairly consistent with Ma's investigation [24].

Table 2. Summary of the maximums of relative error and absolute error in different location. The locations of (a–i) were introduced in Figure 4.

		a	b	c	d	e	f	g	h	i
L = 6 m	**Absolute error** (m/s)	0.03	0.05	0.06	0.04	0.06	0.05	0.01	0.06	0.07
	Relative error (%)	12	10.8	13.9	15.4	12	12.5	4.6	13.3	14.6
L = 4 m	**Absolute error** (m/s)	0.03	0.04	0.05	0.04	0.07	0.06	0.02	0.06	0.06
	Relative error (%)	12	8.7	11.6	15.4	14	15	7.6	13.3	12.2

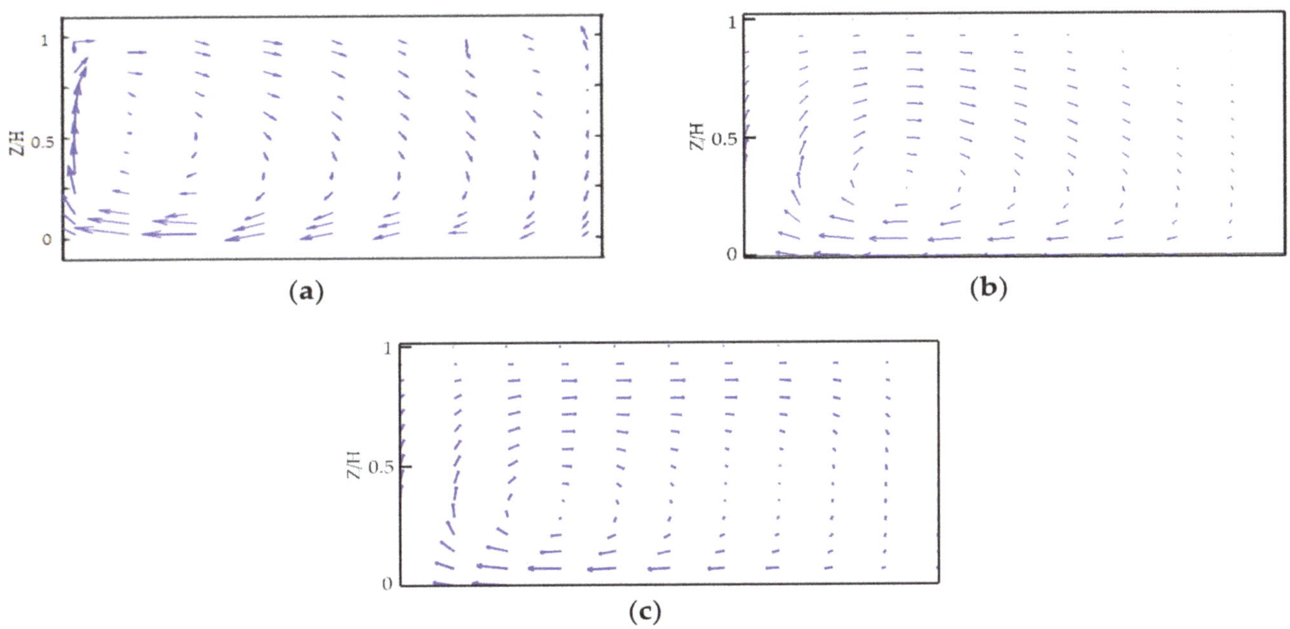

Figure 5. Comparison of the secondary flow between the numerical simulation and experiment, (**a**) experiment [24], (**b**) numerical simulation L = 6 m, (**c**) numerical simulation L = 4 m.

Consequently, the RNG k-ε model could be used for the numerical investigation of the U-shaped channel and ensure the accuracy of the data. To optimize the computational time, the length of 4m were selected for the straight sections. Analyses were performed from the aspects of the WTS-BA, longitudinal velocity, secondary flow, shear stress and TKE in next section.

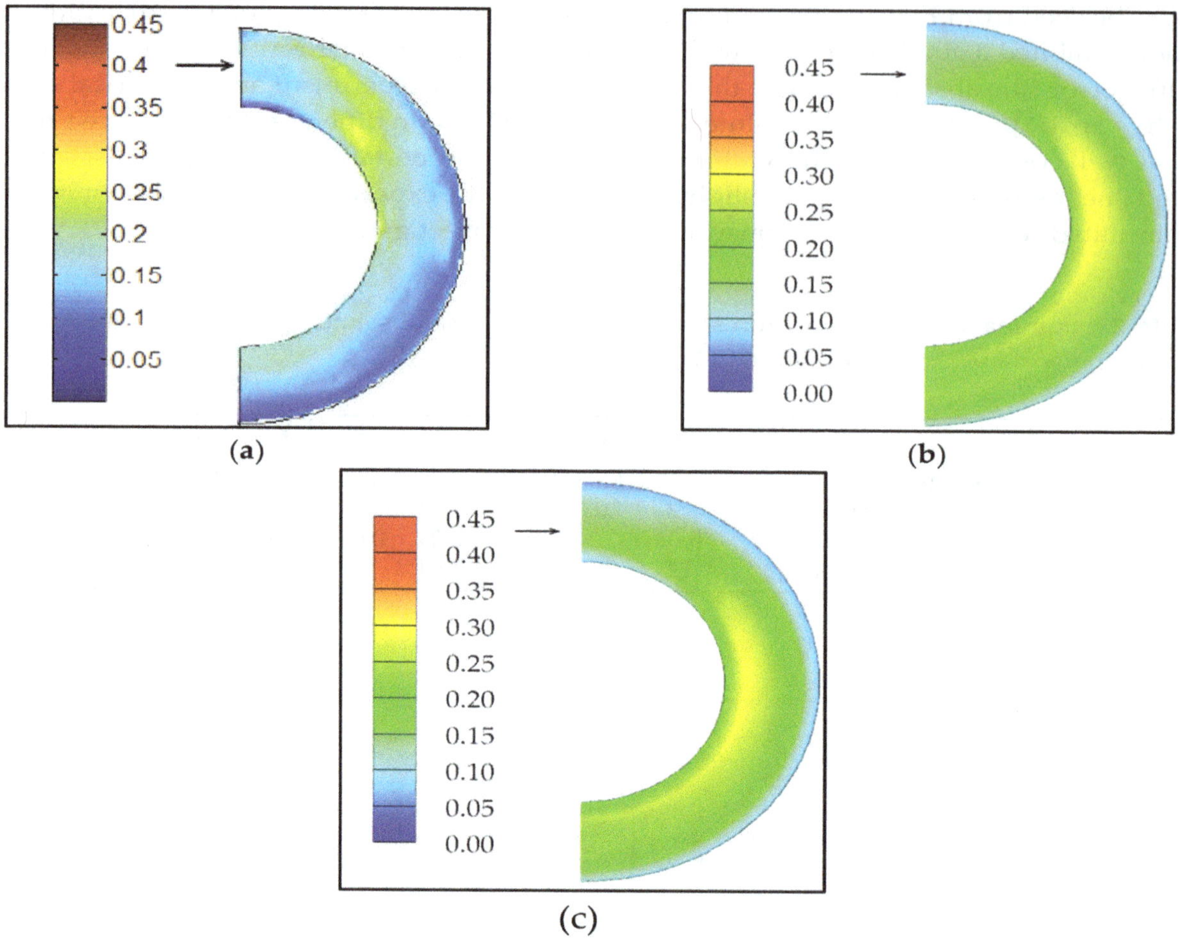

Figure 6. Comparison of the shear stress between the numerical simulation and experiment, (**a**) experiment [24], (**b**) numerical simulation L = 6 m, (**c**) numerical simulation L = 4 m.

3. Results and Discussion

3.1. Analysis of the Water Surface Transverse Slope in Bend Apex

The centrifugal motion of the water in bend is a result of centrifugal force, gravity and the friction with surrounding water. The transverse slope is presented on the water free surface, due to the descending water level in convex bank and rising water level in the concave bank [23]. The WTS varies with radius-to-width ratio, slope coefficient and cross-section shape [31]. The equation for calculating the WTS of cross-section of the bend apex (C90) is [31]:

$$J_r = \left(1 + \frac{g}{\kappa^2 C^2}\right)\frac{u^2}{gr} \tag{21}$$

where C is the Chezy coefficient, κ is the Carmen constant, g is the acceleration of gravity, r is the radius and u is the corresponding velocity. For this investigation, the empirical values $\frac{1}{\kappa^2}$ calculated by Zhang [32], Liu [31] and Rozovskii [32] were 5.75, 6.25 and 4, respectively. The cross-section in the bend apex (C90) was selected for further investigations, r and u were the corresponding centerline radius and the average longitudinal velocity of the cross-section of the bend apex, respectively.

The free surface level across the across-section in the bend apex with varying radius-to-width ratios and slope coefficients were shown in Figure 7a,b, respectively. The vertical coordinate is the water level and the horizontal coordinate is the relative distance from a point on the cross-section to the convex bank on the surface. The WTS-BA with varying radius-to-width ratios and slope coefficients were shown in Figure 8a,b, respectively. The vertical coordinate is a dimensionless WTS and the horizontal ordinate is radius-to-width ratio or slope coefficient.

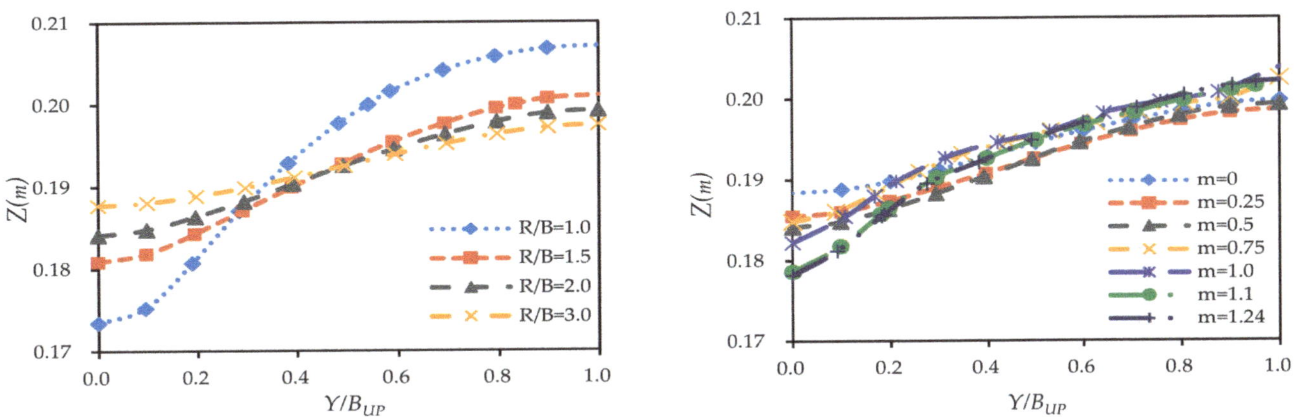

(a) (b)

Figure 7. The free surface level across the across-section in the bend apex (C90), (**a**) the radius-to-width ratios (λ) of 1.0, 1.5, 2.0 and 3.0, (**b**) the slope coefficients (*m*) of 0, 0.25, 0.5, 0.75, 1.0, 1.10 and 1.24.

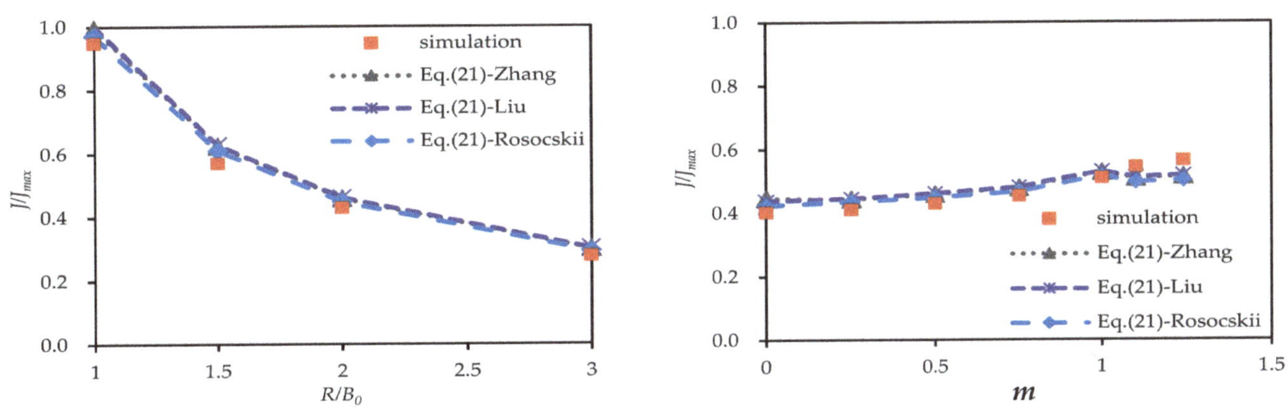

(a) (b)

Figure 8. The WTS-BA (C90), (a) the radius-to-width ratios (λ) of 1.0, 1.5, 2.0 and 3.0; (**b**) the slope coefficients (*m*) of 0, 0.25, 0.5, 0.75, 1.0, 1.10 and 1.24.

The WTS decreased with increasing radius-to-width ratios, which can be explained by the decrease in the water level in the concave bank and the increased water level in the convex bank, shown in Figures 7a and 8a. It can be considered that the numerical simulation exhibited a good agreement with the theoretical Equation (21) for the maximum error is 8.2%, in spite of the fact that the data obtained via simulation was smaller than calculated by Equation (21) with the varying Carmen constants.

The WTS increased with the increasing slope coefficients, for the water level in the concave bank slightly decreased first and then increased, while decreased in a relative larger extent in the concave bank, shown in Figures 7b and 8b. As can be seen from Figure 8b, the data obtained via numerical simulation was smaller than calculated by Equation (21) except the compound cross-section, the maximum errors of which were 7.2% and 10.1%, respectively.

The reason was that Equation (21) was derived on the basis of a single trapezoidal cross-section, which could not be applicable to the compound cross-section. In the whole, the numerical simulation was consistent with the Equation (21) for the varying slope coefficients. As can be seen from Figures 7b and 8a, the effect of the radius-to-width ratio was greater than the slope coefficient on the WTS in the bend apex.

3.2. Analysis of the Longitudinal Velocity

Chang [33] believes that the vertical distribution of the longitudinal velocity in the bend conforms to the Equation (22):

$$\frac{u}{u_{ave}} = \left(1 + \frac{1}{n}\right)\left(\frac{Z}{H}\right)^{\frac{1}{n}} \tag{22}$$

where, n is obtained by equation $n = \kappa\sqrt{8/f}$, κ is the Carmen constant, f is the friction constant; Z is the instantaneous height and H is the corresponding water depth and u and u_{ave} are the longitudinal velocity and the corresponding average in the cross-section. The vertical distributions of the longitudinal velocity calculated by Equation (22) were shown in Figures 9 and 10.

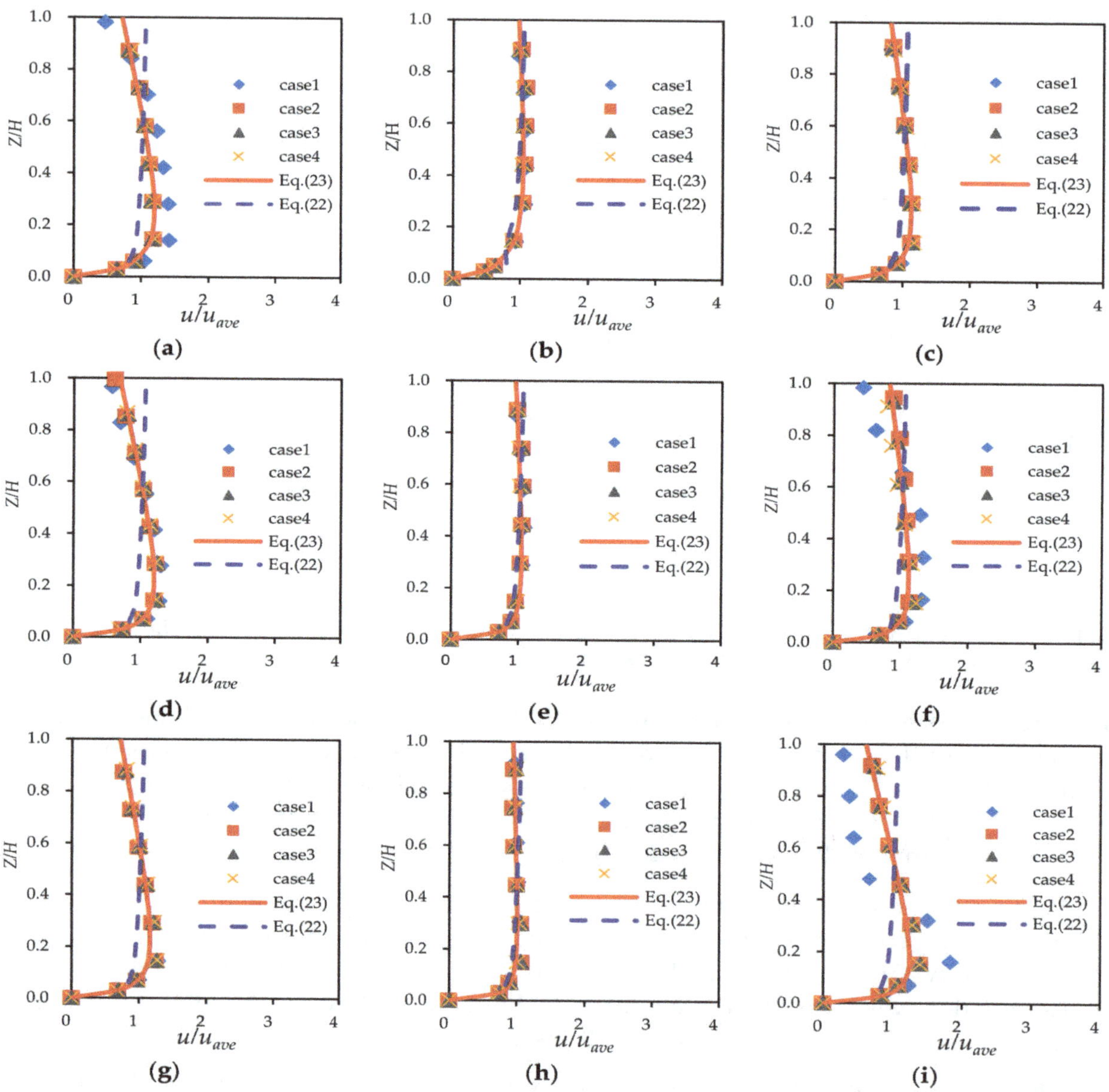

Figure 9. Vertical distribution of the dimensionless longitudinal velocity of the bend in different positions with varying radius-to-width ratios, **(a–c)** the concave bank, centerline and convex bank of the inlet of the bend, respectively; **(d–f)** the concave bank, centerline and convex bank of the bend apex, respectively; **(g–i)** the concave bank, centerline and convex bank of the outlet of the bend, respectively.

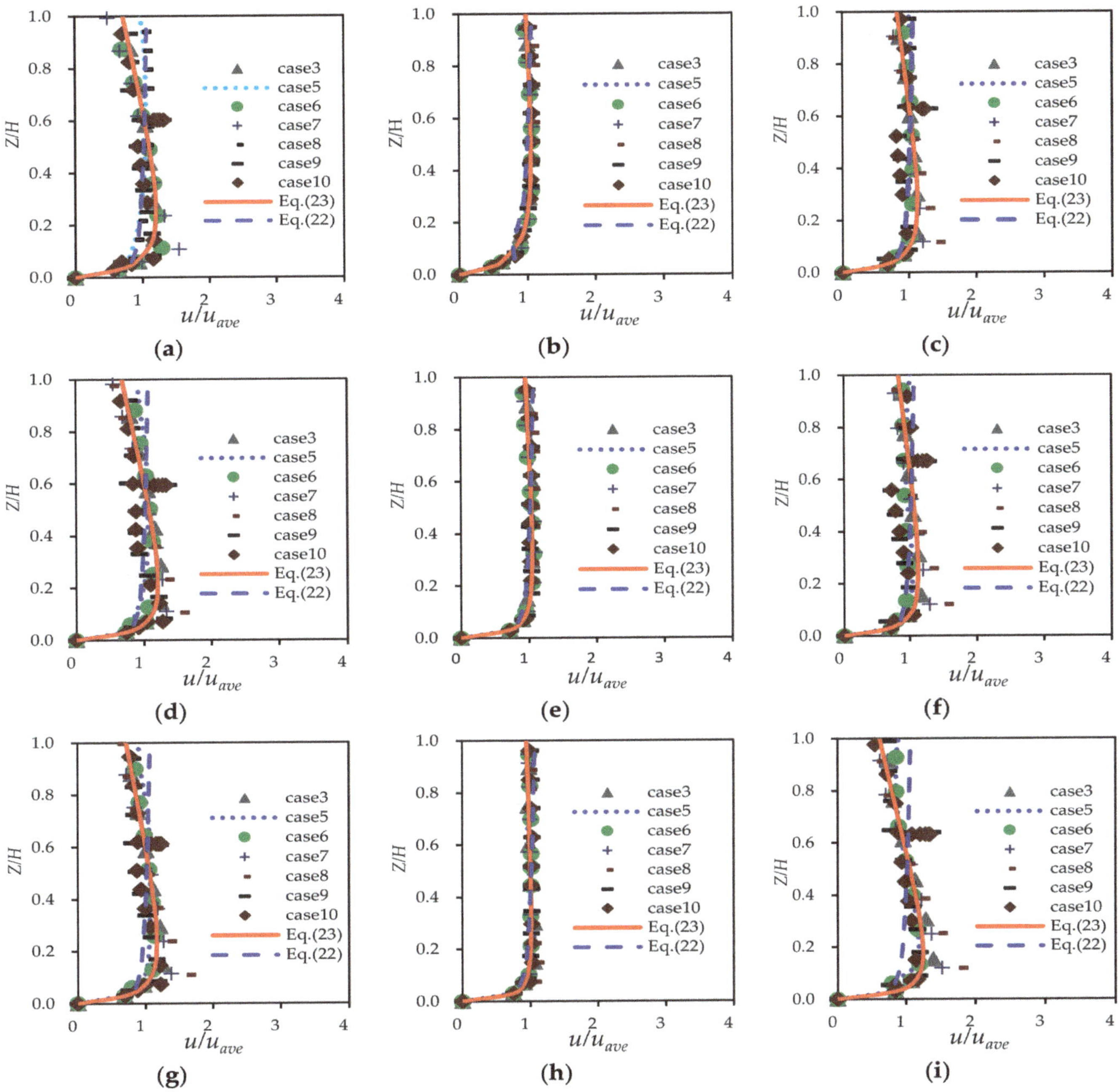

Figure 10. Vertical distribution of the dimensionless longitudinal velocity of the curved channel in different positions with varying slope coefficients, (**a**–**c**) the concave bank, center line and convex bank of the inlet of the bend, respectively; (**d**–**f**) the concave bank, center line and convex bank of the bend apex, respectively; (**g**–**i**) the concave bank, center line and convex bank of the outlet of the bend, respectively.

Figures 9 and 10 showed the vertical distribution of the dimensionless longitudinal velocity of the bend in different positions with the varying radius-to-width ratios and slope coefficients, respectively. In Figures 9 and 10, the vertical coordinate is the relative depth, where Z is instantaneous height, H is the corresponding water depth and the horizontal coordinate is the ratio of the longitudinal velocity (u) to the corresponding average velocity (u_{ave}). The velocity locating close to the sidewall at $5\%B_0$ was extracted to represent the velocity at the sidewall, as the velocity at the sidewall is zero.

As shown in Figure 10, the vertical distribution of longitudinal velocity calculated by Equation (22) was relatively consistent with the distribution of case5 (whose cross-section was rectangular).

The longitudinal velocity in the bend was affected by the shape of cross-section, radius-to-width ratio and other factors [34–37]. The differences of the velocity distribution in the both banks were bigger than the center line, which was shown in Figures 9 and 10. The longitudinal velocity extracted via the numerical simulation conformed to the Equation (13) in the center line, yet the obvious errors were exhibited in the both banks. In this study, Equation (23) was proposed, which was consistent with the vertical distribution of dimensionless longitudinal velocity for the varying operating conditions in any position:

$$\frac{u}{u_{ave}} = A\left(\frac{Z}{H}\right) + \frac{B}{\left(\frac{Z}{H}\right)} + C \tag{23}$$

where, A is equal to -0.09 or -0.03 for the sidewall or the center line, respectively, B is equal to -0.05 and the range of C is from 1 to 2.

Corresponding optimal fitting curves (the red continuous line) calculated by Equation (16) in different positions were shown in Figures 9 and 10. It can be seen that the vertical distribution of the longitudinal velocity in the bend with varying operating conditions were consistent with the optimal fitting curves in the center line but deviations are presented in the both banks. The deviations for the $\lambda = 1.0$ were conspicuous for the obvious gradient of the velocity in both banks and the vertical distribution for the $\lambda > 1.0$ were consistent with the optimal fitting curve. The deviations were lesser than others when the slope coefficient was 0.5. As shown in Figures 9 and 10, the influence of the slope coefficient on the distribution of longitudinal velocity was more conspicuous than the radius-to-width ratio.

The vertical distribution of the longitudinal velocity in both banks at the inlet of the bend were different for the varying operating conditions and demarcation were each $Z/H = 0.5$. As shown in Figure 9, the longitudinal velocity in the concave bank gradually decreased or increased for the $Z/H < 0.5$ or $Z/H > 0.5$ with increasing radius-to-width ratios but did not show the obvious trend for the varying radius-to-width ratios in the convex bank. As shown in Figure 10, the longitudinal velocity gradually increased or decreased for the $Z/H < 0.5$ or $Z/H > 0.5$ with increasing slope coefficients in the both banks. The difference of the velocity distribution between two types of compound cross-sections was slight.

The vertical distributions in the concave bank of the bend apex failed to show regular law for the varying operating conditions and were similar to the distributions in the convex bank of the inlet of the bend in the convex bank, whose demarcations were each $Z/H = 0.6$. The distributions of the longitudinal velocity of the outlet in the bend were similar to the distribution of the cross-section in the bend apex and the demarcations were $Z/H = 0.4$ and $Z/H = 0.5$.

3.3. Analysis of the Secondary Flow

The secondary flow is occurred for the momentum exchange [7]. The force of weeny water column are as follow [32]:

$$\Delta F = \underset{\substack{gravity\\force}}{g} + \underset{\substack{pressure\\force}}{\rho g h J_r} - \underset{\substack{centrifugal\\force}}{\alpha_0 \frac{u_r^2}{gr}} + \underset{resistance}{F_0} \tag{24}$$

where, g is the acceleration of gravity, r is the radius, h and u is the corresponding height and velocity, J_r is the transverse slope, α_0 is coefficient and F_0 is the summation of other resistance; in this investigation, gravity force is neglected in horizontal direction.

The structure of secondary flow of the cross-section in the bend apex for varying operating conditions was shown in Figure 11. In the present study, the clockwise vortex was defined as positive and counterclockwise vortex is reverse. As shown in Figure 11a–d, a reverse vortex is captured in case1 and the vortex structure do not change significantly from case2 to case4. The reason was that the WTS decreased sharply though the velocity increase in the concave bank with increasing radius-to-width ratios (shown in Figure 8a). The ΔF was positive for the convex bank and negative for the concave bank only in case1 in horizontal direction.

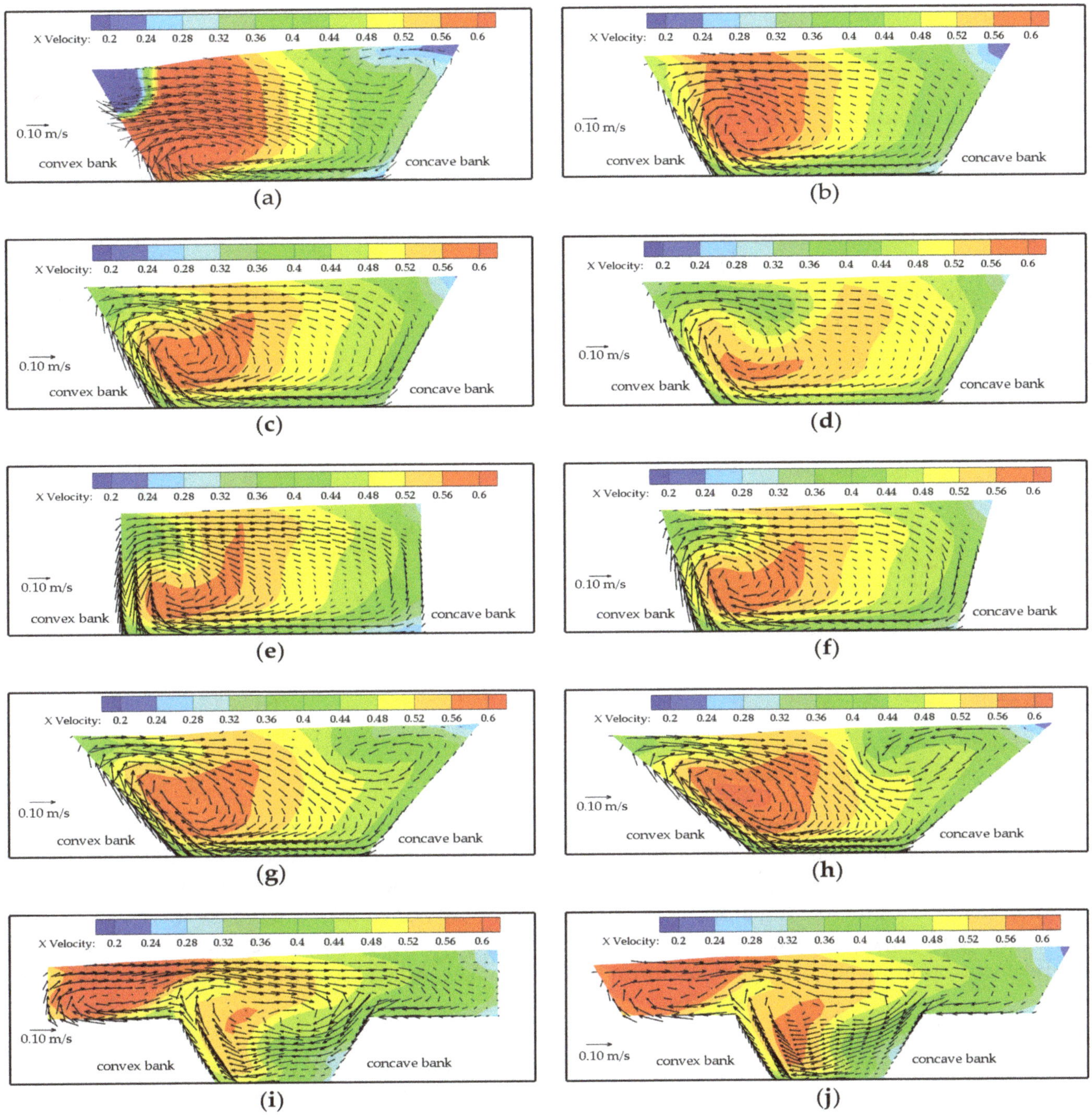

Figure 11. Structure of secondary flow of the cross-section in the bend apex for varying operating conditions, (**a–j**) case1–case10, respectively.

As shown in Figure 11c,e–h, a reverse vortex are gradually obvious in the cross-section of the bend apex with increasing slope coefficients. The obvious double vortexes were captured in Figure 11g,h, the positive vortex near the convex bank was close to the bed but the reverse vortex was generated near the water surface in the concave bank. The reason was that velocity decreased in the concave bank (shown in Figure 11), yet the transverse slope increased (shown in Figure 8b) with increasing slope coefficients and the reverse vortex occur for the ΔF was negative at a certain position. Obvious multi-vortex was captured in Figure 10i,j in the present study, due to the irregular velocity distribution caused by the compound section.

In this study, the equation of the intensity of secondary flow [38] is as follows:

$$\omega = \frac{u_w^2 + u_v^2}{u_l^2 + u_w^2 + u_v^2} \tag{25}$$

where, ω is the intensity of secondary flow, u_w is the streamwise velocity, u_w is the spanwise velocity and u_V is the vertical velocity.

Figure 12a,b showed the graph of dimensionless intensity of secondary flow for varying radius-to-width ratios and slope coefficients. The vertical or horizontal coordinate were the dimensionless intensity of secondary flow or the angle of the bend, respectively. The distribution of the intensity of secondary flow for m = 0 is consistent with the investigation of Ma [24], and Vaghefi et al. [39] also presented that the maximum intensity of secondary flow appears at approximately 60°~80° via various analysis methods, which demonstrated good agreement with the above conclusion.

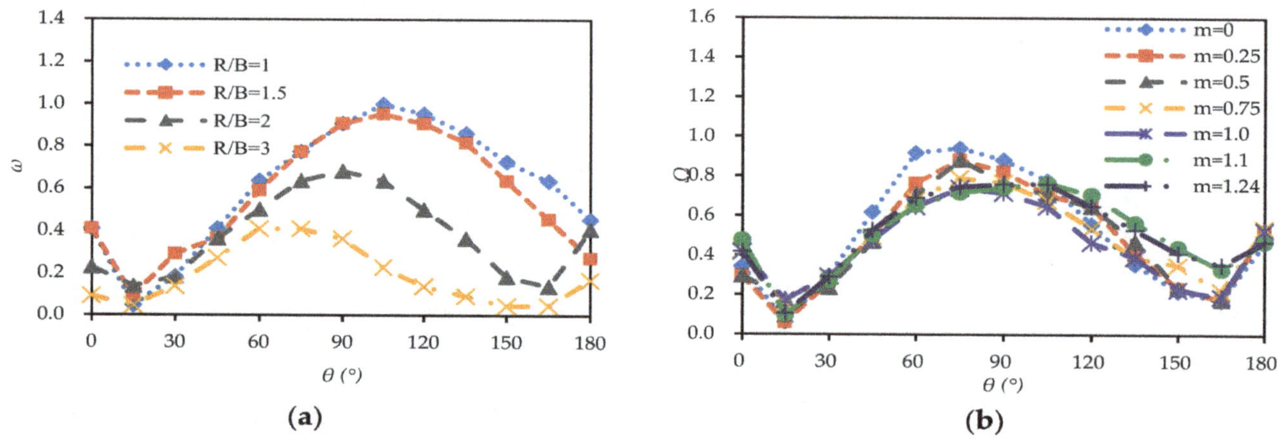

Figure 12. Graph of the dimensionless intensity of secondary flow for varying radius-to-width ratios and slope coefficients, (**a**) radius-to-width ratio, (**b**) slope coefficient.

The intensity of secondary flow decreased firstly, then increased and finally decreased with increasing slope coefficients. The reason why sudden increase was occurred was that the cross-section changes from trapezoid to compound cross-section. It can be considered that the intensity of secondary flow decreased with increasing slope coefficients but the intensity was the weakest when the slope coefficient was 1.0. As the radius-to-width ratio increased, the intensity of secondary flow decreased and the corresponding angle of the maximum intensity of secondary flow transforms from 105° to 60°.

3.4. Analysis of the Shear Stress

The higher the shear stress on the bed and sidewall generated by the water in the bend, the more obvious erosion or siltation in channel and shear stress was investigated by many scholars [8,9,38,39]. The shear stress cannot be measured directly in actual research and various calculation methods were adopted to calculate the shear stress via other physical parameters, for example, energy slope and hydraulic radius [40]. The average velocity and correlation coefficient were employed via linear calculation to determine the shear stress by Williams et al. but this calculation method was not accurate in local location [41] Fluctuating velocity [42] and TKE [43] can be used to calculate the shear stress. The TKE method less affected by local streamline is more suitable for complex areas in the bend [44]. In this study, TKE method was adopted to calculate the shear stress and the calculation equation is as follows:

$$\tau = C_1\left[0.5\rho\left(\overline{u'^2} + \overline{v'^2} + \overline{\omega'^2}\right)\right] \tag{26}$$

where C_1 is 0.19 [43], ρ is corresponding density and u', v' *and* ω' are the streamwise, spanwise and vertical fluctuating velocity, respectively.

The distributions of the shear stress in the bend were shown in Figure 13. The maximum shear stress and the corresponding angle for varying operating conditions were shown in Figures 14 and 15, respectively.

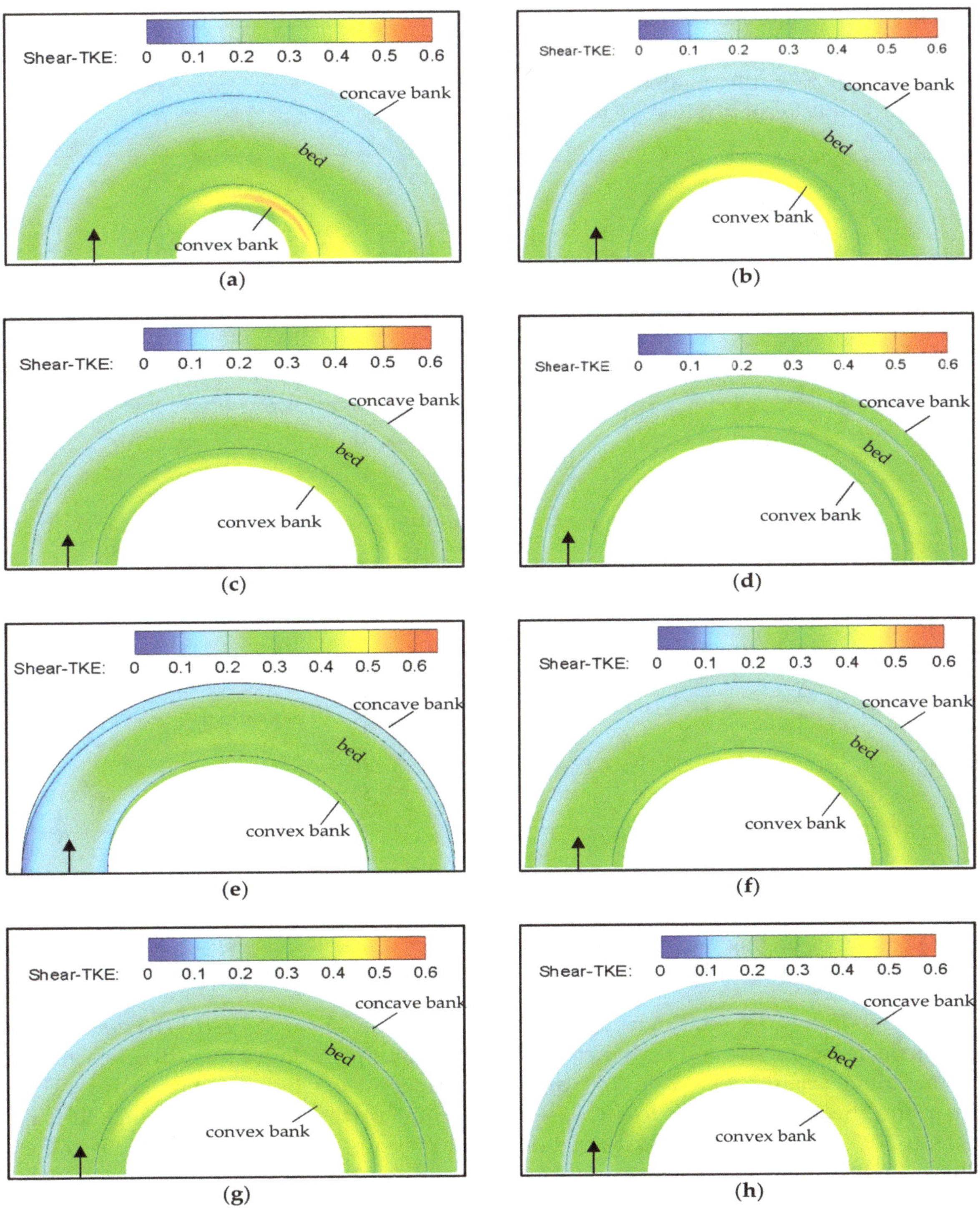

Figure 13. *Cont.*

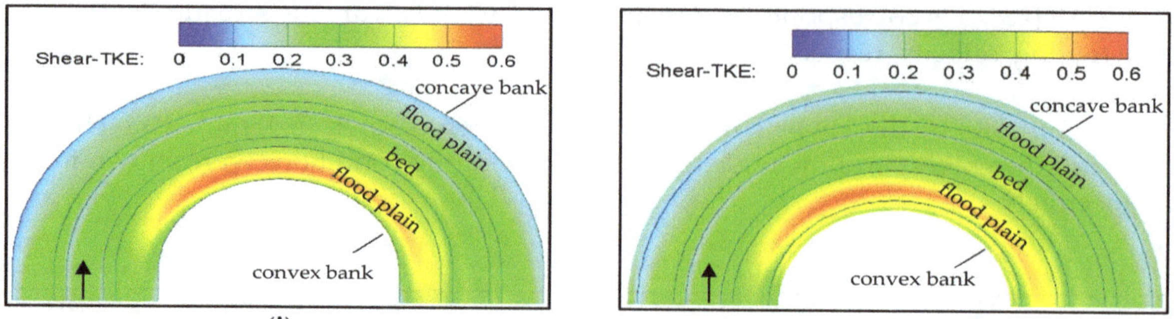

(i) (j)

Figure 13. Shear stress distribution of the riverbed and sidewall for varying operating conditions, (**a**)–(**j**) case1–case10, respectively.

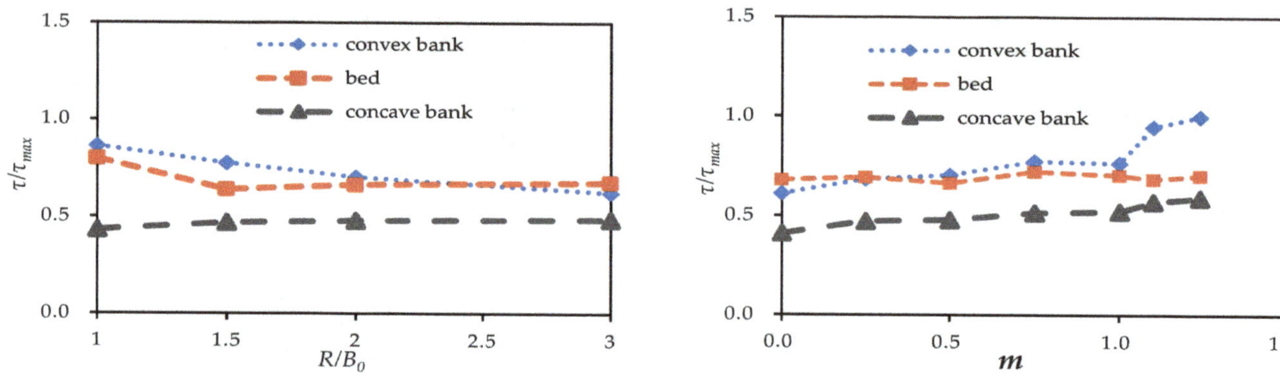

(a) (b)

Figure 14. Maximum shear stress for varying operating conditions, (**a**) radius-to-width ratio, (**b**) slope coefficient.

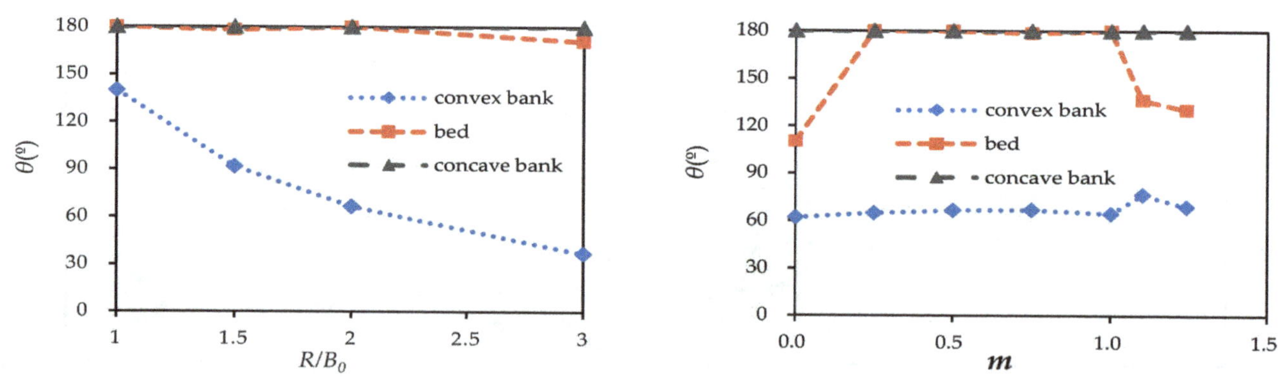

(a) (b)

Figure 15. Corresponding angle of the maximum shear stress for varying operating conditions, (**a**) radius-to-width ratio, (**b**) slope coefficient.

According to Figure 13a–d, the transformation law of the shear stress in the convex bank, bed, concave bank decreased firstly and then increased with increasing radius-to-width ratios and the extent of maximum value variation were 27%, 15% and 12%, respectively (shown in Figure 14a). The position of the maximum shear stress moves to the upstream and the corresponding angle moved from 140° to 37° or 180° to 170° in the convex bank or bed (shown in Figure 15a) but did not move in the concave bank. Why shear stress was significantly higher than others for $\lambda = 1$ was that the momentum exchange and turbulence transfer violently for the dramatic change of the water in the bend.

The shear stress increased in the convex bank and concave bank with the increasing slope coefficients, which was consistent with the conclusion inferred by Ansar et al. [9], who revealed that the shear stress of sidewall increased with increasing slope coefficients in straight trapezoidal channel. Hence, the law of shear stress in both banks increased with increasing slope coefficients can be adopted in both straight and cured trapezoidal channels. The shear stress in the bed failed to present the regular variations with increasing slope coefficients and was higher than the trapezoidal cross-section in the convex bank of the compound cross-section in the same position. The position of maximum shear stress did not show the obvious law in both banks and bed with increasing slope coefficients.

3.5. Analysis of the TKE

Blanckaert K. [45] and Ma [24] investigated the distribution of TKE in the cross-section of the bend. In this study, the equation of TKE calculation is as follows:

$$K = \frac{1}{2}\left(u'^2 + v'^2 + \omega'^2\right) \tag{27}$$

where K presents the TKE, u', v', ω' are the streamwise, spanwise and vertical fluctuating velocity, respectively.

The graphs of the dimensionless TKE with the radius-to-width ratios and slope coefficients at various locations of the bend were shown in Figure 16. The relational graphs of the change extent of the dimensionless TKE with the radius-to-width ratios and the slope coefficients were shown in Figures 17 and 18, respectively. The TKE located at $5\%B_0$ was extracted to represent the sidewall. The turbulent kinetic energy was nondimensionalized via the mean turbulent kinetic energy (TKE_{ave}).

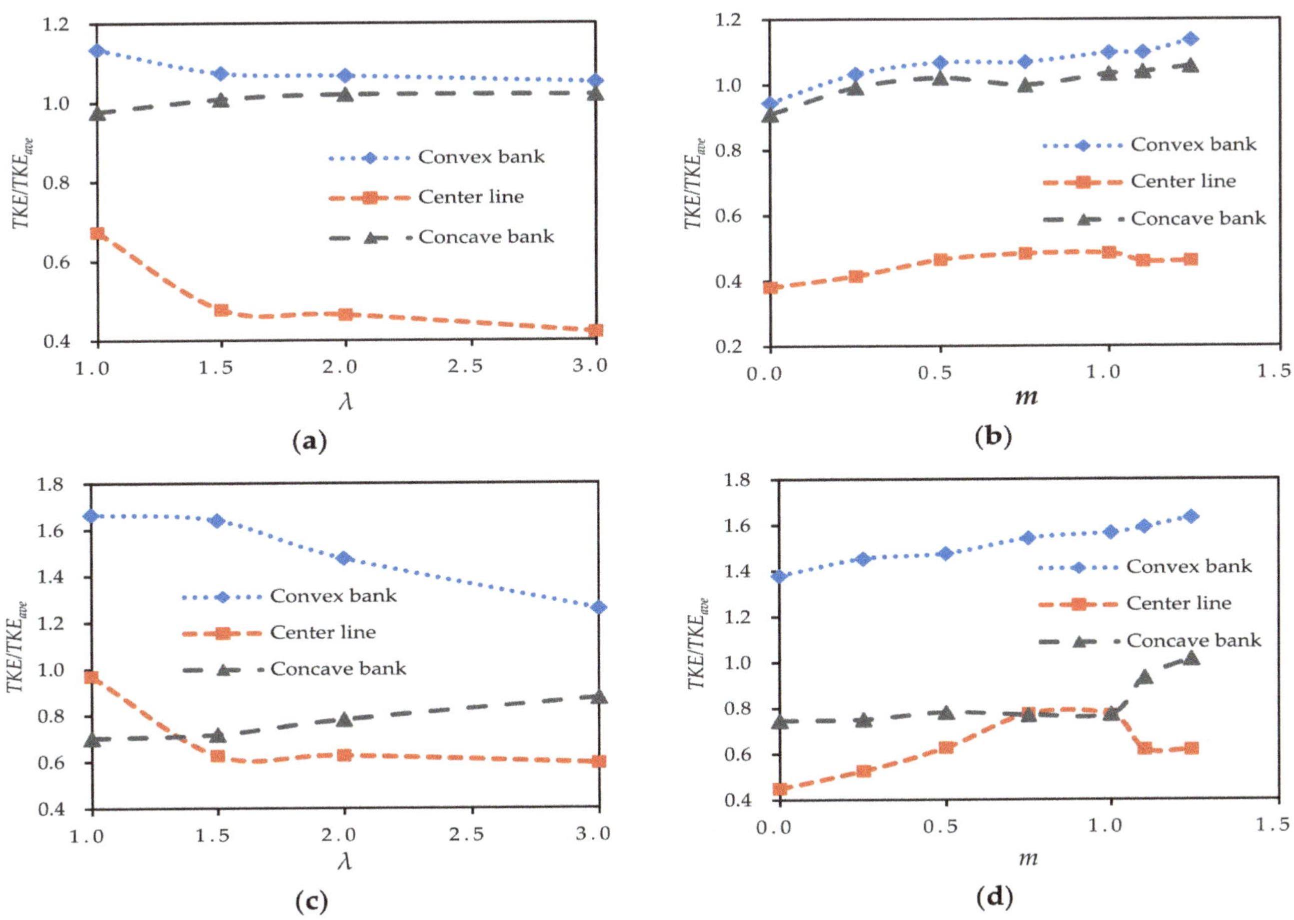

Figure 16. *Cont.*

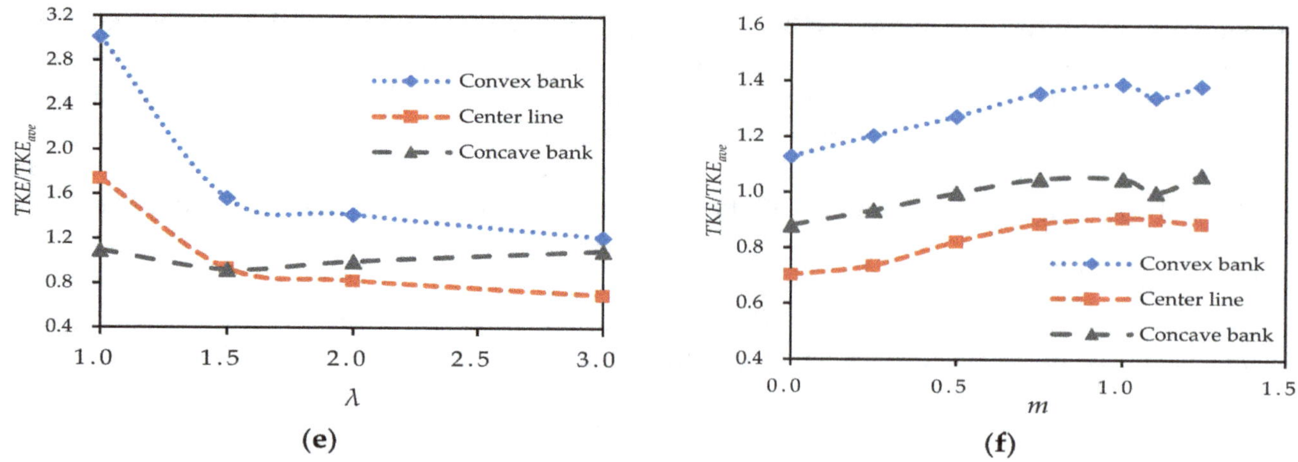

Figure 16. Graphs of the dimensionless TKE with the radius-to-width ratios and slope coefficients in different positions of the bend, (**a**,**b**) the inlet of the bend; (**c**,**d**) the apex of the bend; (**e**,**f**) the outlet of the bend.

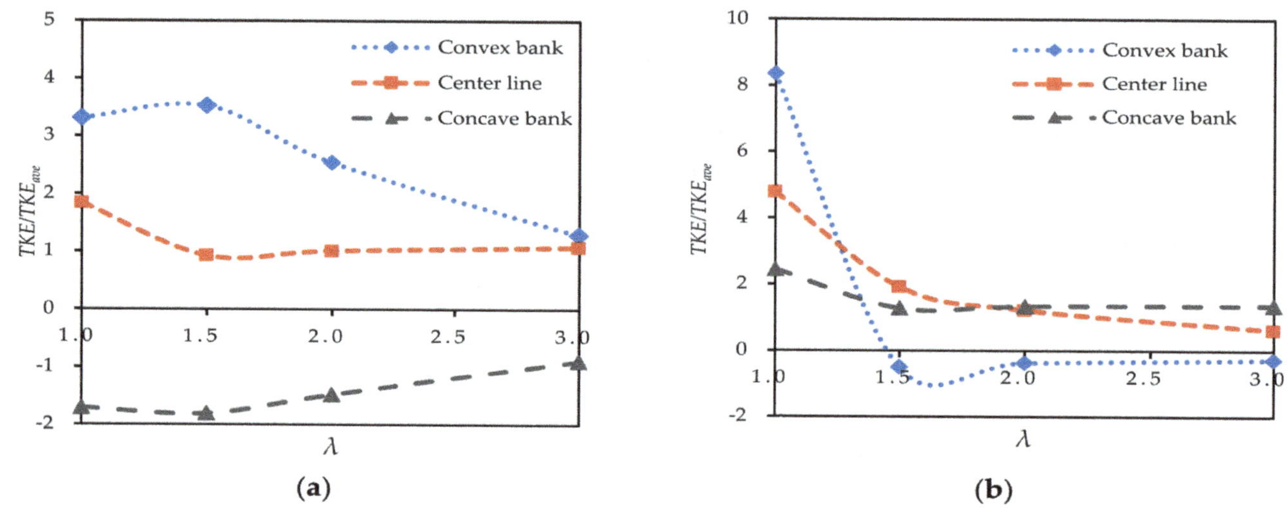

Figure 17. Relation graphs of the change extent of the dimensionless TKE with radius-to-width ratios, (**a**) the difference between inlet and apex of the bend, (**b**) the difference between apex and outlet of the bend.

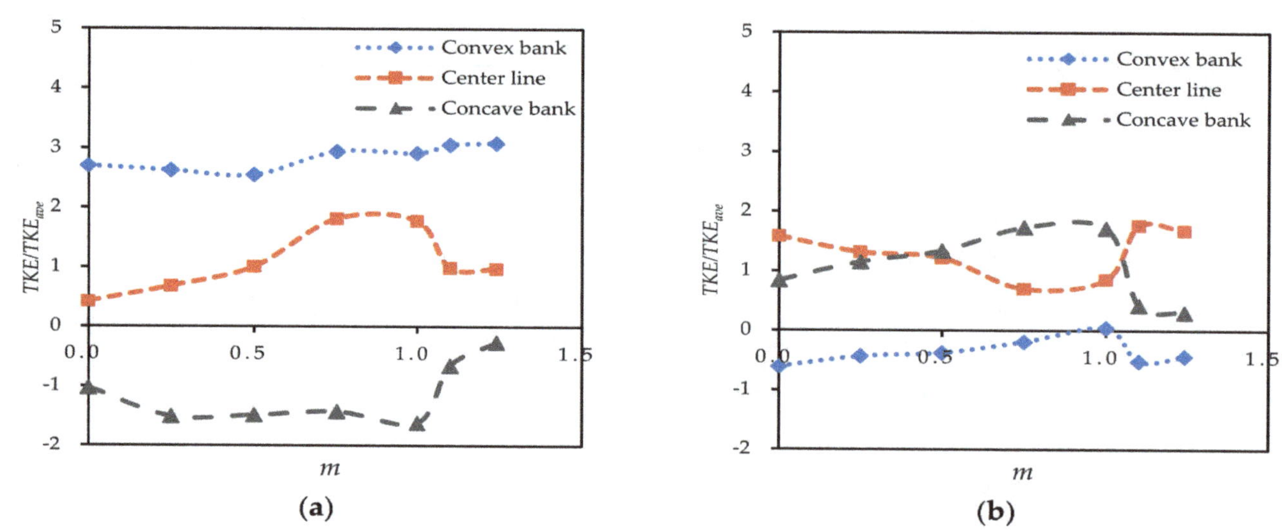

Figure 18. Relation graphs of the change extent of the dimensionless TKE with slope coefficients, (**a**) the difference between inlet and apex of the bend, (**b**) the difference between apex and outlet of the bend.

According to Figure 16, the TKE on both sidewalls was larger than the center line in the bend. The TKE in the convex bank and central line gradually decreased and increased slightly in the concave bank with increasing radius-to-width ratios. The reason was that the turbulence and pulsation were weaker in the convex bank and central line and more violent in the concave bank with the increasing radius-to-width ratios. The TKE of all single trapezoidal cross-section increased with increasing slope coefficients.

As the flow of water in the bend, the TKE obviously increased in the convex bank and center line and obviously reduced for the concave bank in the cross-section of the bend apex, as shown in Figure 16a–d. The amplitudes decreased with increasing radius-to-width ratios and increased with increasing slope coefficients, which were shown in Figures 17a and 18a.

In the outlet of the bend, the TKE significantly increased in the concave bank and center line and decreased in the convex bank, as shown in Figure 16c–f. The amplitudes decreased with increasing radius-to-width ratios and decreased in the convex bank and center line with increasing slope coefficients. An increase was observed in the concave bank for the single trapezoidal cross-section but an opposite trend was presented for the compound cross-section with the increasing slope coefficients.

4. Conclusions

In this study, the numerical model was verified via the existing experimental data and then numerical simulation was performed to investigate the hydraulic characteristics of the U-shaped channel with trapezoidal cross-section under various conditions, that is, four types of radius-to-width ratio and seven types of slope coefficient with the aspect ratio being constant. The conclusions are as follow:

(1) The transverse slope decreased with increasing radius-to-width ratios and increased with increasing slope coefficients. The effect of the radius-to-width ratio on the WTS in the bend apex was greater than that of the slope coefficient.

(2) A new equation concerning the vertical distribution of the longitudinal velocity in trapezoidal cross-section channel was proposed. The influence of the slope coefficient on the distribution of the longitudinal velocity was higher than that of the radius-to-width ratio.

(3) The slope coefficient had an effect on the structure of secondary flow in the bend. When the slope coefficient was large enough ($m = 0.75$ in the present study), obvious double vortexes were captured in the bend apex. The corresponding angle of the maximum intensity of secondary flow in the bend gradually moved to the upstream with increasing radius-to-width ratios. The intensity of secondary flow in the bend gradually decreased with the increase in slope coefficients.

(4) The maximum of the shear stress of the bend was observed in the convex bank, which decreased gradually and the corresponding angle decreased with increasing radius-to-width ratios. As the slope coefficient increased, the maximum of the shear stress in the convex bank gradually increased.

(5) The maximum TKE of the section in the bend decreased with increasing radius-to-width ratios and increased with increasing slope coefficients.

Author Contributions: Conceptualization and Methodology, J.Z.; Numerical Simulation, R.H.; Investigation, R.H.; Writing-Original Draft Preparation, R.H.; Writing-Review & Editing, J.Z.; Supervision, J.Z.; Project administration, J.Z.; Funding acquisition, J.Z.

References

1. Blanckaert, K.; Graf, W.H. Momentum transport in sharp open-channel bends. *J. Hydraul. Eng.* **2004**, *130*, 186–198. [CrossRef]
2. Hey, R.D.; Thorne, C.R. Secondary flows in river channels. *Area* **1975**, *7*, 191–195.
3. Kabiri, S.A.; Farshi, F.; Chamani, M.R. Boundary shear stress in smooth trapezoidal open channel flows. *J. Hydraul. Eng.* **2013**, *139*, 205–212. [CrossRef]
4. Knight, D.W.; Omran, M.; Tang, X.N. Modeling depth-averaged velocity and boundary shear in trapezoidal channels with secondary flows. *J. Hydraul. Eng.* **2007**, *133*, 39–47. [CrossRef]
5. Shiono, K.; Muto, Y. Complex flow mechanisms in compound meandering channels with overbank flow. *J. Fluid Mech.* **1998**, *376*, 221–261. [CrossRef]
6. Knight, D.W.; Shiono, K. Turbulence measurements in a shear layer region of a compound channel. *J. Hydraul. Res.* **1990**, *28*, 175–196. [CrossRef]
7. Tominaga, A.; Nezu, L.; Ezaki, K.; Nakagawa, H. Three-dimensional turbulent structure in straight open channel flows. *J. Hydraul. Res.* **1989**, *27*, 149–173. [CrossRef]
8. Yang, S.Q.; Lim, S.Y. Boundary shear stress distributions in trapezoidal channels. *J. Hydraul. Res.* **2005**, *43*, 98–102. [CrossRef]
9. Ansar, K.; Morvan, H.P.; Hargreaves, D.M. Numerical Investigation into Secondary flow and Wall Shear in Trapezoidal Channels. *J. Hydraul. Eng.* **2011**, *137*, 432–440. [CrossRef]
10. Farhadi, A.; Sindelar, C.; Tritthart, M.; Glas, M.; Blanckaert, K.; Habersack, H. An investigation on the outer bank cell of secondary flow in channel bends. *J. Hydro-Envrion. Res.* **2018**, *18*, 1–11. [CrossRef]
11. Termini, D.; Piraino, M. Experimental analysis of cross-sectional flow motion in a large amplitude meandering bend. *Earth Surf. Process. Landf.* **2011**, *36*, 244–256. [CrossRef]
12. Shukla, D.R.; Shiono, K. CFD modelling of meandering channel during floods. *Water Manag.* **2008**, *161*, 1–12. [CrossRef]
13. Mosalman, A.; Mosalman, M.; Yazdi, H.M. Equations of unsteady flow in curved trapezoidal channels. *Int. J. Phys. Sci.* **2011**, *6*, 671–676. [CrossRef]
14. Shao, S.D. Simulation of breaking wave by SPH method coupled with k-epsilon model. *Int. J. Numer. Methods Fluids* **2006**, *44*, 338–349. [CrossRef]
15. Shao, S.D.; Ji, C.M. SPH computation of plunging waves using a 2-D sub-particle scale (SPS) turbulence model. *J. Hydraul. Res.* **2010**, *51*, 913–936. [CrossRef]
16. Peng, Y.; Zhou, J.G.; Burrows, R. Modelling the free surface flow in rectangular shallow basins by lattice Boltzmann method. *J. Hydraul. Eng.* **2011**, *137*, 1680–1685. [CrossRef]
17. Peng, Y.; Zhou, J.G.; Burrows, R. Modelling solute transport in shallow water with the lattice Boltzmann method. *Comput. Fluids* **2011**, *50*, 181–188. [CrossRef]
18. Peng, Y.; Mao, Y.F.; Wang, B.; Xie, B. Study on C-S and P-R EOS in pseudo-potential lattice Boltzmann model for two-phase flows. *Int. J. Mod. Phys. C* **2017**, *28*, 1750120. [CrossRef]
19. Bai, R.; Liu, S.; Tian, Z.; Wang, W.; Zhang, F. Experimental investigations of air-water flow properties of offse-aerator. *J. Hydraul. Eng.* **2017**, *144*, 04017059. [CrossRef]
20. Bai, R.; Zhang, F.; Liu, S.; Wang, W. Air concentration and bubble characteristics downstream of a chute aerator. *Int. J. Multiph. Flow* **2016**, *87*, 156–166. [CrossRef]
21. Ducoin, A.; Shadloo, M.S.; Roy, S. Direct Numerical Simulation of flow instabilities over Savonius style wind turbine blades. *Renew. Energy* **2017**, *105*, 374–385. [CrossRef]
22. Mendez, M.; Shadloo, M.S.; Hadjadj, A.; Ducoin, A. Boundary layer transition over a concave surface caused by centrifugal instabilities. *Comput. Fluids* **2018**, *171*, 135–153. [CrossRef]
23. Shadloo, M.S.; Hadjadj, A.; Hussain, F. Statistical behavior of supersonic turbulent boundary layers with heat transfer at $M_\infty = 2$. *Int. J. Heat Fluid Flow* **2015**, *53*, 113–134. [CrossRef]
24. Ma, M. Flow Characteristics and Influence of Geometric Factors in a Bend. Ph.D. Thesis, Xi'an University of Technology, Xi'an, China, 2017. (In Chinese)
25. Bai, Z.L.; Peng, Y.; Zhang, J.M. Three-Dimensional Turbulence Simulation of Flow in a V-Shaped Stepped Spillway. *J. Hydraul. Eng.* **2017**, *143*, 06017011. [CrossRef]
26. Sun, X.; Shiono, K.; Rameshwaran, P. Modelling vegetation effects in irregular meandering river. *J. Hydraul. Res.* **2010**, *48*, 775–783. [CrossRef]

27. Yakhot, V.; Orszag, S.A. Renormalization group analysis of turbulence: I. Basic Theory. *J. Sci. Comput.* **1986**, *1*, 3–51. [CrossRef]
28. Hirt, C.W.; Nichols, B.D. Volume of Fluid (VOF) Method for the Dynamics of Free Boundaries. *J. Comput. Phys.* **1981**, *39*, 201–225. [CrossRef]
29. Han, Z.; Reitz, R.D. Turbulence modeling of internal combustion engines using RNG k~ε models. *Combust. Sci. Technol.* **1995**, *106*, 267–295. [CrossRef]
30. Celik, I.B.; Ghia, U.; Roache, P.J.; Freitas, C.J. Procedure of estimation and reporting of uncertainty due to discretization in CFD applications. *J. Fluids Eng.* **2008**, *130*, 078001–078004. [CrossRef]
31. Liu, H.F. The study of water surface in curved channel. *J. Hydraul. Eng.* **1990**, *4*, 46–50. [CrossRef]
32. Zhang, H.W.; Lv, X. *Hydraulics in the Curved Channel*, 1st ed.; China Water Power Press: Beijing, China, 1993; pp. 48–49. ISBN 7–120-02333-0/TV•835. (In Chinese)
33. Chang, H.H. *Fluvial Processes in River Engineering*; John Wiley and Sons: New York, NY, USA, 1988.
34. Sarma, K.V.N.; Lakshminatayana, P.; Rao, N.S.L. Velocity distribution in smooth rectangular open channels. *J. Hydraul. Eng.* **1988**, *109*, 270–289. [CrossRef]
35. Yan, J.; Tang, H.W.; Xiao, Y.; Li, K.J.; Tian, Z.J. Experimental study on influence of boundary on location of maximum velocity in open channel flows. *Water Sci. Eng.* **2011**, *4*, 185–191. [CrossRef]
36. Song, T.; Graf, W.H. Velocity and turbulence distribution in unsteady open-channel flows. *J. Hydraul. Eng.* **1996**, *122*, 141–154. [CrossRef]
37. Kirkgoz, M.S. Turbulent Velocity Profiles for Smooth and Rough Open Channel Flow. *J. Hydraul. Eng.* **1989**, *115*, 1543–1561. [CrossRef]
38. Abhari, M.N.; Ghodsian, M.; Vaghefi, M.; Panahpur, N. Experimental and numerical simulation of flow in a 90° bend. *Flow Meas. Instrum.* **2010**, *21*, 292–298. [CrossRef]
39. Vaghefi, M.; Akbari, M.; Fiouz, A.R. An Experimental Study of Mean and Turbulent Flow in a 180 Degree Sharp Open Channel Bend: Secondary Flow and Bed Shear Stress. *KSCE J. Civ. Eng.* **2016**, *20*, 1582–1593. [CrossRef]
40. Koopaei, K.B.; Ervine, D.A.; Carling, P.A.; Cao, Z. Velocity and turbulence measurements for two overbank flow events in River Severn. *J. Hydraul. Eng.* **2002**, *128*, 891–900. [CrossRef]
41. Williams, J.J.; Rose, C.P.; Thorne, P.D.; O'connor, B.A.; Humphery, J.D.; Hardcastle, P.J.; Moores, S.P.; Cooke, J.A.; Wilson, D.J. Field observations and predictions of bed shear stresses and vertical suspended sediment concentration profiles in wave-current conditions. *Cont. Shelf Res.* **1999**, *19*, 507–536. [CrossRef]
42. Pope, S.B. *Turbulent Flow*, 1st ed.; Cambridge University Press: Cambridge, UK, 2000.
43. Kim, S.C.; Friedrichs, C.T.; Ma, J.P.Y.; Wright, L.D. Estimating bottom stress in tidal boundary layer from acoustic Doppler velocimeter data. *J. Hydraul. Eng.* **2000**, *126*, 399–406. [CrossRef]
44. Biron, P.M.; Robson, C.; Lapointe, M.F.; Gaskin, S.J. Comparing various methods of bed shear stress estimates in simple and complex flow fields. *Earth Surf. Process. Landf.* **2010**, *29*, 1403–1415. [CrossRef]
45. Blanckaert, K.; Vriend, H.J. Turbulence characteristics in sharp open-channel bends. *Phys. Fluids* **2005**, *17*, 717–731. [CrossRef]

3

Numerical Analysis on the Hydrodynamic Performance of an Artificially Ventilated Surface-Piercing Propeller

Dongmei Yang *, Zhen Ren, Zhiqun Guo and Zeyang Gao

College of Shipbuilding Engineering, Harbin Engineering University, Harbin 150001, China; forfanqiang@126.com (Z.R.); guozhiqun@hrbeu.edu.cn (Z.G.); gaozeyangpc@hotmail.com (Z.G.)
* Correspondence: yangdongmei@hrbeu.edu.cn.

Abstract: When operated under large water immersion, surface piercing propellers are prone to be in heavy load conditions. To improve the hydrodynamic performance of the surface piercing propellers, engineers usually artificially ventilate the blades by equipping a vent pipe in front of the propeller disc. In this paper, the influence of artificial ventilation on the hydrodynamic performance of surface piercing propellers under full immersion conditions was investigated using the Computational Fluid Dynamics (CFD) method. The numerical results suggest that the effect of artificial ventilation on the pressure distribution on the blades decreases along the radial direction. And at low advancing speed, the thrust, torque as well as the efficiency of the propeller are smaller than those without ventilation. However, with the increase of the advancing speed, the efficiency of the propeller rapidly increases and can be greater than the without-ventilation case. The numerical results demonstrates the effectiveness of the artificial ventilation approach for improving the hydrodynamic performance of the surface piercing propellers for high speed planning crafts.

Keywords: surface-piercing propeller; artificial ventilation; hydrodynamic performance; numerical simulation

1. Introduction

Surface-piercing propellers (SPPs) are also known as Surface penetrating propellers. The SPP is so named because part of the propeller is above the water surface and the rest is under the water during normal operation. Compared with the conventional propellers, the SPPs mainly have the following three advantages (Ding et al. [1]): (1) the resistance of the appendages, such as the paddle shaft and the shaft bracket, is minimized; (2) propeller diameter is no longer limited by some parameters such as soak depth and stern frame; and (3) the cavitation erosion on the blade surface is substantially reduced. Due to these advantages, the SPPs become an optimal choice for high-speed boats and some shallow draft ships, and thus have really good application prospects.

However, in some cases the SPPs might be operated under the large water immersion, e.g., the boats are at the bow-up status, which makes the SPPs be in heavy load conditions. One of the practical approaches to solve this problem is setting an aeration pipe in front of SPPs to ventilate the blades. This approach has been adopted in some actual boats and achieved significant results. Nonetheless, few attentions have been paid on how the ventilation of blades improves the hydrodynamic characteristics of SPPs. To this end, this paper is dedicated to investigate the influence of the ventilation of blades on the hydrodynamic performance of SPPs using numerical methods.

In some simplified studies, the surface-piercing process of an SPP blade can be deemed as the water-entry process of a 2D profile, which is easy to be understood. Zhao et al. [2] used the non-linear boundary element method to investigate the characteristics of the sprays generated by the water-entry of wedges with different bottom-angles. It was found that the increase of the wetted area directly leads to a great slamming pressure on the wetted surface. And when the slope angle of the bottom is larger than 30°, there is no typical slamming pressure concentration phenomenon due to the smaller wetted surface. Young et al. [3], in his doctoral dissertation, carried out similar studies on the water-entry of wedges, and analyzed the pressure distribution on the wetted surface of a flat plate under different entry angles.

Yari et al. [4] conducted a preliminary study on a wedge entering the water under different inclination angles. The obtained pressure distribution, free surface deformation, and so on agree well with the experimental data. Yu et al. [5] carried out a numerical study on the water entry problem of a 2D cross-section profile, and found that the cup shape of the trailing edge can enhance the hydrodynamic load on the edge, while reducing the lift-to-drag ratio as well as the efficiency of the propeller.

Ghassemi et al. [6] investigated the hydrodynamics of SPPs named SPP-1 and SPP-2 under full and half immersed conditions using a boundary element method (BEM). The numerical results agree well with experimental ones. In the study, they found that the Weber number has significant influence on the ventilation status of the SPP. Kinnas et al. [7] studied the bubble flow around the hydrofoil and SPPs using a BEM, in which they investigated the hydrodynamic effects of the bubble flow on the hydrofoil and SPPs with various blade profiles. Young et al. [8,9] carried out a series of studies on large-scale SPPs and super-vacuum paddles also by using the BEM, which contributed to the knowledge of the hydrodynamics of these propellers.

With the development of the computer technology, the CFD methods become popular and play the role of benchmark in the numerical studies on surface-piercing propellers. Young et al. [3] predicted the hydrodynamic performance of large-scale surface paddles using a coupled boundary element method–finite element method (BEM–FEM), and found that the numerical results compared well with those obtained by the Reynolds Averaged Navier–Stokes (RANS) solver (Fluent software), in which the Reynolds stress tensor was modeled using the SST form of the $k-\omega$ turbulence model. Shi et al. [10] investigated the wake of surface-piercing propellers, as well as the pulsating pressure on the paddles and the deformation of the free surface after the paddles. Alimirzazadeh et al. [11] employed the OpenFOAM software to study the hydrodynamic performance of SPP-841B surface paddle under different sway angles and immersion depths. In the study, the $k-\omega$ based Shear Stress Transport (SST) model with automatic wall functions (mixed formulation) was employed. The numerical results agree with experimental ones well under different immersion depths, while poorly under different sway angles. They found that increasing the yaw angle would decrease the thrust coefficient and torque coefficient, while the efficiency was improved. In addition, the maximum efficiency point has not been achieved at the zero shaft yaw angle, due to the implementation of SPPs.

Obviously, in the existing works the most attention has been paid to investigating the hydrodynamic performance of SPPs under the natural operation conditions, while very little attention has been paid on the artificial ventilated SPPs. It is believed that the hydrodynamic characteristics of the artificial ventilated SPPs should be significantly different from the without-ventilation case. Thereby, we were motivated to study the SPPs under artificial ventilation conditions. In this paper, a right-handed three-blade paddle was proposed for the investigation, and an aeration pipe was installed in front of the paddle disk for the ventilation purpose. The numerical simulation was carried out based on the commercial CFD software Star-CCM+ [12]. The rotation of the blades was realized by using the overlapped mesh technique.

2. SPP Model and Numerical Setup

The submergence ratio of SPPs is defined as $It = h/D$, where h denotes the immersion depth of the propeller disc, D is the diameter of the propeller, as shown in Figure 1.

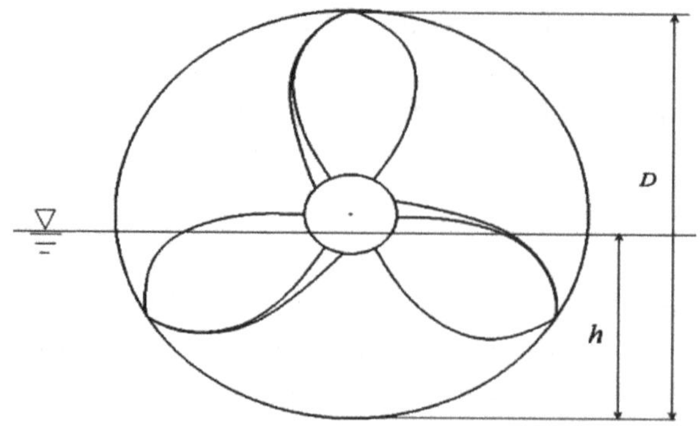

Figure 1. The submergence ratio of surface-piercing propellers (SPPs), which is defined as the ratio of the immersion depth h to the propeller diameter D.

Generally, the SPPs work on the water-air interface, and the blades rush into and out of the water in a staggered manner. In this process, the air is sucked into the water forming an air cavity, which connects the atmosphere and the paddle. This phenomenon is known as the ventilation of SPPs.

When the submergence ratio is sufficient large, the torque on the paddle will dramatically increase. For the adjustable SPPs, the torque can be reduced by decreasing the immersion depth. For the fixed SPPs, however, it is commonly difficult to adjust the immersion depth. Although the paddle may pump some air into water when it rotates, the torque can still be very large. For the purpose of torque reduction, an aeration pipe is generally installed in front of the paddle to ventilate the blades and thus reduce the torque.

2.1. SPP Model

The SPP proposed for the calculation is the SPP-1 type surface-piercing propeller model given in Table 1 [6], with the rotation direction being right-handed and the profile being S-C type. The diameter of the aeration pipe is 50 mm, and the distance from the outlet of the aeration pipe to the paddle surface is 100 mm. The sketch of the propeller and the vent pipe are shown in Figure 2.

Table 1. SPP-1 propeller geometric parameters [6].

Parameters	
Diameter D	0.2
Tilt angle/°	10
Number of leaves Z	3
Pitch ratio P/D	1.6
Disk ratio AE/AO	0.5
Hub diameter ratio d/D	0.2
Side angle/°	0
Profile	S-C

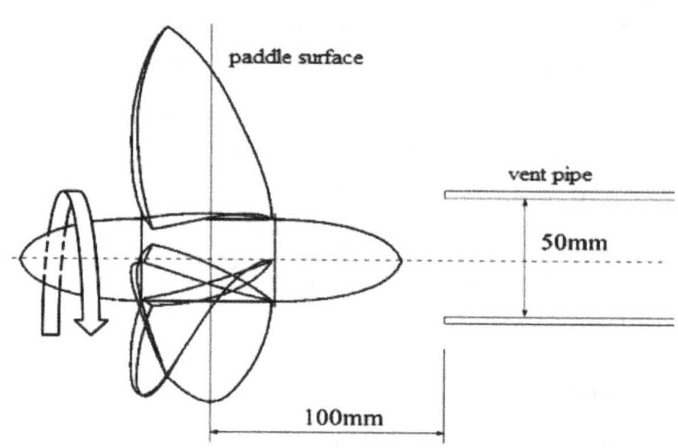

Figure 2. The sketch of the propeller and the vent pipe.

2.2. Control Equation

Under the full immersion condition, ventilation through the aeration pipe allows the paddle always being in an unsteady air-water two-phase flow field. The VOF (volume of fluid) model [13] was selected for capturing the interface between water and air. Air and water in the flow field are deemed incompressible.

Mass conservation equation can be written as:

$$\frac{\partial p}{\partial t} + \nabla(\rho \vec{\mu}) = S_m \tag{1}$$

The above equation is a general expression of the mass conservation equation, where the term S_m can be any user-defined source term added to the continuity term.

The momentum equations are given as:

$$\frac{\partial(\rho\mu_i)}{\partial t} + \frac{\partial(\rho\mu_i\mu_j)}{\partial x_j} = -\frac{\partial p}{\partial x_i} + \frac{\partial}{\partial x_j}\left(\mu\frac{\partial \mu_i}{\partial x_j} - \rho\overline{\mu_i'\mu_j'}\right) + S_i \tag{2}$$

where μ_i is the velocity component in the xi direction ($i, j = 1,\ 2,\ 3$) of the Cartesian coordinate system, p the fluid pressure, ρ the fluid density, μ the dynamic viscosity coefficient, t the time, and S_i the volume force.

In this paper, the $k - \omega$ SST [14,15] was employed as the turbulence model, which could take the transport characteristics of turbulent shear forces into account, and make possible to obtain more accurately results of the flow separation in the counter pressure gradient region [16].

2.3. Computational Domain and Grid Division

In order to make the flow field around the propeller be fully developed, and the numerical results be accurate and reliable, the size of the flow domain in the simulation should be large enough. In practice, the entire flow domain is divided into two parts: a cylindrical rotating zone containing the paddle and an outer cylindrical stationary zone. The outer diameter of the stationary zone is 5.0D, the inlet is 3.5D from the paddle plane, and the outlet is 7.0D from the paddle plane. The diameter of the rotation domain is 1.2D, and both end surfaces of the cylinder are 0.45D from the paddle surface. Figure 3 shows an oblique view of the propeller's computational domain.

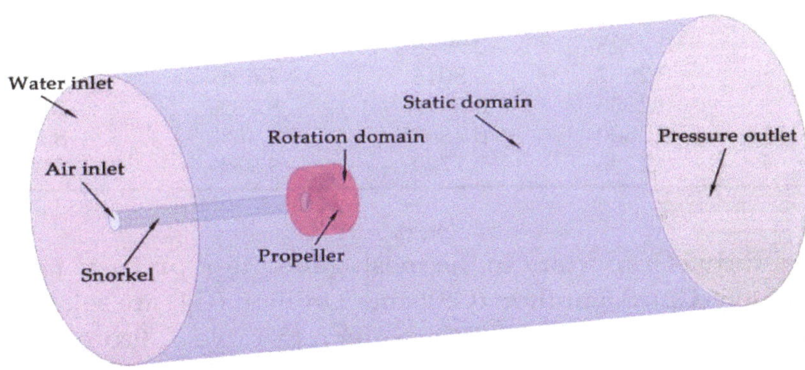

Figure 3. Oblique view of propeller and computational field.

To simulate the rotation of the paddle in the flow field, the overlapped mesh method was employed. At the overlapping domain between the stationary and the rotational domains, the information of the flow field is exchanged through the overlapping grids. The mesh in the overlapping domain should be the same size as much as possible. We set the polyhedral mesh in the inner rotational domain, while the cutting body mesh in the outer stationary domain. Moreover, local mesh densification was performed around the overlapping domain. Finally, the total number of grids is 3.5 million, in which the rotation domain is 1.7 million, and the stationary domain is 1.8 million. The min and max size of computational grids are 7.65×10^{-4} mm and 11 mm, respectively. The surface mesh is shown in Figure 4.

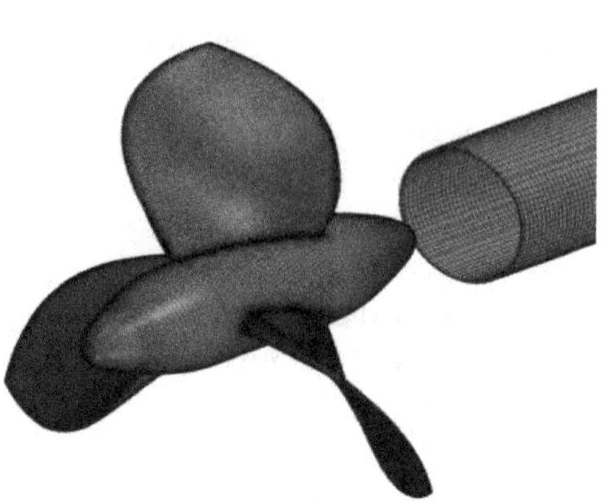

Figure 4. Mesh on the propeller and aeration pipe.

2.4. Grid Convergence

Let $Y+$ be the non-dimensional wall distance defined as $Y+ \equiv u_* y/\nu$, where u_* is the friction velocity at the nearest wall, y the distance to the nearest wall, and ν the local kinematic viscosity of the fluid. In this sub-section the sensitivity of numerical results with respect to $Y+$ is studied. Generally, the range of dimensionless wall distance $Y+ \approx 30 \sim 100$ is acceptable for blade surfaces. Table 2 lists the error of the thrust coefficient K_t and torque coefficient $10K_q$ compared with the test value with different $Y+$ value. It can be seen that three errors reach minimum when the $Y+$ is equal to 60. Therefore, we set $Y+ = 60$ in the meshing setup.

Table 2. Results for the $Y+$ for selection.

Y+	K_t	$10K_q$	**Error**K_t	**Error10**K_q
Exp.	0.3093	0.8579		
30	0.2759	0.7758	−10.81%	−9.57%
40	0.2791	0.7840	−9.77%	−8.61%
50	0.2825	0.7956	−8.69%	−7.26%
60	0.2857	0.8016	−7.65%	−6.56%
70	0.2836	0.7977	−8.33%	−7.01%
80	0.2800	0.7843	−9.48%	−8.57%
90	0.2786	0.7779	−9.95%	−9.32%

Grid convergence study is important in the mesh generation process. Here four grid numbers, 1.77 million (G1), 2.5 million (G2), 3.5 million (G3), and 5 million (G4) are selected for the convergence study. The grid number and simulation results are given in Table 3. From these results, it is clearly noticed that with the growth of grid number, the results get closer to the experimental ones, i.e., the grid convergence can be guaranteed in these numbers. However, the greater the grid number, the lower computational efficiency for the numerical simulation. It is found that the accuracy was only slightly improved in the G4 case at the expense of a lot of calculation time. Hence, when determining the grid number, one should make a compromise between accuracy and efficiency. In this paper, the fine mesh of G3 was adopted in the present simulations, which achieves both computational efficiency and accuracy.

Table 3. Results from the grid convergence.

Grid	Control Volumes	K_t	εK_t	$10K_q$	εK_q
G1	1,770,000	0.2756	10.89%	0.7761	9.53%
G2	2,500,000	0.2806	9.28%	0.7874	8.21%
G3	3,500,000	0.2857	7.63%	0.8016	6.65%
G4	5,000,000	0.2884	6.77%	0.8070	5.93%

2.5. Boundary and Initial Conditions

Due to the air-suck phenomenon, the flow around the surface-piercing paddle is usually an air-water two-phase unsteady one. The density of water and air in the simulation is 997.56 kg/m^3 and 1.18 kg/m^3, respectively. The dynamic viscosity of water and air is 8.89×10^{-4} Pa·s and 1.86×10^{-5} Pa·s, respectively. The VOF approach was employed to track the air-water interface. The principle of this approach is to determine the interface by counting the volume ratio function of the fluid in each cell rather than the motion of the particle on the free surface.

The fluid density can be expressed as:

$$m\frac{dv}{dt} = \sum \alpha_m \rho_m \tag{3}$$

where α_m is the volume fraction of the mth type of fluid in each cell, which satisfies the following condition:

$$\sum_{m=1}^{n} \alpha_m = 1 \tag{4}$$

In this paper, velocity inlet was divided into air velocity inlet and water velocity inlet. Both inlets have the same velocity vector. The volume fraction of the overall inlet was set as the mixture of air and water fractions, where the volume fraction of air at the air velocity inlet was set to 1, and the water volume fraction was set to 0. The outlet is set as a pressure outlet and the standard atmospheric pressure is made as the reference pressure. The outer wall of the stationary zone is set as a plane of symmetry, and the surfaces of the surface-piercing paddle and the vent pipe are set to be non-slip and non-penetrable. The outer boundary of the rotation domain is set to overlapped grids.

The second-order schemes were applied for spatial discretization and linear interpolation, as well as the time integration. The time step is 6.94×10^{-5} s, during which the propeller rotates 1° under the given rotational speed. As suggested by Blocken and Gualtieri [17], this time step makes the condition CFL ≤ 1 satisfied. The numerical calculation was performed on a PC with an Intel Xeon CPU X5690 (6 cores, 3.46 GHz), and one case simulation approximately costs 40 h.

2.6. Validation of the Numerical Setup

To verify the aforementioned numerical setup, the hydrodynamic performance of a propeller in the open water was investigated and compared with experimental results from Ghassemi et al. [6]. The main parameters of the SPP-1 propeller are listed in Table 1. The hydrodynamic characteristics are plotted using solid lines (see Figure 5). Let T, Q, n, and V_A be the thrust, torque, rotational speed, and forward speed of the propeller, respectively. The relevant parameters for the propeller can be nondimensionalized as follows.

Speed coefficient:

$$J = \frac{V_A}{nD} \tag{5}$$

Thrust coefficient:

$$K_t = \frac{T}{\rho n^2 D^4} \tag{6}$$

Torque coefficient:

$$m\frac{dv}{dt} = f \tag{7}$$

Promote efficiency:

$$\eta = \frac{K_t}{K_q}\frac{J}{2\pi} \tag{8}$$

In the numerical and experimental setup, the submergence ratio for the propeller is $It = 1/3$, and the rotational speed is $n = 2400$ rpm. The speed coefficient J ranges from 0.85 to 1.45, which was obtained by varying the magnitude of the incoming flow velocity V_A.

By using the CFD solver, we obtained the thrust coefficient K_t, the torque coefficient $10K_q$ of the propeller, and the open water efficiency η. $10K_q$ is used due to the fact that K_q is one order smaller than K_t and η. Figure 5 compares the CFD results with the experimental ones given in Table 1 [6].

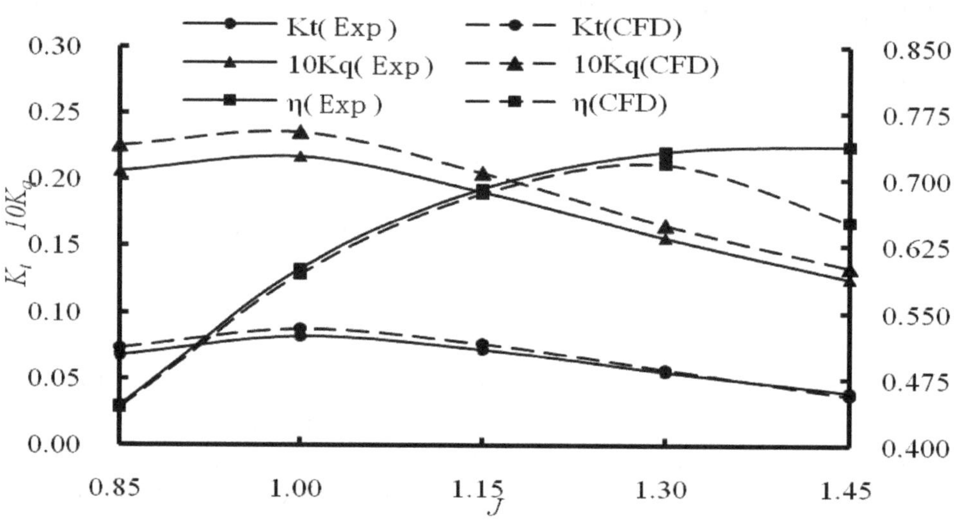

Figure 5. Comparison of CFD and experimental results at $It = 1/3$.

As shown in Figure 5, the thrust coefficient K_t obtained by the CFD is slightly larger than the experimental one when $J = 0.85$–1.30, and then the discrepancy gradually decreases with the growth of J. The torque coefficient $10K_q$ has a similar trend but the discrepancy between CFD and experimental results is slightly larger than that in the K_t case. In contrast, the open water efficiency η obtained by the CFD is always slightly smaller than the experimental one when $0.85 < J < 1.30$, but the error rapidly grows with J when $1.30 < J < 1.45$ and reaches its maximum 12% at $J = 1.45$. This study demonstrates that the numerical setup in the CFD solver is effective and can provide the required precision for the hydrodynamic simulation of the propeller with the existence of air-water interface.

3. Numerical Results of the Ventilated SPP

3.1. Calculation of Hydrodynamic Coefficient of Wigley Ship

Figure 6 depicts the wake of the ventilated SPP at $J = 1.15$, from which it can be seen that the majority of the air stream flow bypasses the hub, while the rest of the flow is swung away by the rotating blades. Figure 6 displays a snapshot of the air volume fraction (green color) flowing around the rotating propeller. One can find that the trailing edges are enclosed by air bubbles due to the relatively low pressure around them. The air bubbles on the trailing edges generally stretch from the root to the tip of the blades, though on the leading edges the bubbles may only concentrate around the root domain of the blades. In the wake, three helical air bubbles are forming and rotating about the centric air stream. In the near propeller F field, the helical air bubbles and the centric air stream are connected to each other. With the flow moving downstream, the helical air bubbles gradually separate from the centric air stream, and then both helical air bubbles and centric air stream keep breaking into smaller bubbles until all of them disappear in the far field. Through setting up the interface and defining two-phase flow displayable distribution, it also shows the air and water distribution on the two cross sections $x/D = -0.5$ and $x/D = -1.0$ after the paddle plane. The red and blue colors refer to air and water, respectively. It can be seen that the air-content ratio gradually decreases along the downstream direction of the wake.

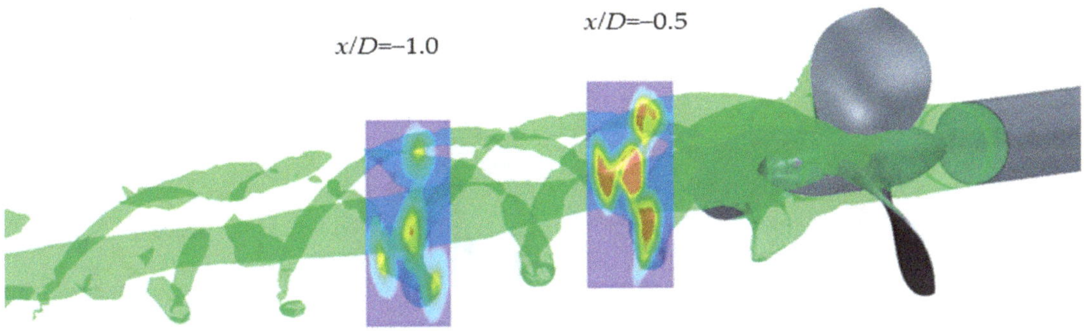

Figure 6. The wake after the ventilated propeller at $J = 1.15$ when $x/D = -1.0$ (left section) and $x/D = -0.5$ (right section).

3.2. Effect of Ventilation on the Hydrodynamic Performance of the SPP

Table 4 compares the numerical results for the ventilated SPP with the unventilated one under full submergence condition.

Table 4. Numerical results for the SPP under full submergence conditions.

Speed Coefficient J		Unventilated SPP			Ventilated SPP		
		K_t	$10K_q$	η	K_t	$10K_q$	η
0.85	Value	0.3763	0.9679	0.5259	0.3199	0.8452	0.5120
	Variation/%				−14.99	−12.68	−2.64
1.00	Value	0.3093	0.8579	0.5738	0.2749	0.7441	0.5880
	Variation/%				−11.12	−13.26	2.47
1.15	Value	0.2495	0.7502	0.6087	0.2308	0.6462	0.6537
	Variation/%				−7.49	−13.86	7.39
1.30	Value	0.1897	0.6426	0.6108	0.1867	0.5389	0.7168
	Variation/%				−1.58	−16.14	17.35

From Table 4, it can be seen that the thrust coefficient K_t and torque coefficient $10K_q$ of the propeller decrease after ventilation. With the growth of the speed coefficient J (from 0.85 to 1.30), the influence of ventilation on the thrust coefficient K_t gradually diminishes, while the effect on the torque coefficient $10K_q$ is reinforced. On the other hand, the efficiency η of the ventilated propeller is slightly less than the unventilated one under low speed coefficient J. However, with the growth of the speed coefficient J, the efficiency η of the ventilated propeller quickly increases as compared to the unventilated one.

As one sees, the thrust T and the torque Q decrease after ventilation. This is mainly due to the fact that the air bubbles attached to the surfaces of the blades after ventilation. However, with the growth of the speed factor, the effect of the ventilation on the thrust coefficient K_t is getting weaker, while the effect on the torque coefficient $10K_q$ becomes stronger, which makes the efficiency η of the ventilated propeller greater than the unventilated one.

3.3. Effect of Ventilation on the Pressure Distribution

Figure 7 portrays the pressure distribution cloud on the blades of the unventilated SPP at $J = 0.85$. Figure 7a,b show the pressure distribution on the pressure surface and the suction surface, respectively. The arrows in the figures indicate the rotational direction of the propeller. From Figure 7a, one notes that on the pressure surface the high pressure is mainly concentrated on the region near the leading edge of the blades. Obviously, the closer the leading edge, the higher the pressure. While the closer the trailing edge, the lower the pressure. On the thick trailing edge, the pressure may even be negative. In contrast, as shown in Figure 7b, the lower pressure mainly locates on the region near the leading edge of the blades.

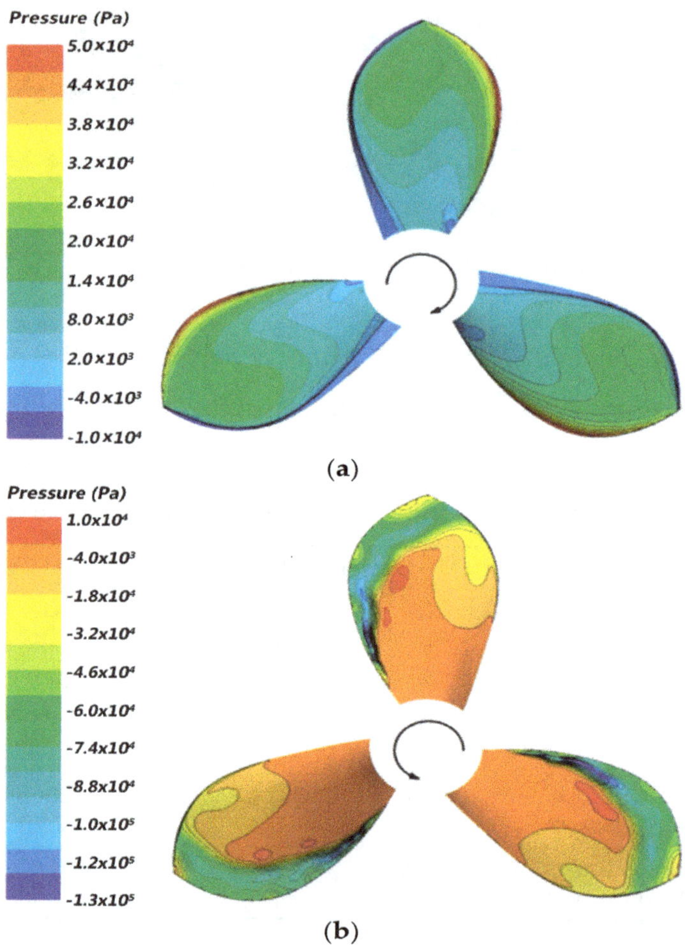

Figure 7. The pressure distribution on the blades at $J = 0.85$ under unventilated condition. (**a**) The pressure distribution on the pressure surface; (**b**) The pressure distribution on the suction surface.

As shown in Figure 7b, there exists a pressure jump near the leading edge, which is caused by the geometric characteristics of the SPP, i.e., the relative large pitch of the SPP induces a vortex formed after the leading edge, as shown in Figure 8. Figure 8a,b depict a non-dimensional velocity distribution cloud at the $0.5R$ section and a non-dimensional tangential velocity vector near the leading edge, respectively. The flow velocity is nondimensionalized with respect to the forward speed of the propeller V_A. From Figure 8a,b, one can find that there exists a high flow velocity zone on the suction surface near the leading edge. The local high velocity produces a sudden pressure drop on the suction surface. Moreover, it can be seen from Figure 8a that around the root of the thick trailing edge, the flow velocity is also very high, which makes the pressure in this zone be low and thus the air bubbles can adhere to this zone.

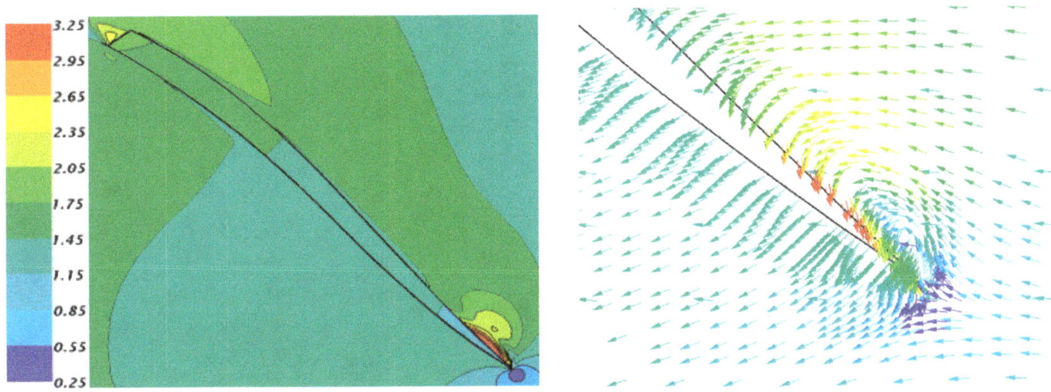

(a) (b)

Figure 8. Non-dimensional velocity distribution cloud at 0.5R section. (**a**) The non-dimensional velocity cloud; (**b**) The non-dimensional velocity vector cloud near the leading edge.

Figure 9 shows the pressure distribution on the surfaces of the ventilated SPP. Figure 9a,b are the pressure distribution on the pressure surface and the suction surface, respectively. Comparing Figure 9a with Figure 7a, it can be found that the pressure on the trailing edge of the pressure surface increases after ventilating due to the fact that air bubbles adhere to it. On the other hand, comparing Figure 9b with Figure 7b, it can be found that the area of the high pressure zone on the suction surface is significantly increased. This is because of the adsorption of air on the suction surface after ventilation. The increase in the area of the high pressure zone on the suction surface results in a reduction in the thrust and torque of the SPP.

Figure 10 plots the pressure distribution at $J = 0.85$ on various profiles (0.24R, 0.50R and 0.70R) of the ventilated/unventilated SPP. C_p is the pressure coefficient. Blue and red lines refer to the pressure distribution profiles of the unventilated and ventilated SPP, respectively. From Figure 10a one can see that on the profile near the root (0.24R), when there is no ventilation, the difference between the pressure on the pressure surface and the one on the suction surface is very large, especially near the leading edge. After ventilating, the pressure on the suction surface of the blade significantly increases and evenly distributes along the entire chord direction. The pressure almost remains unchanged because the air bubbles completely cover the root region of the blades, and in the vicinity of the leading edge ($x/C \leq 0.1$), the pressure on the pressure surface and the suction surface are almost the same. This is because both the pressure surface and suction surface near the leading edge are covered by the air bubble. The same phenomenon can be observed around the trailing edge.

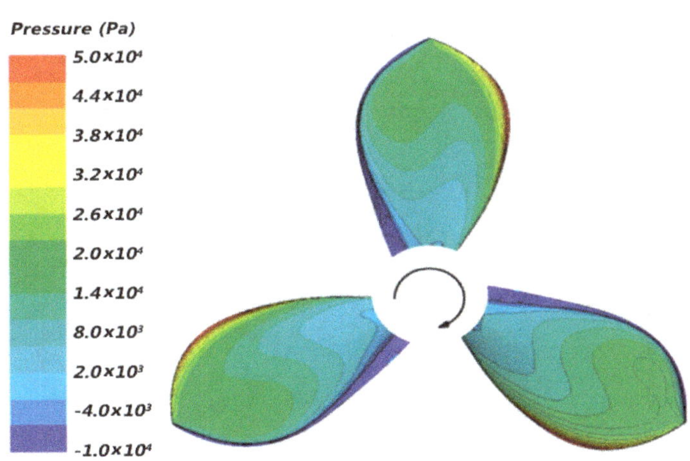

(a)

Figure 9. *Cont.*

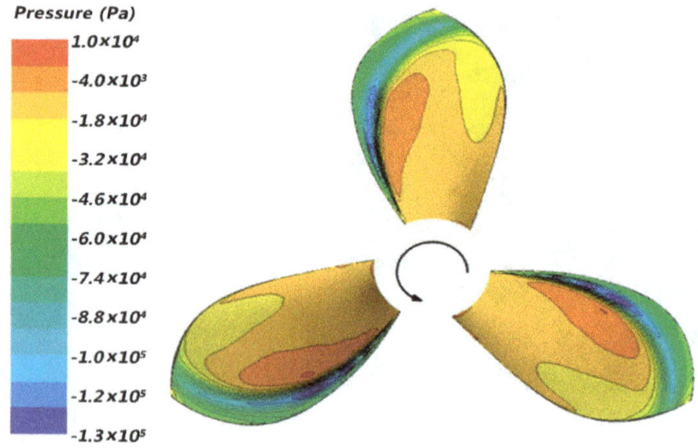

(**b**)

Figure 9. The pressure distribution on the blades at $J = 0.85$ under ventilated condition. (**a**) The pressure distribution on the pressure surface; (**b**) The pressure distribution on the suction surface.

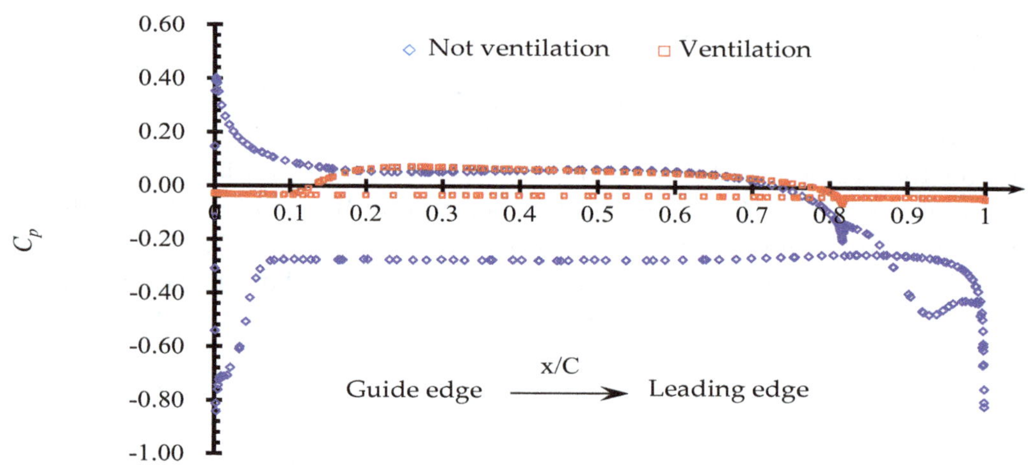

(**a**)

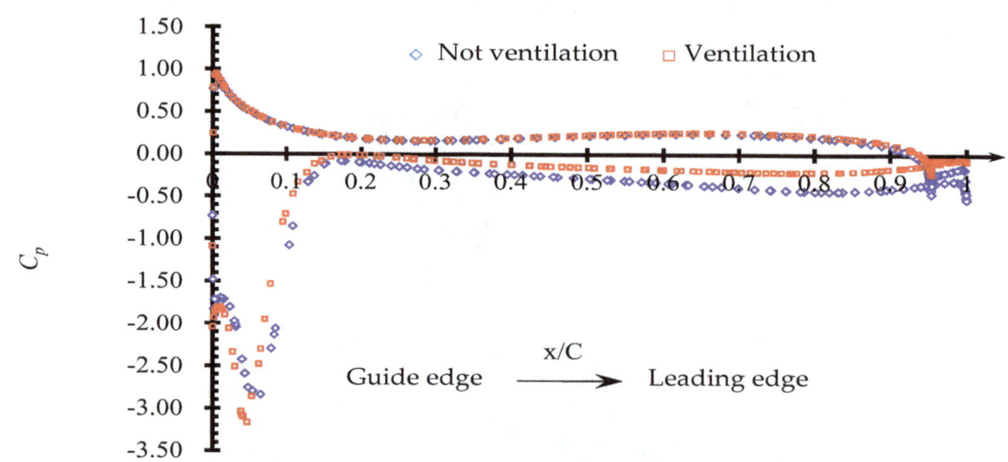

(**b**)

Figure 10. *Cont.*

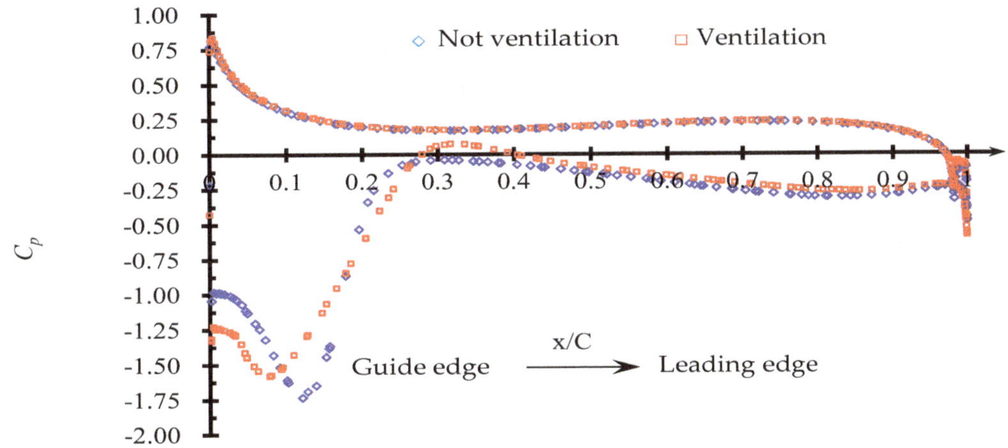

(c)

Figure 10. The pressure distribution profiles at $J = 0.85$. (**a**) The pressure distribution profile at 0.24R; (**b**) The pressure distribution profile at 0.50R; (**c**) The pressure distribution profile at 0.70R.

As shown in Figure 10b, on the profile 0.50R, the pressures on the pressure surface of the ventilated and unventilated SPP are almost the same. And only on the suction surface the pressure from the ventilated SPP is slightly bigger than the unventilated one due to the ventilation.

On the profile closer to the tip of the blade (0.70R), see Figure 10c, the pressure distribution on the pressure surface of the ventilated and unventilated SPP are completely the same. Only in the vicinity of the guide edge were the pressures on the suction surface of the ventilated and unventilated SPP slightly different, which is due to the non-uniformity of the flow field.

Comparing Figure 10a–c, one observes that the effect of ventilation on the pressure distribution of the SPP is limited to a zone in the vicinity of the blade root. In this zone, the ventilation has significant effect on the pressure. And along the outward radius direction, the effect of the ventilation gradually diminishes.

4. Conclusions

In this paper, the effect of artificial ventilation on the hydrodynamic performance of a surface-piercing propeller (SPP) was investigated using the CFD technique. The present work is different from previous works as follows. In previous works, such as Ghassemi et al. [6,7], Young et al. [9–11], and Alimirzazadeh et al. [13], the SPPs work under half immersed conditions with the propeller naturally ventilated. In such conditions, all blades experience periodic water-entry and water-exit processes. While in the present work, the SPP works under full immersion conditions with artificial ventilation through a vent pipe in front of the propeller disc. In such condition, all blades are partially surrounded by air bubbles, in which the phenomenon as well as the hydrodynamic performance of the SPP are significantly different from those given in literatures. The main conclusions are summarized as follows:

(1) The turbulence model SST and the overlapped mesh method can make faithful simulation for the ventilated SPP in unsteady flow field, and precisely forecast its hydrodynamic performance.

(2) In the wake of the ventilated SPP, there are three helical air bubbles that rotate about the centric air stream. With the flow moving downstream, the helical air bubbles and the centric air stream gradually break into smaller bubbles until all of them disappear in the far field.

(3) The effect of ventilation on the pressure distribution of the SPP is limited to a zone in the vicinity of the blade root (less than 0.5R). In this zone, the ventilation has significant effect on the pressure, and along the outward radius direction, the effect gradually diminishes.

(4) The thrust coefficient (K_t) and torque coefficient ($10K_q$) of the propeller decrease after ventilation. However, with the growth of the forward speed of the SPP, the influence of the ventilation on

the thrust coefficient is smaller than on the torque coefficient, which makes the efficiency (η) of the ventilated SPP, be significantly greater than that of unventilated SPP. The numerical results demonstrate the effectiveness of the ventilation approach for improving the hydrodynamic performance of the SPPs for high speed planning crafts.

Author Contributions: D.Y. performed the numerical calculations; Z.R. and Z.G. (Zhiqun Guo) completed the geometric modeling and mesh generation; Z.G. (Zhiqun Guo) and Z.G. (Zeyang Gao) analyzed the post-processing data; D.Y. wrote the paper.

References

1. Ding, E.B.; Tang, D.H.; Zhou, W.X. A review of research on semi-submerged propellers. *J. Ship Mech.* **2002**, *6*, 75–84.
2. Zhao, R.; Faltinsen, O. Water entry of two-dimensional bodies. *J. Fluid Mech.* **1993**, *246*, 593–612. [CrossRef]
3. Young, Y.L.; Savander, B.R. Numerical analysis of large-scale surface-piercing propellers. *Ocean Eng.* **2011**, *38*, 1368–1381. [CrossRef]
4. Yari, E.; Ghassemi, H. Numerical analysis of surface piercing propeller in unsteady conditions and cupped effect on ventilation pattern of blade cross-section. *J. Mar. Sci. Technol.* **2016**, *21*, 501–516. [CrossRef]
5. Yu, Y.Q.; Yu, J.X.; Ding, E.B. Numerical study of two-dimensional "cup" shaped supercavitation cross-section inflow. *J. Ship Mech.* **2008**, *12*, 539–544.
6. Ghassemi, H.; Shademani, R. Hydrodynamic characteristics of the surface-piercing propellers for the planing craft. *J. Mar. Sci.* **2009**, *8*, 267–274.
7. Kinnas, S.A.; Young, Y.L. Modeling of cavitating or ventilated flows using BEM. *Int. J. Numer. Methods Heat Fluid Flow* **2003**, *13*, 672–697. [CrossRef]
8. Young, Y.L.; Kinnas, S.A. Numerical analysis of surface piercing Propellers. In *2003 Propeller and Shaft Symposium*; Society of Naval Architects and Marine Engineers: Virginia Beach, VA, USA, 2003.
9. Young, Y.L.; Kinnas, S.A. Analysis of supercavitating and surface piercing propeller flows via BEM. *J. Comput. Mech.* **2003**, *32*, 269–280. [CrossRef]
10. Shi, Y.X.; Zhang, L.X.; Shao, X.M. Numerical simulation of hydrodynamic performance of semi-submerged propeller. *Mechatron. Eng.* **2014**, *31*, 985–990.
11. Alimirzazadeh, S.; Roshan, S.Z.; Seif, M.S. Unsteady RANS simulation of a surface piercing propeller in oblique flow. *Appl. Ocean Res.* **2016**, *56*, 79–91. [CrossRef]
12. CD-adapco Group. *STAR-CCM+ User Guide Version 9.06*; CD-adapco Group: Melville, NY, USA, 2014.
13. Zhang, J.; Fang, J.; Fan, B.Q. A review of theories and applications of VOF methods. *Adv. Sci. Technol. Water Resour.* **2005**, *4*, 67–70.
14. Menter, F.R. Zonal two-equation k-ω turbulence model for aerodynamics flows. In Proceedings of the 23rd Fluid Dynamics, Plasmadynamics, and Lasers Conference, Orlando, FL, USA, 6–9 July 1993; AIAA Paper 93-2906. The American Institute of Aeronautics and Astronautics (AIAA): Reston, VA, USA, 1993.
15. Calomino, F.; Alfonsi, G.; Gaudio, R.; D'lppplito, A.; Lauria, A.; Tafarojnoruz, A.; Artese, S. Experimental and Numerical Study of Free-Surface Flows in a Corrugated Pipe. *Water* **2018**, *10*, 638. [CrossRef]
16. Huang, S.; Wang, C.; Wang, S.Y. Application and comparison of different turbulence models in the calculation of propeller hydrodynamic performance. *J. Harbin Eng. Univ.* **2009**, *5*, 481–485.
17. Blocken, B.; Gualtieri, C. Ten iterative steps for model development and evaluation applied to Computational Fluid Dynamics for Environmental Fluid Mechanics. *Environ. Model. Softw.* **2012**, *33*, 1–22. [CrossRef]

4

Consistent Particle Method Simulation of Solitary Wave Interaction with a Submerged Breakwater

Yaru Ren [1], Min Luo [2,*] and Pengzhi Lin [1]

1. State Key Laboratory of Hydraulics and Mountain River Engineering, Sichuan University, Chengdu 610065, China; 2017223060054@stu.scu.edu.cn (Y.R.); cvelinpz@126.com (P.L.)
2. Zienkiewicz Centre for Computational Engineering, College of Engineering, Swansea University, Swansea SA1 8EN, UK

* Correspondence: min.luo@swansea.ac.uk.

Abstract: This paper presents a numerical study of the solitary wave interaction with a submerged breakwater using the Consistent Particle Method (CPM). The distinct feature of CPM is that it computes the spatial derivatives by using the Taylor series expansion directly and without the use of the kernel or weighting functions. This achieves good numerical consistency and hence better accuracy. Validated by published experiment data, the CPM model is shown to be able to predict the wave elevations, profiles and velocities when a solitary wave interacts with a submerged breakwater. Using the validated model, the detailed physics of the wave breaking process, the vortex generation and evolution and the water particle trajectories are investigated. The influence of the breakwater dimension on the wave characteristics is parametrically studied.

Keywords: consistent particle method; solitary wave; submerged breakwater; breaking wave; vortex

---

1. Introduction

Tsunamis possess tremendous destructive power and are among the most horrible natural hazards in coastal regions. For example, the catastrophic tsunami disaster in Indonesia in September 2018 caused more than 2100 deaths, more than 4400 people seriously wounded and more than 67,000 houses destroyed or damaged [1]. The most recent tsunami disaster triggered by the collapse of the volcano in Indonesia has also caused heavy casualties and damaged thousands of properties [2]. To protect the coastal communities, breakwaters are extensively constructed along the coast for reflecting waves and/or dissipating wave energy. Among the various types of breakwaters, the submerged breakwater has been frequently used because it has less effect on the ecosystem and is cheaper to construct [3]. However, the wave-structure interaction induces complex flows, which can be dangerous for underwater craft navigations and swimmers as well as transport excessive sediments and hence cause the erosion of the structure foundation. In this context, the present study aims to investigate the detailed physics of the tsunami wave-submerged breakwater interaction. The wave breaking process, vortex generation and evolution and fluid particle trajectory will be focused.

Solitary waves have resemblances to tsunami waves and hence are frequently used as the substitute in tsunami studies [4,5], although some studies looked into the differences between a solitary wave and a tsunami wave [6,7]. The review here discusses the works on both solitary and tsunami waves without emphasizing their differences. An early study of the solitary wave-structure interaction is Grilli et al. [8] who investigated the breaking characteristics of solitary waves over submerged and emerged trapezoidal breakwaters by experimental measurements of the wave profiles and a fully nonlinear potential model. Lin et al [9] studied the solitary wave runup and rundown on sloping beaches by using a RANS (Reynolds Average Navier-Stokes) model that was able to reproduce the breaking wave and the associated turbulence. Chang et al [10] studied the solitary wave interaction with a submerged rectangular obstacle and particularly measured the velocity field near the obstacle by the Particle Image Velocimetry (PIV). Huang and Dong [11] numerically studied the viscous interaction between a solitary wave and a submerged rectangular dike by a finite-analytical method and found that the primary vortex generated at the leeward of the dike and the secondary vortex at the right toe of the dike may scour the bottom of the dike. Based on laboratory experiments and the numerical model developed in Lin et al [9], Liu and Al-Banaa [12] studied the action of non-breaking solitary waves on a vertical surface-piercing barrier with the emphasis on wave runup and forces. Lin [13] systematically explored the wave reflection, transmission and dissipation of a solitary wave interacting with submerged rectangular obstacles using his RANS model. More recently, Hsiao and Lin [14] experimentally and numerically studied the solitary wave impinging and overtopping on a seawall, and Wu et al [15] investigated the propagation of solitary waves over a rectangular breakwater. In both studies, the characteristics of breaking waves and the accompanied vortexes were analyzed numerically based on the COBRAS model (COrnell BReaking And Structure), which is a cognate of Lin's numerical model.

All the numerical studies mentioned above are based on the mesh-based methods. In recent years, another category of numerical method has obtained significant developments, i.e., the particle method. Two typical methods belonging to this category are the Smoothed Particle Hydrodynamics (SPH) [16,17] and Moving Particle Semi-implicit (MPS) method [18]. Without predefined meshes, the particle methods are advantageous in dealing with the large deformations (e.g., fluid merging and splitting) and tracking the free surface or fluid interface, and hence have been extensively applied to study free surface flows [19,20] and wave-structure interaction problems [21–23]. The most up-to-date developments of particle methods are reviewed in [24]. The pioneer particle method study of solitary wave-structure interaction is Monaghan and Kos [25], who studied the run-up and return of a solitary wave on a beach. Shao [26] is the pioneer of using the incompressible SPH (ISPH) for solitary wave interaction with structure. Specifically, this work investigated the characteristics of wave reflection, transmission, and dissipation of solitary waves impinging on a vertical surface-piercing barrier. Zhang and Tang [27] employed the MPS method to study the interaction between the solitary wave and a flat plate. Huang and Zhu [28] investigated the tsunami actions on a seawall.

Although a large number of researches have investigated the solitary wave interaction with submerged structures, some key phenomena associated with the breaking wave generation and evolution in the wave-structure interaction process are not fully understood. The present study aims to reveal the physical process of these phenomena. Particularly for the particle methods such as SPH and MPS, although having the advantage of being able to capture large deformations, they suffer from spurious pressure fluctuations [29]. One major cause is that the derivative approximation schemes introduce numerical errors particularly for irregular particle distributions [30]. To overcome this issue, the Consistent Particle Method (CPM) has been developed [30]. Different from the kernel approximation schemes in SPH and the weighted-average particle interaction models in MPS, the CPM computes the first- and second-order derivatives simultaneously based on the Taylor series expansion [30]. In this way, the numerical consistency that is a key issue in the derivative approximation of particle methods can be achieved and hence the numerical accuracy is improved. The performance of CPM in alleviating spurious pressure fluctuation has been well demonstrated by the cases of violent free surface flow [30,31] and fluid-structure interaction [32]. Another issue that has restricted the application of particle methods in large-scale problems is the relatively low computational efficiency. CPM adopts the OpenMP parallel computing to accelerate the computation [33], and the Graphics Processing Unit parallel computing is ongoing.

In this paper, the CPM is used to study the solitary wave interaction with a submerged breakwater. The numerical accuracy is validated by the published experimental results of wave elevation, wave profile and wave velocity distributions at three sections. Using the validated model, the characteristics of the breaking wave process and the associated vortex structure will be discussed in detail. The water particle trajectories will be analyzed to get insights into the possible sediment transport. And a parametric study will be conducted to explore the influences of the breakwater dimension on the fluid fields. It is expected that the research findings can provide some guidance for the engineering design.

2. CPM Methodology

2.1. Governing Equations

The governing equations of CPM are the conservations of mass and momentum, i.e., the Navier-Stokes equations, as follows:

$$\frac{1}{\rho}\frac{D\rho}{Dt} + \nabla \cdot \mathbf{v} = 0 \tag{1}$$

$$\frac{D\mathbf{v}}{Dt} = -\frac{1}{\rho}\nabla p + \nu\nabla^2\mathbf{v} + \mathbf{g}, \tag{2}$$

where ρ is the density of a fluid, $\mathbf{v}$ the particle velocity vector, p the fluid pressure, $\mathbf{v}$ the kinematic viscosity of a fluid, $\mathbf{g}$ the gravity acceleration and t the time. The fluid domain is discretized by non-connecting particles. Each particle has a fixed mass and moves under the external forces arising from the gravity, pressure difference and viscosity as in Equation (2).

2.2. Two-Step Projection Method

CPM solves the governing equations by a two-step projection method [34]. In the predictor step, the intermediate particle velocities and positions are computed by neglecting the pressure gradient term as follows:

$$\mathbf{v}^* = \mathbf{v}^{(k)} + \left[\nu\nabla^2\mathbf{v}^{(k)} + \mathbf{g}\right]\Delta t \tag{3}$$

$$\mathbf{r}^* = \mathbf{r}^{(k)} + \mathbf{v}^*\Delta t, \tag{4}$$

where $\mathbf{v}^{(k)}$ and $\mathbf{r}^{(k)}$ are the particle velocity and position in the k-th (previous) time step respectively, $\mathbf{v}^*$ and $\mathbf{r}^*$ are the temporary particle velocity and position, respectively, and Δt is the time step size.

In the corrector step, a pressure Poisson equation (PPE) is derived as:

$$\nabla \cdot \left(\frac{1}{\rho^*}\nabla p^{(k+1)}\right) = \frac{1}{\Delta t^2}\frac{\rho^{(k+1)} - \rho^*}{\rho^{(k+1)}}. \tag{5}$$

The fluid incompressibility condition is enforced by setting the fluid density at the current time step ($\rho^{(k+1)}$) to the initial value (ρ_0). The intermediate fluid density ($\rho*$) of a particle is computed by evaluating the relative distances of this particle to its neighbor particles within the influence domain (the influence radius $r_e = 2.1L_0$, where L_0 the initial particle spacing) [31].

Applying a proper derivative computation scheme to the left-hand side of Equation (5), a linear equation system with sparse and non-symmetric coefficients can be obtained and solved efficiently. With the solved pressure, particle velocities and positions in the entire computational domain are corrected as:

$$\mathbf{v}^{(k+1)} = \mathbf{v}^* - \left(\frac{\nabla p}{\rho}\right)^{(k+1)}\Delta t \tag{6}$$

and

$$\mathbf{r}^{(k+1)} = \mathbf{r}^{(k)} + 0.5\left(\mathbf{v}^{(k)} + \mathbf{v}^{(k+1)}\right)\Delta t. \tag{7}$$

The computational time step size Δt is governed by the Courant condition as

$$\frac{v_{\max}\Delta t}{L_0} \le C_{\max}, \tag{8}$$

where $v_{\max}$ is the maximum particle velocity at the k-th time step and the coefficient C_{max} is selected to be 0.25.

This above is a standard solving scheme that has been used in other particle methods such as MPS [18] and ISPH [26]. The distinct feature of CPM compared to other particle methods lies in the computation of the spatial derivatives in the governing equations, as elaborated in the following section.

2.3. Derivative Computation Based on Taylor Series Expansion

The Taylor series expansion for a smooth function $f(x, y)$ in the vicinity of a reference particle (x_0, y_0) can be expressed as:

$$f(x,y) = f_0 + hf_{,x0} + kf_{,y0} + \frac{1}{2}h^2 f_{,xx0} + hkf_{,xy0} + \frac{1}{2}k^2 f_{,yy0} + O(r^3), \tag{9}$$

where $h = x - x_0$, $k = y - y_0$, $f_0 = f(x_0, y_0)$, $f_{,x0}$ is the first order derivative of function f with respect to x at (x_0, y_0), and $f_{,xy0}$ the second-order derivative of function f with respect to x and y at (x_0, y_0). By writing Equation (9) for each of the neighboring particles, the following equation system can be obtained:

$$[\mathbf{A}]\{\mathbf{Df}\} - \{\mathbf{f}\} = 0, \tag{10}$$

where $[\mathbf{A}]$ is a function of relative particle positions (i.e., h and k), $\{\mathbf{f}\}$ is a combination of the variable differences between the reference particle and its neighboring particles, (i.e., $f - f_0$), and $\{\mathbf{Df}\}$ is a vector including the five derivatives in Equation (9). Theoretically, five neighboring particles are enough to solve Equation (10). In practice, more particles, (typically 12–18 for 2D problems), are involved and hence Equation (10). is over-determined. The over-determined equation system is solved by the weighted-least-squares scheme [35], giving the first- and second-order derivatives simultaneously as follows [30]:

$$\frac{\partial p_i}{\partial x} = \sum_{j \neq i} \left[w_j^2 \left(a_1 h_j + a_2 k_j + 0.5 a_3 h_j^2 + a_4 h_j k_j + 0.5 a_5 k_j^2 \right) (p_j - p_i) \right] \tag{11}$$

and

$$\frac{\partial^2 p_i}{\partial x^2} = \sum_{j \neq i} \left[w_j^2 \left(c_1 h_j + c_2 k_j + 0.5 c_3 h_j^2 + c_4 h_j k_j + 0.5 c_5 k_j^2 \right) (p_j - p_i) \right], \tag{12}$$

where w_j is the weighting function used in the weighted-least-squares approximation to solve an overdetermined equation system. Note that this weighting function is essentially different from the kernel in SPH and the weighting function in the particle interaction model of MPS, both of which serve as the weighting in the weighted average calculation of function value or derivatives. p_i in Equations (11) and (12) is the pressure on particle i, and a and c the coefficients generated by the weighted-least-squares scheme (refer to Equation (21) in [30]).

2.4. Free Surface and Solid Boundaries

In CPM, the free surface particles are recognized by the "arc" method [35]. If any arcs of a circle around a reference particle are not covered by the circles of its neighbors, this reference particle is a free surface particle, (the reader is referred to [30] for more details). The essential boundary condition, i.e., $p = 0$, is enforced on the free surface particles. The impermeable walls are modeled by the mirror particle approach [26,36], i.e., generating fictitious particles outside the physical boundary by mirroring real fluid particles along the boundary line. Particularly, a moving wall can be simulated by updating the axis (in accordance with the physical wall motion) along which mirror particles are generated. This is an intrinsic advantage compared to the mesh-based methods. The submerged structure is simulated by fixed particles. On all the solid boundaries, the Neumann boundary condition $\partial p/\partial n = -\rho g \times n$ is applied, where n is the outward unit vector of the solid boundary.

3. Parameters of the Studied Problem

The CPM model is used to study the experimental case of solitary wave interaction with a submerged bottom-mounted breakwater in Wu et al. [15]. Figure 1 is the schematic view of this example, in which the length of the wave flume is 6.5 m, the water depth $h = 0.14$ m and the rectangular structure (dimension 0.1 m $\times$ 0.02 m) located 2.5 m from the home position of the wave maker. Wu's experimental study was conducted in the wave flume at Tainan Hydraulics Laboratory, National Cheng Kung University. Wave heights at E1 ($x = -0.657$ m), E2 ($x = 0.010$ m) and E3 ($x = 0.357$ m) were measured by wave gauges. The wave profile and velocity filed in a certain area at the leeward of the breakwater were measured by a PIV system. The solitary wave was produced by a translational wave maker, whose motion was generated by the procedure presented in Goring [37]. In CPM simulations, stationary particles of initial particle distance $L_0 = 0.0025$ m are generated in the whole domain. Using the approach introduced in Section 2.3 to move the wave maker in the described way, a solitary wave can be produced. A fixed time step $\Delta t = 0.0005$ s is adopted to satisfy the Courant condition. The CPM results of wave elevations at E1, E2 and E3, wave profiles and wave velocities at three sections (V1-$x = 0.06$ m, V2-$x = 0.10$ m and V3-$x = 0.14$ m, as shown in Figure 1) are compared with the experimental results of Wu et al. [15].

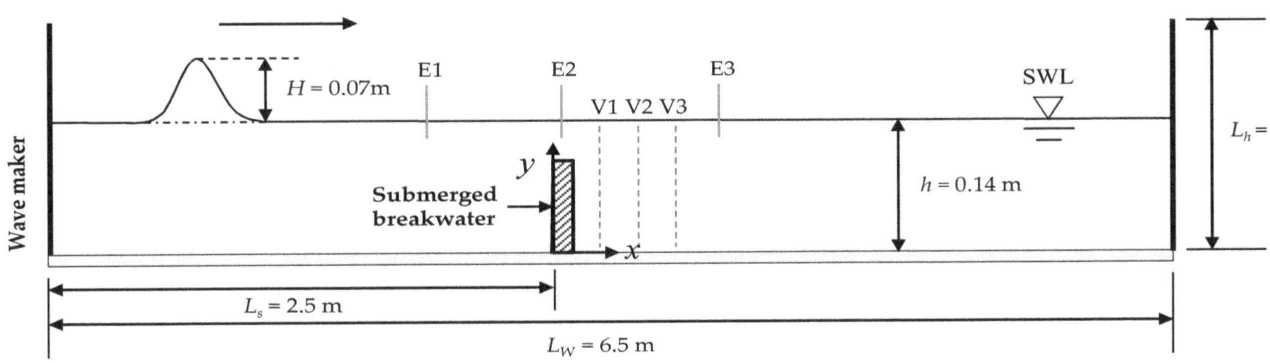

E1: x = -0.657 m E2: x = 0.010 m E3: x = 0.357 m V1: x = 0.060 m V2: x = 0.100 m V3: x = 0.140 m

Figure 1. Schematic view of the solitary wave propagating over a submerged breakwater.

4. Results and Discussions

4.1. Wave Transmission and Reflection

The wavemaker generates a solitary wave of wave height of $H = 0.07$ m, which propagates forward. The wave elevations just in front of (E1), at the location of (E2) and just behind of (E3) the breakwater is presented in Figure 2. The time when the solitary wave crest occurs at E1 is $t = 0$ s. In general, the CPM results and the COBRAS numerical results of Wu et al. [15] match quite well and are in a good agreement with Wu's experimental data [15]. The relative differences of these two sets of numerical wave crests against the experimental data are presented in Table 1. The maximum difference is 3.1% for COBRAS and 0.4% for CPM. Both numerical models are very accurate in predicting the wave elevations. From the next section on, the capability of CPM in reproducing the wave breaking process and tracking the particle trajectories will be demonstrated.

The solitary wave-breakwater interaction happens when the wave approaches the structure. The wave crests at E1 and E3 are treated as the amplitudes of the incident and transmitted waves. Since most of the energy of a solitary wave concentrates on the bulge part of the wave, the ratio of the transmitted and incident wave heights, i.e., Ht/Hi, is used as an indicator of the wave transmission. Based on this, the wave transmission coefficient of this case is 92.6%. The breakwater also reflects part of the wave, which is manifested in the second crest of wave elevation at E1 that occurs at around $t = 1$ s. With the same argument for the wave transmission, the wave reflection coefficient is evaluated by the ratio of the reflected and incident wave heights, i.e., Hr/Hi. It is computed to be 17.7% for this case. Analytical solutions have been derived for estimating the wave transmission (Equation 4.8 in [38]) and reflection (Equation (11) in [13]) coefficients of a solitary wave propagating rectangular submerged obstacles. Based on these equations, the wave transmission and reflection coefficients of this case are 95.3% and 18.1%, the relative differences between which and the CPM solutions are 2.8% and 2.2%, respectively.

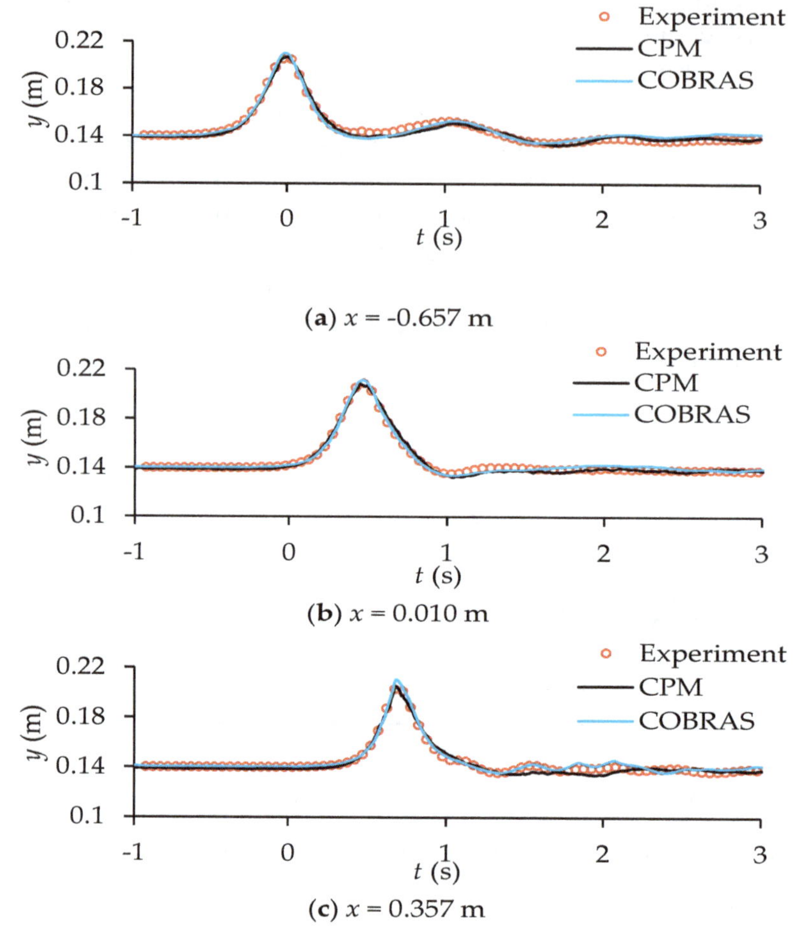

Figure 2. Wave elevations at E1, E2 and E3: (**a**) $x = -0.657$ m; (**b**) $x = 0.010$ m; (**c**) $x = 0.357$ m.

Table 1. Relative differences between the numerical and experimental results of wave crests at E1, E2 and E3.

Numerical models	Relative Errors Against Experimental Data in [15]		
	E1	E2	E3
COBRAS in [15]	2.0%	1.4%	3.1%
CPM	0.2%	0.4%	0.2%

4.2. Fluid Characteristics of Wave-Structure Interaction

When the wave approaches the breakwater, the wave accelerates due to the blockage of part of the flume section. This leads to a jet flow toward the leeward side of the breakwater. Accompanied is a clockwise vortex behind and at an elevation near the crest of the breakwater. These phenomena can be clearly seen from the velocity profile at $t = 0.46$ s as shown in Figure 3a. The jet flow impinges the main water body behind the breakwater. Due to the velocity of the main water body being smaller than that of the impinging flow, the elevation of the water body is raised (see Figure 3b). The raised water body and the crest of the incoming jet flow form the double crests. This is the so-called crest-crest exchange phenomenon [15,39]. As the main crest of the solitary wave propagates downstream, the jet flow impingement increases the steepness of the tail of the transmitted wave (see the wave profile at t = 0.74 s). The impinging flow continuously reduces the water elevation of the impinging location and causes the tail surface to break in the direction opposite to that of the main wave direction (Figure 3d). In this process, the vortex develops in size and moves downstream with the wave. Specifically, the vortex occupies the area of 0.2 m × 0.1 m behind the breakwater at t = 0.88 s. It is noteworthy that the CPM model has accurately reproduced the experimental wave profiles at the time instants discussed above. And the predicted velocity distributions at sections V1, V2 and V3 are in a reasonably good agreement with the experimental results as shown in Figure 4. Although some discrepancies exist near the center of the vortex region, the numerical model is able to capture the key phenomena during this process. This shows the satisfactory accuracy of CPM.

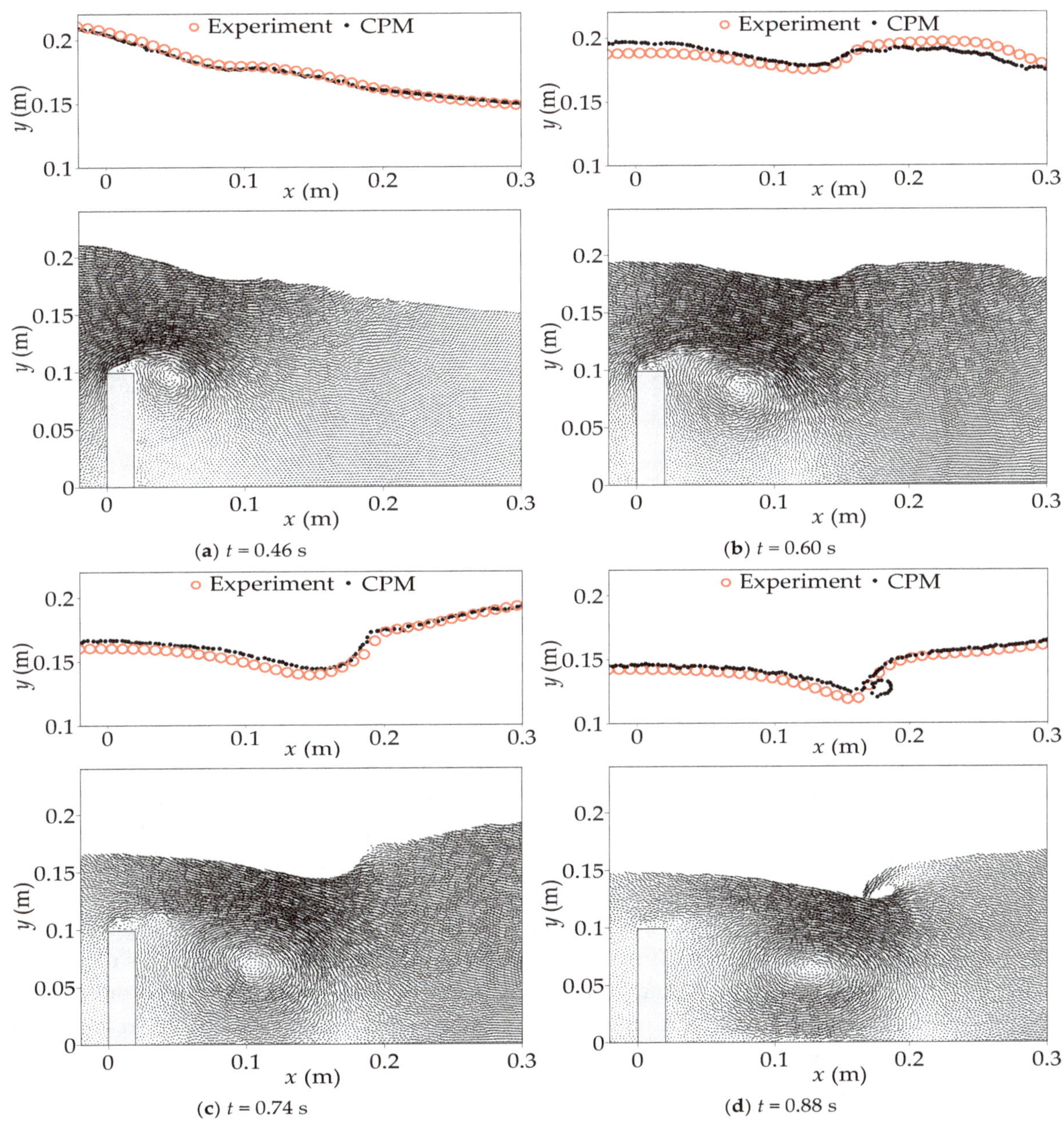

Figure 3. Wave profiles and velocity fields at typical time instants: (**a**) t = 0.46 s; (**b**) t = 0.60 s; (**c**) t = 0.74 s; (**d**) t = 0.88 s.

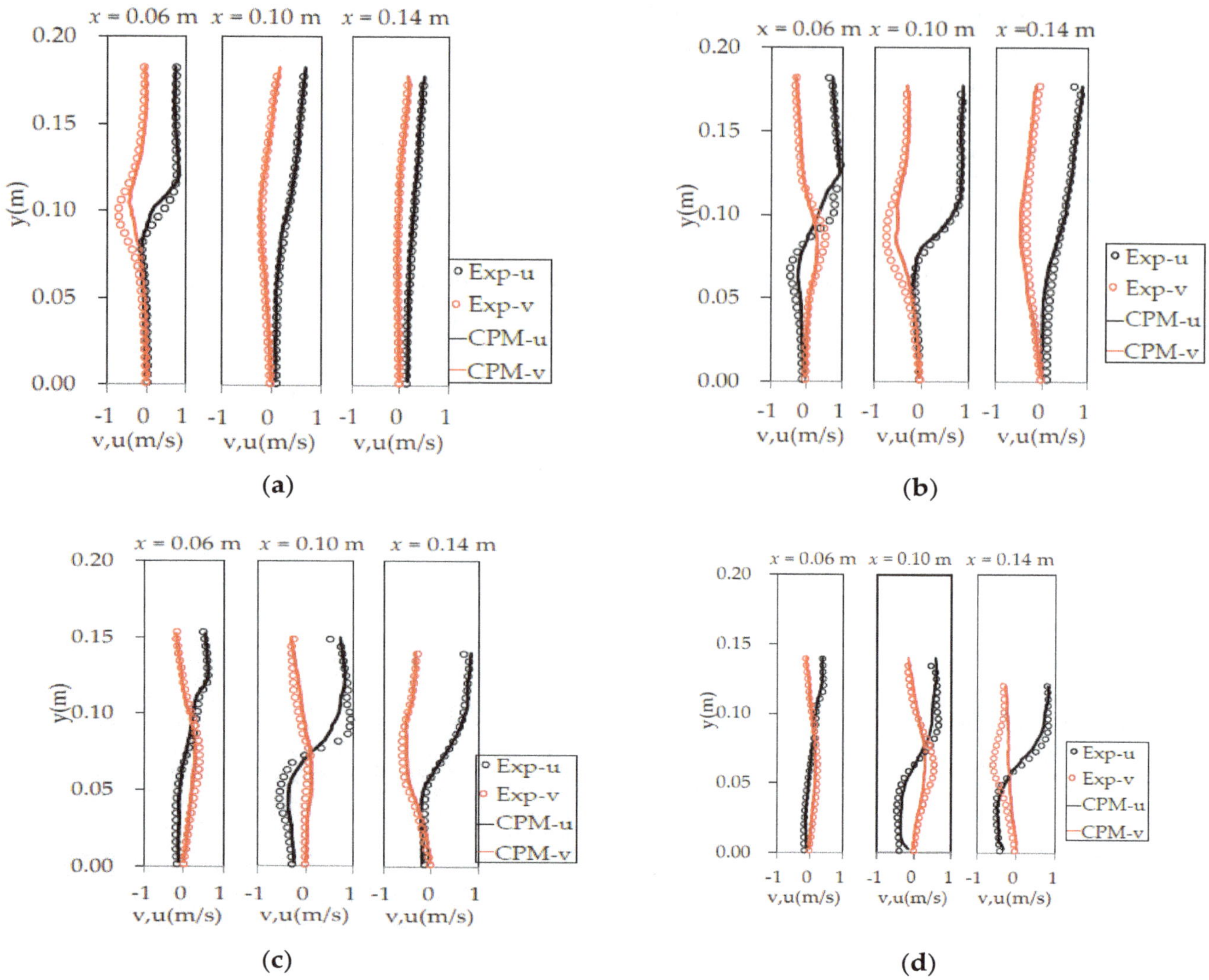

Figure 4. Horizontal and vertical velocity distributions at the sections of $x = 0.06$ m, $x = 0.10$ m and $x = 0.14$ m at typical time instants: (**a**) $t = 0.46$ s; (**b**) $t = 0.60$ s; (**c**) $t = 0.74$ s; (**d**) $t = 0.88$ s.

The backward (in terms of the main wave direction) plunging breaker interacts with the forward main jet flow, inducing energetic and complicated local fluid motion. Specifically, the impingement of the plunging breaker on the main jet causes a velocity bifurcation, as shown by the bent arrows in Figure 5a. The fluid with an upward velocity component forms a second jet, while the majority of the main jet maintains the original moving direction and merges into the main water body. The second jet rises to an elevation higher than the crest of the plunging wave because the wave kinetic energy is converted to the potential energy, and then splits into two parts (see Figure 5b). The leftward part overturns and impinges on the main jet flow again, and the rightward part impinges on the plunging breaker. All violent wave-breaking scenarios during this process dissipate wave energy. Another interesting phenomenon accompanying this process is the air entrapment as shown in Figure 5a. Since the present CPM model does not include the air phase, the effects of the air entrapment on the local fluid characteristics are not simulated. To capture this in detail, a two-phase simulation that allows for the compressible air necessitates and is left for the future work. The air entrapment region is occupied by water very quickly because of gravity. The fluid that surrounds this region forms an anti-clockwise vortex as shown in Figure 5b.

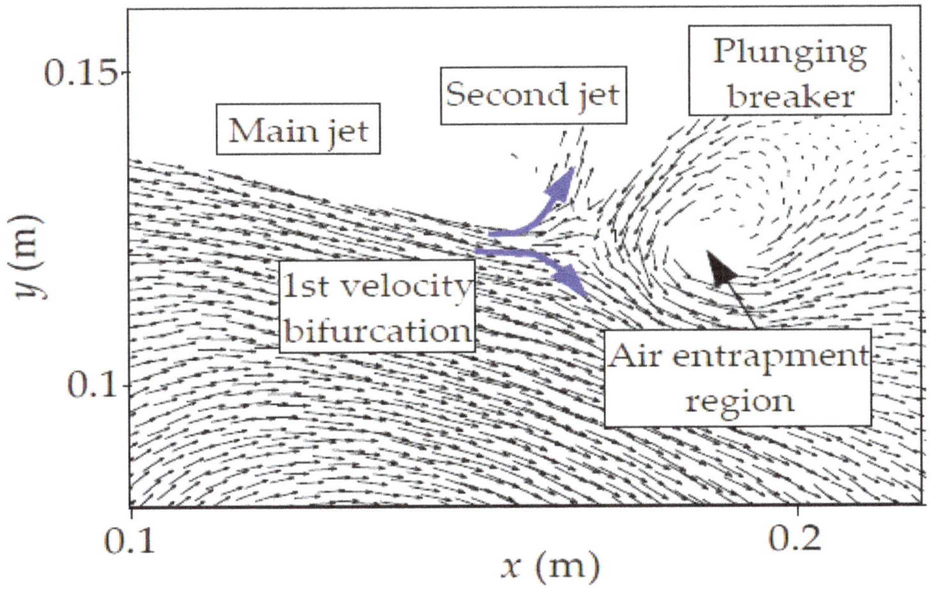

(**a**) $t = 0.92$ s

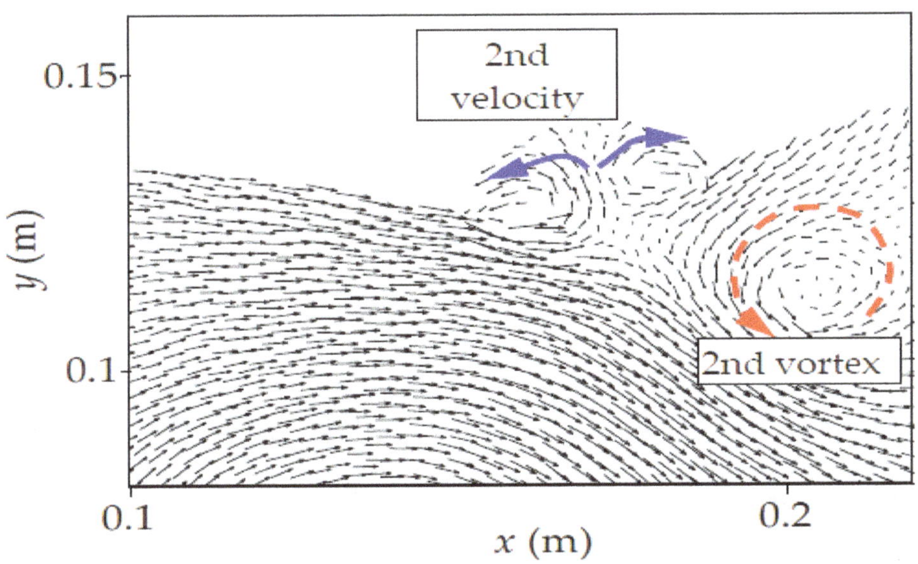

(**b**) $t = 0.99$ s

Figure 5. The interaction between the backward plunging breaker and the main jet flow: (**a**) $t = 0.92$ s; (**b**) $t = 0.99$ s.

The main vortex induced by the main jet flow and the second vortex initiated by the plunging breaker keep on developing after the main body of the solitary wave passes by this region. While the second vortex is rotating itself, it rotates around the main vortex, liking the movement relationship between a satellite and a planet. When the water at the leeward starts to flow back to the seaward at around $t = 1.2$ s (manifested in the wave elevation at E2 in Figure 2), a third vortex is initiated near the leeward toe of the breakwater (see Figure 6a). By rotating around the main vortex, this vortex subsequently goes up to the free surface (Figure 6b,c). Due to the energy dissipation induced by fluid viscosity, the third and second vortex disappear at $t = 2.55$ s (Figure 6d) and $t = 3.59$ s (Figure 6e), respectively. The backflow goes to the wave paddle and is reflected again. This new incident wave that has a much smaller amplitude pushes the remaining main vortex onshore and generates a small vortex near the crest of the breakwater at the leeward (Figure 6f). These vortexes will disappear eventually under the action of fluid viscous force.

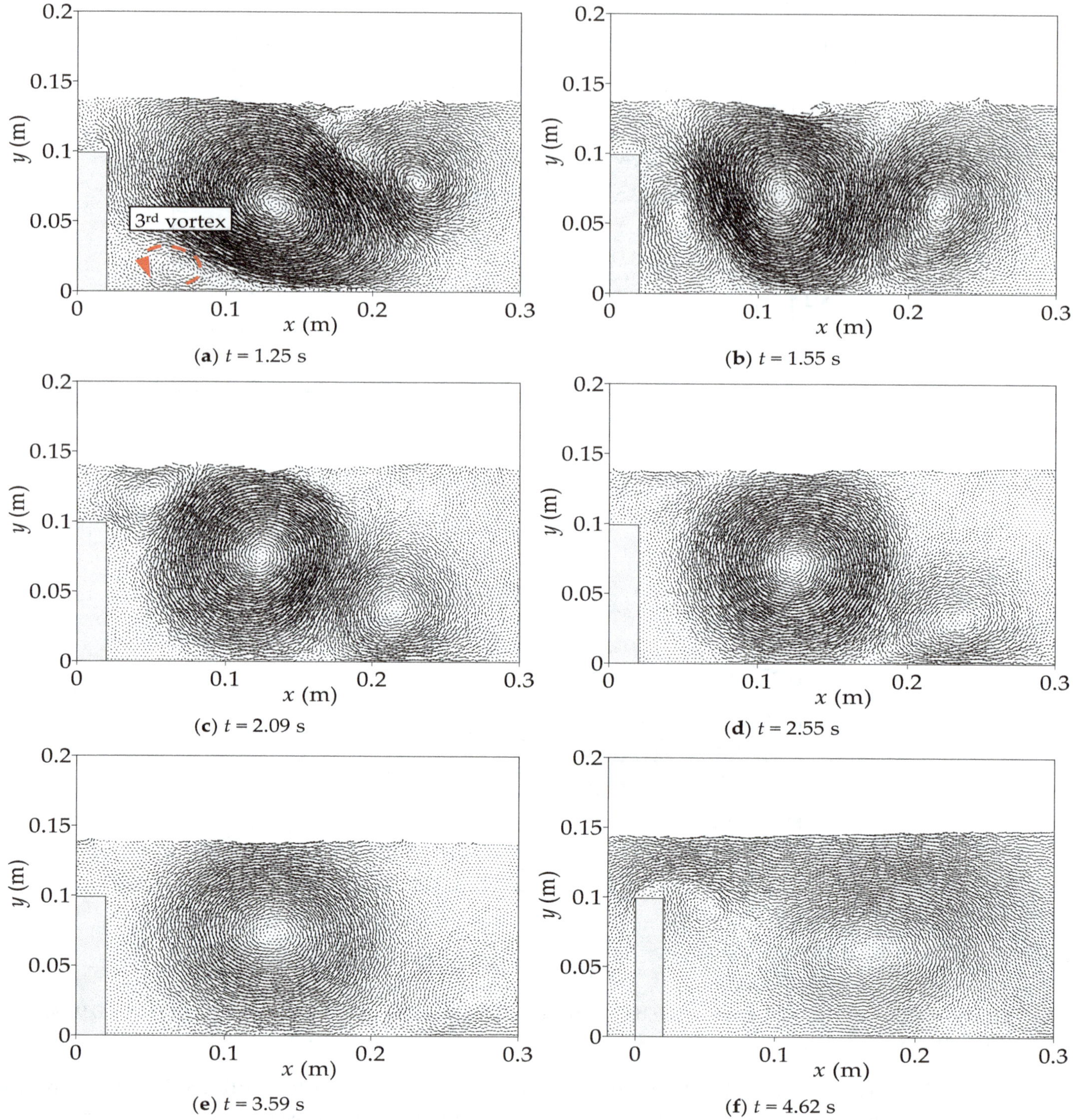

Figure 6. Snapshots of backward plunging breaker and forward jet flow interaction: (**a**) t = 1.25 s; (**b**) t = 1.55 s; (**c**) t = 2.09 s; (**d**) t = 2.55 s; (**e**) t = 3.59 s; (**f**) t = 4.62 s.

4.3. Trajectories of the Fluid Particles

The vortex transports sediment particles and can cause the failure of structure foundations or excessive sediment deposition, both of which will affect the proper functioning of a submerged breakwater. Since the sediment motion is closely related to fluid motion, this section discusses the trajectories of fluid particles at typical locations, to provide some insights into the possible sediment transportations. Note that, because of the Lagrangian description, the particle method offers convenience for this analysis.

Eight particles (the initial positions shown in Figure 7) at the seaward and leeward sides of the breakwater are traced and the trajectories are shown in Figure 8. Particles 1 and 2, which are initially located at the seaward and at an elevation of half of the breakwater height, go along the onshore direction with the main jet flow because of the vortex and their trajectories display spirals, as shown in Figure 8a,b. Particle 3 moves from a position near the flume bed to an elevation more than half of the breakwater (Figure 8c). The seaward bottom corner of the structure is a stagnation point. Hence, particle 4 that is near to structure toe has a small velocity during this period. However, this particle does move up a distance (Figure 8d). It can be anticipated that, with the actions of some more waves, particles 3 and 4 will go up further and eventually move to the leeward of the breakwater. The trajectories of particles 1 to 4 imply that the running-up component of the wave upon the interaction with the breakwater has the power to move the sediment particles near the seaward toe to the leeward. This will induce the erosion of the breakwater foundation.

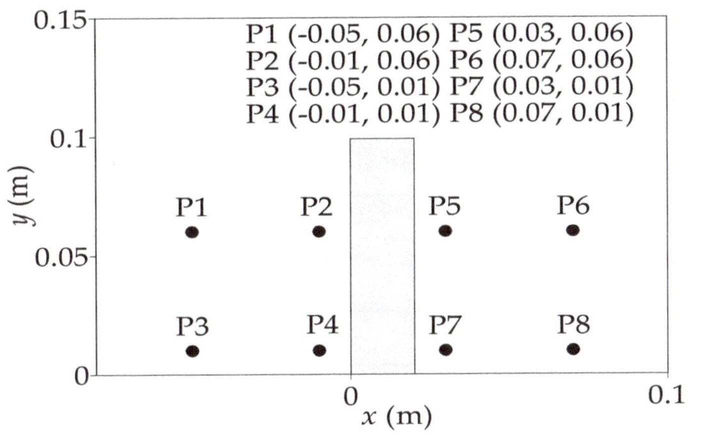

Figure 7. Initial positions of the traced particles.

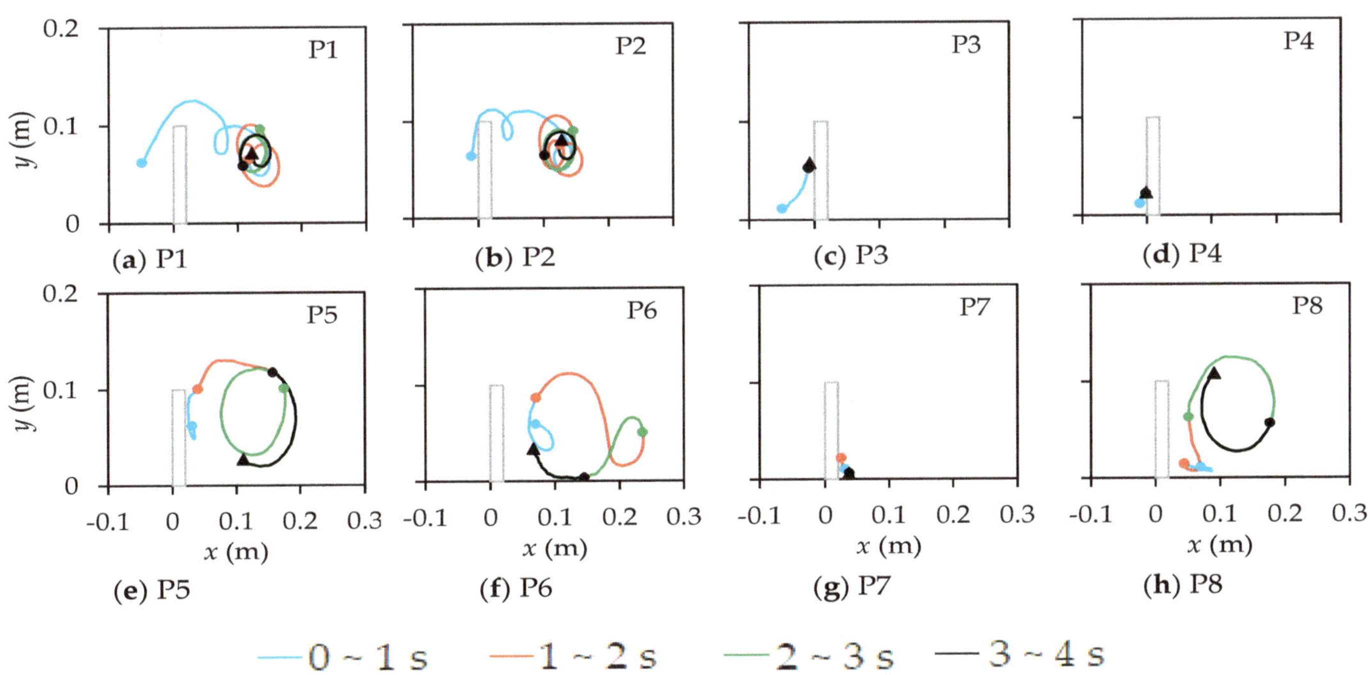

Figure 8. Trajectories of eight particles from 0 ~ 4 s: Blue line (0 ~ 1 s), red line (1 ~ 2 s), green line (2 ~ 3 s) and black line (3 ~ 4 s)

The particles originally located at the leeward and at an elevation of half of the breakwater height (i.e., particles 5 and 6) move with the main vortex, and the trajectories are spirals (Figure 8e,f). Particle

8 that initially locates near the flume bed is easily "transported" by the main vortex (Figure 8h). Similar to particle 4, particle 7 (near to the leeward toe of the breakwater) does not have a long trajectory as shown in Figure 8g. The main feature of the water particle trajectories at the lee side of the breakwater is that the vortex flow initiates the sediment particles. With the combined actions of vortexes and incident waves, the sediment particles will be transported onshore and the foundation of the breakwater will be eroded. Therefore, these factors should be considered carefully in practical designs.

5. Influence of Breakwater Dimension on Fluid Characteristics

In practical applications, the dimension of a breakwater and the wave conditions are important factors that affect the effectiveness of the breakwater in preventing incident waves and dissipating wave energy. To provide some insights, this section conducts a parametric study by analyzing the wave characteristics of the abovementioned solitary wave interacting with six rectangular submerged breakwaters, whose dimensions are shown in Table 2. The wave transmission (K_T) and reflection coefficients (K_R) with breakwaters of different dimensions are shown in Figure 9. As can be seen, increasing the breakwater height increases the wave reflection and reduces the wave transmission, while the breakwater width has little effect on wave reflection and transmission.

Table 2. Dimensions of six rectangular breakwaters.

Dimension	Case 1	Case 2	Case 3	Case 4	Case 5	Case 6
Width a (m)	0.02	0.02	0.02	0.1	0.1	0.1
Height b (m)	0.08	0.1	0.12	0.08	0.1	0.12

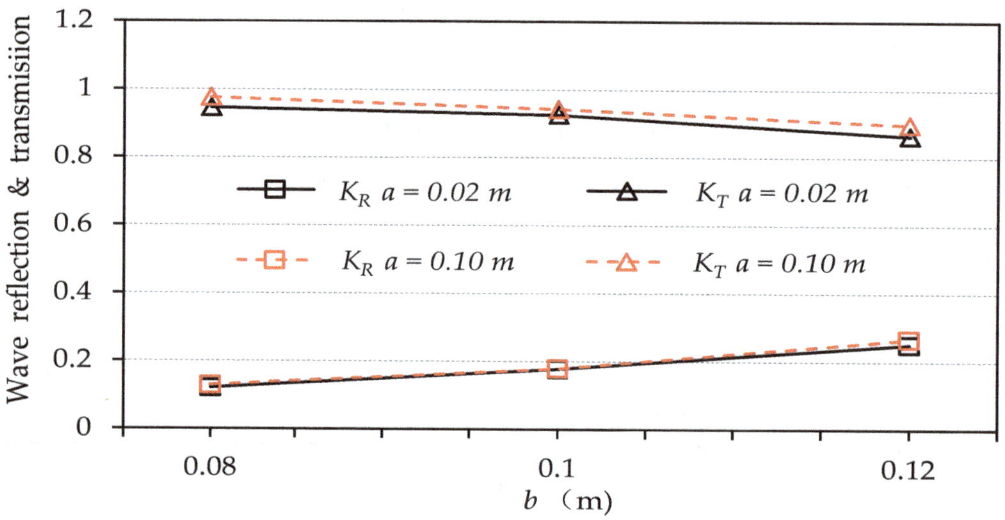

Figure 9. Wave transmission and reflection coefficients with breakwaters of difference dimensions.

When the breakwater width is fixed, a higher breakwater causes more wave reflections and hence less transmitted waves. However, because of the reduction of the section that the wave goes through, the jet flow that impinges towards the lee side has a larger velocity. This induces breaking waves and vortexes of higher intensity. Besides, the main vortex is further away from the structure because the jet flow ejects further from the crest of the breakwater. These phenomena are manifested by the velocity fields of cases 1 (Figure 10a,b), 2 (Figure 3b,d) and 3 (Figure 10c,d) at $t = 0.60$ s and 0.88 s. The strong vortex induces more sediment transportations and hence erosions. In practical designs, therefore, one needs to consider the balance between reflecting more waves and the severity of the erosion at the lee side of the breakwater. The effect of the breakwater width on the flow field is demonstrated by comparing cases 2 and 5 (Figure 10e,f). The key difference is that the main vortex in the case with a wider breakwater is closer to the lee side toe of the breakwater. The practical implementation is that more sediment particles near the structure toe will be initiated and transported away, causing the attached scour as termed in Young and Testik [40]. For a narrow breakwater, since the main vortexes are relatively far from the toe of the breakwater, the sediment particles underneath of the main vortexes are initiated. Some sediments are transported closer to the breakwater and deposited at the lee side toe. This causes the detached scour [40].

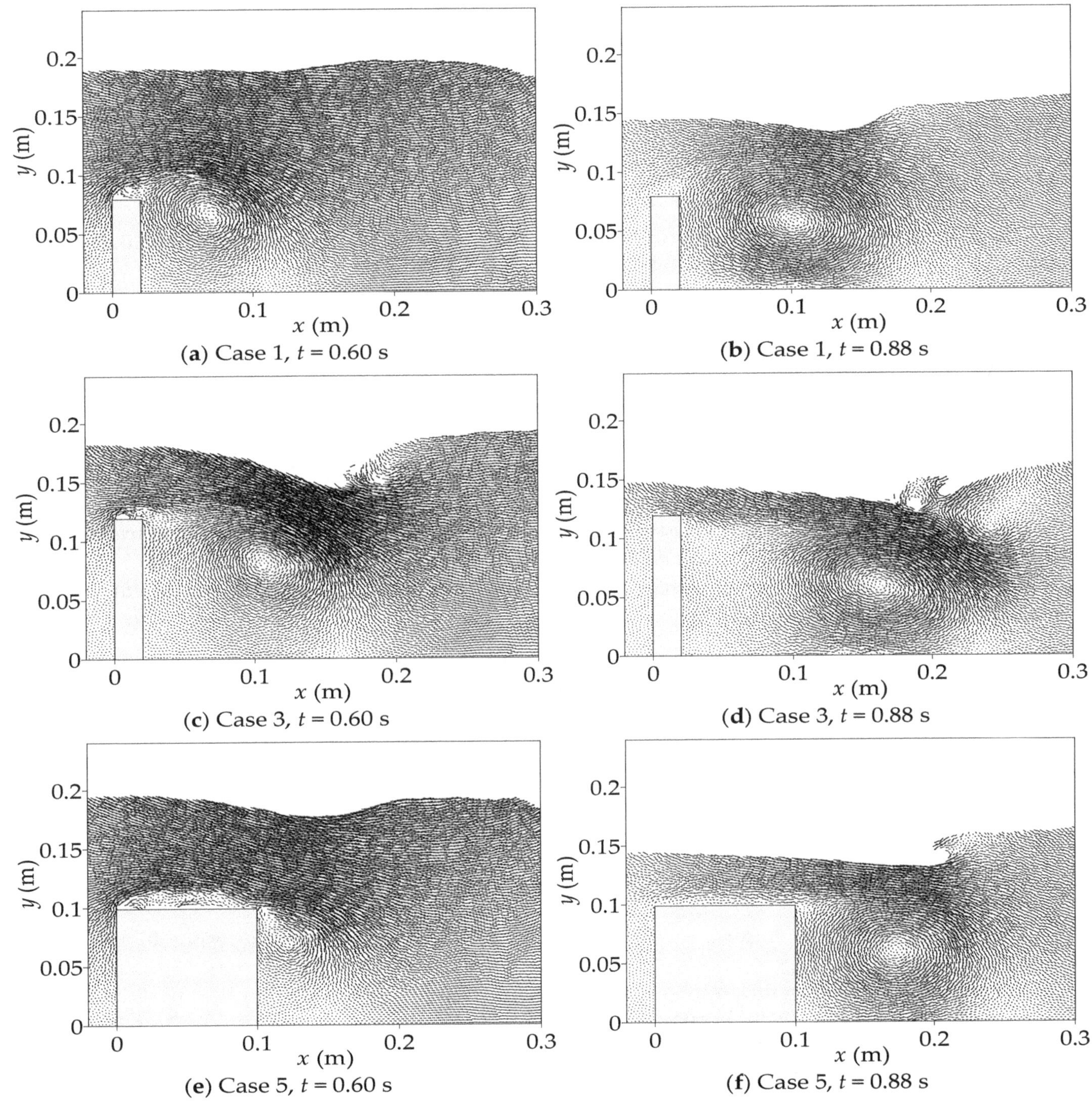

Figure 10. Velocity fields of cases 1, 3 and 5 at $t = 0.60$ s and $t = 0.88$ s: (**a**) Case 1, $t = 0.60$ s; (**b**) Case 1, $t = 0.88$ s; (**c**) Case 3, $t = 0.60$ s; (**d**) Case 3, $t = 0.88$ s; (**e**) Case 5, $t = 0.60$ s; (**f**) Case 5, $t = 0.88$ s.

6. Conclusions

In this work, the CPM is used to study the solitary wave interaction with rectangular submerged breakwaters. The distinct feature of CPM is that it computes the gradient and Laplacian operators in the governing equations in a way consistent with the Taylor series expansion. This addresses the issue of numerical consistency that exists in some other particle methods and hence enhances the numerical accuracy. The CPM model is used to study a documented experimental case. The wave elevation, wave profile and velocity distributions are in good agreement with the experimental data.

Using the validated numerical model, the process of solitary wave-breakwater interaction is studied in detail. The incident solitary wave turns into a jet flow when passing through the breakwater. The jet flow moves onshore and impinges on the water body at the lee side of the breakwater. Subsequently, a backward plunging wave is generated, the interaction between which and the jet flow induces a second jet. The second jet goes up and then splits into two parts. One part re-impinges on the main jet flow and the other part impinges on the plunging breaker, both of which induce splashes. The wave splitting and merging during the wave breaking process cause the dissipation of wave energy.

The main jet flow induces a big vortex behind the breakwater and the plunging breaker induces a secondary vortex. While both vortexes cause the spin of fluid particles, the second vortex, as a whole, rotates around the main vortex. The vortexes play key roles in transporting sediment particles and dissipating wave energy. The potential or capability of the wave flows around the breakwater to transport sediment particles are further studied by analyzing the trajectories of fluid particles. In general, the waves tend to move the seaward sediment particles near the breakwater to the leeward, and transport the sediment leeward particles further onshore.

The influence of the breakwater dimension on the fluid flow around the breakwater is studied. It is found that a higher breakwater causes more wave reflection and hence fewer waves go through the structure. Although the breakwater width has a negligible effect on wave reflection and transmission, the vortex distribution and the associated toe scour pattern are closely related to the width of a breakwater. Specifically, the detached scour is more likely to happen for a narrow breakwater and the attached scour for a wide breakwater.

Author Contributions: Conceptualization, Y.R., M.L. and P.L.; methodology, Y.R., M.L. and P.L.; simulation and data analysis, Y.R. and M.L.; writing—original draft preparation, Y.R. and M.L.; writing—review and editing, Y.R. and M.L.

Acknowledgments: The authors thank Yun-ta Wu at Tamkang University and Shih-Chun Hsiao in National Cheng Kung University for providing the experimental data as presented in Section 4.

References

1. Relief Web. Available online: https://reliefweb.int/disaster/eq-2018-000156-idn (accessed on 20 November 2018).
2. MSF. Available online: https://www.msf.org.uk/country/indonesia-tsunami-response (accessed on 20 November 2018).
3. CoastalWiki. Available online: http://www.coastalwiki.org/wiki/Detached_breakwaters#Reasons_for_selecting_a_submerged.2Flow_breakwater (accessed on 20 November 2018).
4. Synolakis, C.E. The runup of solitary waves. *J. Fluid Mech.* **2006**, *185*, 523–545. [CrossRef]
5. Briggs, M.J.; Synolakis, C.E.; Harkins, G.S.; Green, D.R. Laboratory experiments of tsunami runup on a circular island. *Pure Appl. Geophys.* **1995**, *144*, 569–593. [CrossRef]
6. Madsen, P.A.; Fuhrman, D.R.; Schäffer, H.A. On the solitary wave paradigm for tsunamis. *J. Geophys. Res.* **2008**, *113*. [CrossRef]
7. Qu, K.; Ren, X.Y.; Kraatz, S. Numerical investigation of tsunami-like wave hydrodynamic characteristics and its comparison with solitary wave. *Appl. Ocean Res.* **2017**, *63*, 36–48. [CrossRef]
8. Sté Grilli, P.T.; Losada, M.A. Characteristics of solitary wave breaking induced by breakwaters. *J. Waterway Port Coast. Ocean Eng.* **1994**, *120*, 74–92. [CrossRef]

9. Lin, P.Z.; Chang, K.A.; Liu, L.F. Runup and rundown of solitary waves on sloping beaches. *J. Waterway Port Coast. Ocean Eng.* **1999**, *125*, 247–255. [CrossRef]
10. Chang, K.A.; Hsu, T.J.; Liu, L.F. Vortex generation and evolution in water waves propagating over a submerged rectangular obstacle: Part II: Cnoidal waves. *Coast. Eng.* **2005**, *52*, 257–283. [CrossRef]
11. Huang, C.J. On the interaction of a solitary wave and a submerged dike. *Coast. Eng.* **2001**, *43*, 265–286. [CrossRef]
12. Liu, P.L.F.; Albanaa, K. Solitary wave runup and force on a vertical barrier. *J. Fluid Mech.* **2004**, *505*, 225–233. [CrossRef]
13. Lin, P.Z. A numerical study of solitary wave interaction with rectangular obstacles. *Coast. Eng.* **2004**, *51*, 35–51. [CrossRef]
14. Hsiao, S.C.; Lin, T.C. Tsunami-like solitary waves impinging and overtopping an impermeable seawall: Experiment and RANS modeling. *Coast. Eng.* **2010**, *57*, 1–18. [CrossRef]
15. Wu, Y.T.; Hsiao, S.C.; Huang, Z.C. Propagation of solitary waves over a bottom-mounted barrier. *Coast. Eng.* **2012**, *62*, 31–47. [CrossRef]
16. Monaghan, J.J. Simulating free surface flows with SPH. *J. Comput. Phys.* **1994**, *110*, 399–406. [CrossRef]
17. Liu, G.R.; Liu, M.B. *Smoothed Particle Hydrodynamics: A Meshfree Particle Method*; World Scientific: Singapore city, Singapore, 2003.
18. Koshizuka, S.; Nobe, A.; Oka, Y. Numerical analysis of breaking waves using the moving particle semi-implicit method. *Int. J. Numerical Methods Fluids* **1998**, *26*, 751–769. [CrossRef]
19. Marrone, S.; Antuono, M.; Colagrossi, A.; Colicchio, G.; le Touzé, D.; Graziani, G. δ-SPH model for simulating violent impact flows. *Comput. Methods Appl. Mech. Eng.* **2011**, *13–16*, 1526–1542. [CrossRef]
20. Lind, S.J.; Xu, R.; Stansby, P.K.; Rogers, B.D. Incompressible smoothed particle hydrodynamics for free-surface flows: A generalised diffusion-based algorithm for stability and validations for impulsive flows and propagating waves. *J. Comput. Phys.* **2012**, *231*, 1499–1523. [CrossRef]
21. Liu, X.; Lin, P.; Shao, S. An ISPH simulation of coupled structure interaction with free surface flows. *J. Fluids Struct.* **2014**, *48*, 46–61. [CrossRef]
22. Ren, B.; He, M.; Li, Y.; Dong, P. Application of smoothed particle hydrodynamics for modeling the wave-moored floating breakwater interaction. *Appl. Ocean Res.* **2017**, *67*, 277–290. [CrossRef]
23. Amicarelli, A.; Albano, R.; Mirauda, D.; Agate, G.; Sole, A.; Guandalini, R. A Smoothed Particle Hydrodynamics model for 3D solid body transport in free surface flows. *Comput. Fluids* **2015**, *116*, 205–228. [CrossRef]
24. Gotoh, H.; Khayyer, A. On the state-of-the-art of particle methods for coastal and ocean engineering. *Coast. Eng. J.* **2018**, *60*. [CrossRef]
25. Monaghan, J.J.; Kos, A. Solitary waves on a cretan beach. *J. Waterway Port Coast. Ocean Eng.* **1999**, *125*, 145–155. [CrossRef]
26. Shao, S.; Lo, E.Y.M. Incompressible SPH method for simulating newtonian and non-newtonian flows with a free surface. *Adv. Water Resour.* **2003**, *26*, 787–800. [CrossRef]
27. Zhang, Y.L.; Tang, Z.Y.; Wan, D.C. Simulation of solitary wave interacting with flat plate by MPS method. *Chin. J. Hydrodyn.* **2016**, *31*, 395–401.
28. Huang, Y.; Zhu, C. Numerical analysis of tsunami-structure interaction using a modified MPS method. *Nat. Hazards* **2015**, *75*, 2847–2862. [CrossRef]
29. Khayyer, A.; Gotoh, H.; Shao, S. Enhanced predictions of wave impact pressure by improved incompressible SPH methods. *Appl. Ocean Res.* **2009**, *31*, 111–131. [CrossRef]
30. Koh, C.G.; Gao, M.; Luo, C. A new particle method for simulation of incompressible free surface flow problems. *Int. J. Numerical Methods Eng.* **2012**, *89*, 1582–1604. [CrossRef]
31. Luo, M.; Koh, C.G.; Gao, M.; Bai, W. A particle method for two-phase flows with large density difference. *Int. J. Numerical Methods Eng.* **2015**, *103*, 235–255. [CrossRef]
32. Koh, C.G.; Luo, M.; Gao, M. Modelling of liquid sloshing with constrained floating baffle. *Comput. Struct.* **2013**, *122*, 270–279. [CrossRef]
33. Luo, M.; Koh, C.G. Shared-Memory parallelization of consistent particle method for violent wave impact problems. *Appl. Ocean Res.* **2017**, *69*, 87–99. [CrossRef]
34. Chorin, A.J. The numerical solution of the Navier-Stokes equations for an incompressible fluid. *Bull. Am. Math. Soc.* **1967**, *73*, 928–931. [CrossRef]

35. Dilts, G.A. Moving least-squares particle hydrodynamics II: Conservation and boundaries. *Int. J. Numerical Methods Eng.* **2000**, *48*, 1503–1524. [CrossRef]
36. Liu, X.; Xu, H.; Shao, S. An improved incompressible SPH model for simulation of wave–structure interaction. *Comput. Fluids* **2013**, *71*, 113–123. [CrossRef]
37. Goring, D.G. Tsunamis—The Propagation of Long Waves onto a Shelf. Ph.D. Thesis, California Institute of Technology, Pasadena, CA, USA, 1978.
38. Sugimoto, N.; Nakajima, N.; Kakutani, T. Edge-layer theory for shallow-water waves over a step–reflection and transmission of a soliton. *J. Phys. Soc. Jpn.* **1987**, *56*, 1717–1730. [CrossRef]
39. Cooker, M.J.; Peregrine, D.H.; Vidal, C.; Dold, J.W. The interaction between a solitary wave and a submerged semicircular cylinder. *J. Fluid Mech.* **1990**, *215*, 1–22. [CrossRef]
40. Young, D.M.; Testik, F.Y. Onshore scour characteristics around submerged vertical and semicircular breakwaters. *Coast. Eng.* **2009**, *56*, 868–875. [CrossRef]

5

Experimental Study on the Air Concentration Distribution of Aerated Jet Flows in a Plunge Pool

Weilin Xu *, Chunqi Chen and Wangru Wei

State Key Laboratory of Hydraulics and Mountain River Engineering, Sichuan University, Chengdu 610065, China; ccq-cedric@163.com (C.C.); wangru_wei@hotmail.com (W.W.)

* Correspondence: xuwl@scu.edu.cn

Abstract: There is a lack of knowledge on the air concentration distribution in plunge pools affected by aerated jets. A set of physical experiments was performed on vertical submerged aerated jet flows impinging a plunge pool. The air concentration distribution in the plunge pool was analyzed under different inflow air concentrations, flow velocities, and discharge rate conditions. The experimental results show that the air concentration distribution follows a power-law along the jet axis, and it is independent of the initial flow conditions. A new hypothetical analysis model was proposed for air diffusion in the plunge pool, that is, the air concentration distribution in the plunge pool is superposed by the lateral diffusion of three stages of the aerated jet motion. A set of formulas was proposed to predict the air concentration distribution in the plunge pool, the results of which showed good agreement with the experimental data.

Keywords: submerged jets; aerated flow; air concentration; plunge pool

---

1. Introduction

The impingement of a jet flow on a floor is a common issue that is important for many engineering applications, such as the flood discharging of hydraulic structures, and heat and mass transfer in industrial operations. Due to the presence of the floor being located at a distance from the nozzle exit, the diffusion of jets in a plunge pool is different from that of submerged free jets. Based on how the jet impacts the floor, the flow field can be divided into the free jet region, the impingement region, and the free shear region. The jet flow velocity conforms to the distribution law of submerged free jets in the free jet region. In the impingement region, the jet flow velocity decreases rapidly and reaches zero at the stagnation point [1,2].

Aerated and non-aerated impinging jet flows have been studied by many associated investigators [3–5]. Ervine summarized the characteristics of free turbulent jets, and compared the jet diffusion in the plunge pool with submerged and impinging jets [6]. Ervine collected data for both mean and fluctuating components of the pressure field on a plunge pool floor subjected to jet impingement, and compared the data for circular jets with those for wide rectangular nappes and rectangular slot jets [7]. Castillo analyzed the degree of break-up of a rectangular jet before entering the basin, and its relationship to the pressure fluctuation on the plunge pool floor [8]. Chanson discussed the characteristics and similitude of the air–water flow caused by impinging jets [9]. Wei analyzed the influence of aeration and initial water thickness on the axial velocity attenuation of jet flows using experiments and numerical calculations [10,11]. The research concludes that increasing the air concentration of jet flows and decreasing the initial jet thickness are effective ways to improve the axial velocity attenuation of jet flows. Some researchers focused on the pressure on the floor of the plunge pool by an aerated and non-aerated jet flow [12–15]. The results showed that the decreasing effect of air concentration on the time-averaged impact pressure is minimal, if any. However, the fluctuating pressure rises as the air concentration increases. Duarte compared the jet impingement on the center and side of a block embedded on the floor using plunging and submerged jets [16,17]. Others have inquired into the scouring of the plunge pool floor due to jets [18–20]. In addition to the velocity and pressure fields in the plunge pool, the process of air bubble entrainment at the impinging point of the water surface has also been studied extensively [21,22]. From the interaction of the jet flow and the plunge pool surface, an air sheet forms around the impinging point before air bubbles are entrained into the water surface [23]. Several correlative and predictive models on the air entrainment ratio have been reported, due to plunging jets under different working conditions [24–26]. Some researchers used numerical simulations to study aerated jet flows. Brouillio performed a series of two-dimensional numerical experiments on a translating impacting jet, and studied the dynamics of air entrainment by the jets impacting on the surface of the water [27]. Samanta studied the effect of the permeable wall on secondary cross-stream flow by performing direct numerical simulations of the fully developed turbulent flow through a porous square duct [28]. In model tests, the jet velocity profiles and pressure fluctuations on the plunge floor have been studied extensively, with detailed results available. Numerical simulations can essentially reflect characteristics such as the flow field and air entrainment process on the water surface. However, for mass and heat transfer characteristics at the air–water interface under complex conditions, a large number of systematic tests are still needed.

Few researchers have studied the air concentration distribution in the entire plunge pool, including the free shear region. Dong described the concentration diffusion of an aerated jet in the plunge pool with different plunging angles [29]. The air concentration along the jet axis gradually decreased, following a hyperbolic curve. Chanson conducted separate experiments with a vertical supported jet and a horizontal hydraulic jump [30]. The results showed that the air concentration in both cases exhibits a Gaussian distribution, and the longitudinal maximum air content declines exponentially. Khaled used the Particle Image Velocimetry (PIV) technique to characterize the flow field of both aerated and non-aerated jet flows beneath the water surface [31]. However, analyses and predictions of air concentration distribution in the entire plunge pool, especially where affected by the swirling jet, are still missing.

In the present paper, measurements and analyses of the air concentration distribution are presented for a plunge pool. A new analysis model is proposed to describe the air diffusion, with jet swirling in the entire plunge pool. A set of formulas has been built up to predict the air concentration distribution, and the results are believed to be useful for understanding and predicting the air concentration distribution in both the plunge pool and downstream.

2. Experimental Methods

Experiments were carried out at the State Key Laboratory of Hydraulics and Mountain River Engineering (Sichuan University). An experimental facility was set up, as shown in Figure 1. The jet was aerated before being plunged vertically into the still water in a rectangular glass flume of 4.00 m $\times$ 0.25 m $\times$ 0.80 m. The airflow was provided using an air compressor with a power rating of 7.5 kW at a pressure of 0.8 MPa. The inflow jet was controlled using valves and measured with a TDS-100H ultrasonic flowmeter, of which the measurement range was 0.8–128.0 L/s, and the relative error of measurement was $\pm 1\%$. The air was controlled using valves, and measured with a SLDLB-Y150D vortex flowmeter, of which the measurement range was 2.0–60.0 L/s, and the relative error of measurement was $\pm 1\%$. To form a stable and well-mixed air-water flow, the air was forced through micro-perforated plates with a diameter of 1 mm, and mixed into the water flow in the air entrainment region. The water in the flume flowed out at both ends, where the water level was maintained at $H = 0.55$ m. The incident angle of the jet flow was 90 degrees. The rectangular nozzle, 0.20 m in width and 0.02 m–0.05 m at the opening, was installed slightly below the water surface to ensure a precise, stable initial air concentration.

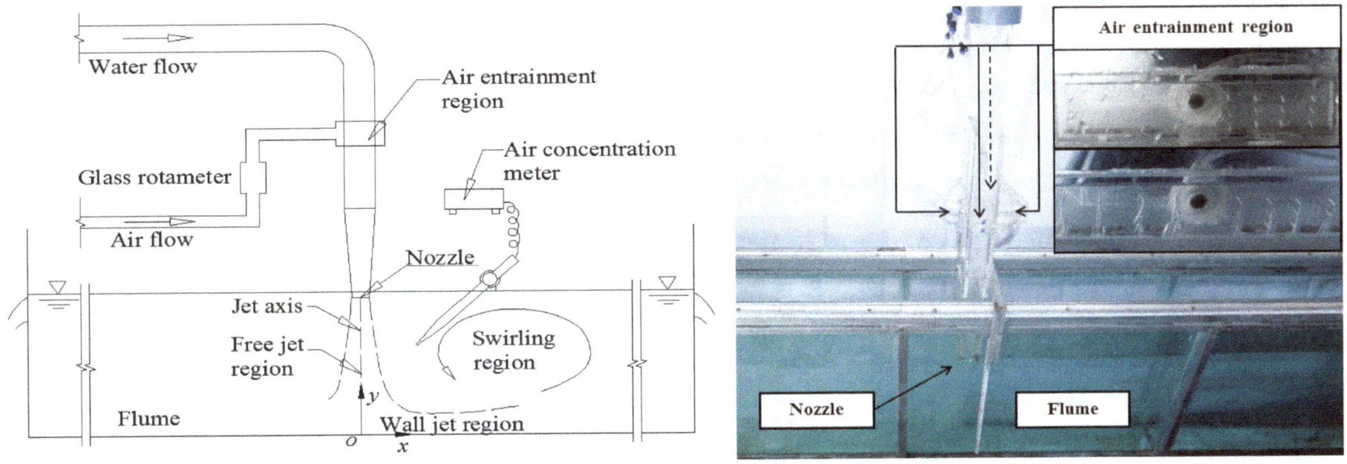

(a) Sketch map (b) Photograph

Figure 1. Experimental facility.

The coordinate origin of an x-0-y coordinate system was located on the flume floor, the horizontal x-axis faced the downstream direction of the plunge pool, and the vertical y-axis coincided with the jet axis. In total, there were 264 measurable points for the air concentration in the plunge. A total of 22 rows were arranged along the x-axis direction. First, six rows were set at an interval of 5 cm (0 cm $\leq x \leq$ 25 cm), and the following 16 rows were set at an interval of 10 cm (30 cm $\leq x \leq$ 180 cm). In total, 12 rows of measuring points were arranged along the y-axis direction, with the first row at y = 1 cm and the last row at y = 54 cm (1 cm from the flume floor). The remaining 10 rows of measuring points along the y-axis were set between y = 5 cm and y = 50 cm at intervals of 5 cm. The air concentration was tested using a CQ6-2004 resistance-type air concentration meter (China Institute of Water Resources and Hydropower Research) with a resolution of 1%. The measurement principle of the resistance-type air concentration meter was to determine the air concentration by detecting the clear water resistance and the aerated water resistance between the two electrodes, and averaging the air concentration values in the selected integration time. Accurate flow measurements can be an important factor in obtaining adequate conclusions [32]. Different integration times for the air concentration measurements were tested beforehand, as shown in Figure 2. It can be seen that the mean air concentration data fall in the measurement accuracy zone if the integration time of the air concentration meter exceeds 60 seconds. Therefore, the sampling period for each measurement point was set to 60 seconds to minimize random error.

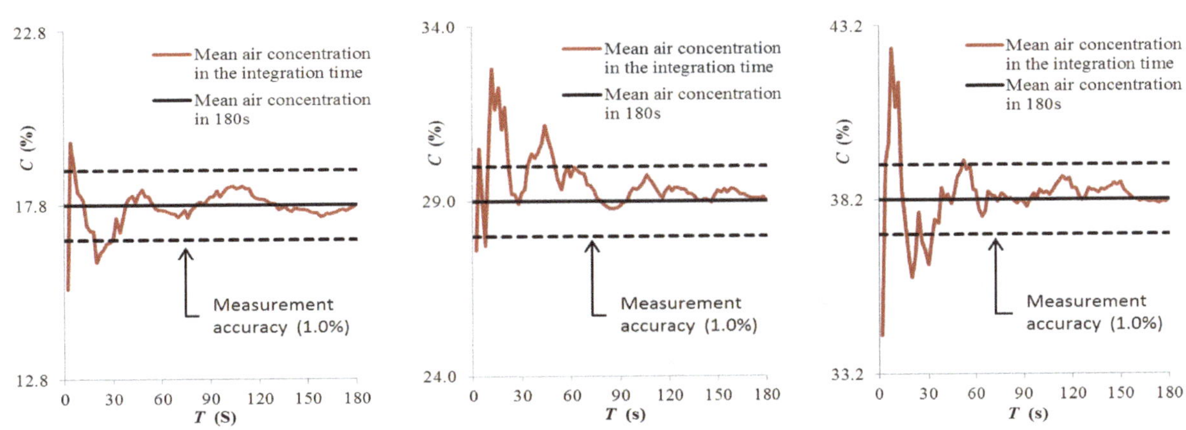

(a) (b) (c)

Figure 2. The uncertainty of air concentration measurements in the plunge pool.

The experiments were performed under different conditions, including the jet flow thickness d_0 (2 cm, 3 cm, 4 cm, 5 cm), the initial air concentration C_0 (10%, 15%, 30%, 40%, 50%, 60%), the clear water flow rate Q_w (14.0 L/s–35.2 L/s), and the air flow rate Q_a (2.1 L/s–42.0 L/s), as listed in Table 1.

Table 1. Experimental conditions.

No.	d_0 (cm)	C_0 (%)	Q_w (L/s)	No.	d_0 (cm)	C_0 (%)	Q_w (L/s)
1	2	10	18.6	16	3	30	21.1
2	2	15	16.1	17	3	40	21.1
3	2	30	14.0	18	3	40	23.9
4	2	30	16.8	19	3	50	23.9
5	2	30	19.6	20	4	15	27.8
6	2	30	22.4	21	4	30	21.1
7	2	40	16.7	22	4	30	28.2
8	2	50	14.7	23	4	40	28.2
9	2	60	14.5	24	4	50	21.1
10	2.3	10	21.1	25	4	50	23.9
11	2.3	15	23.9	26	4	50	27.8
12	2.5	15	21.1	27	5	30	35.2
13	2.5	30	21.1	28	5	30	21.1
14	3	10	21.3	29	5	40	35.2
15	3	15	21.1	30	5	50	23.9

In the experiment, the aerated jet was vertically injected and outflowed simultaneously at both ends of the flume, which was slightly wider than the nozzle exit. Jets could develop on the bottom plate of the flume under all experimental conditions. Thus, the development of the aerated jet in the plunge pool could be simplified as a two-dimensional flow, which is symmetrical to the jet axis (y-axis). For typical working conditions (d_0 = 2 cm, C_0 = 30%, Q_w = 16.8 L/s), an experimental photograph and a contour map of the air concentration distribution on one side of the jet axis are shown in Figure 3. The majority of the air bubbles flow with the jet flow in the free jet region, before being blocked by the flume floor and deflected by 90 degrees into the wall jet region on both sides. Afterwards, the air bubbles flow gradually towards the water surface, with some eventually escaping into the surrounding air, whereas the rest are carried into the swirling region.

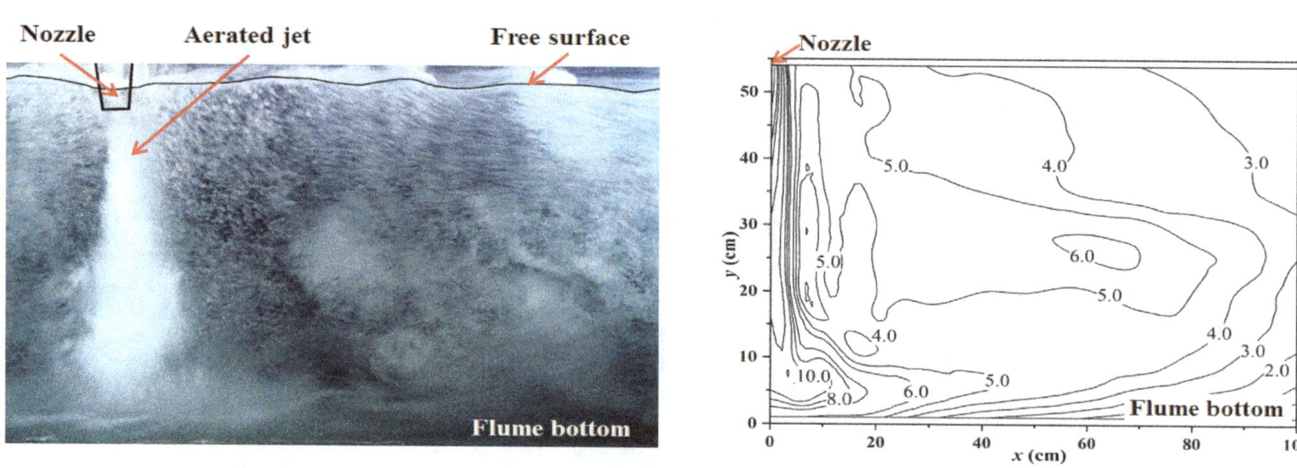

(a) Photograph (b) Air concentration contour map

Figure 3. Air concentration distribution for d_0 = 2 cm, C_0 = 30%, Q_w = 16.8 L/s.

3. Results and Discussion

3.1. Air Concentration Distribution in the Free Jet Region

Along the jet axis, the air concentration distribution should be closely related to the jet velocity distribution. The jet velocity decreases sharply as the aerated jet approaches the plunge pool floor. Due to the pressure gradient, the jet carrying air bubbles drifts toward both sides of the plunge pool, resulting in a steep decrease in air concentration in the impingement region. The region higher than one-fifth of the water depth is considered to be the free jet region, where the attenuation of air concentration was studied in the test.

Considering the influencing factors, the initial water velocity V_W, the initial air concentration C_0, and the clear water flow rate Q_W were adopted, and a series of experiments were performed to test each effect on the air concentration attenuation, as shown in Figure 4. Under a different initial water velocity (3 m/s < V_W < 8 m/s) and a clear water flow rate (21.1 L/s < Q_W < 35.2 L/s), the attenuation of C_m/C_0 varies in a similar pattern along the jet axis. The reason for the results is that the velocity attenuation along the jet axis changes little with different flow conditions, including the initial jet velocity and the clear water flow rate [33].

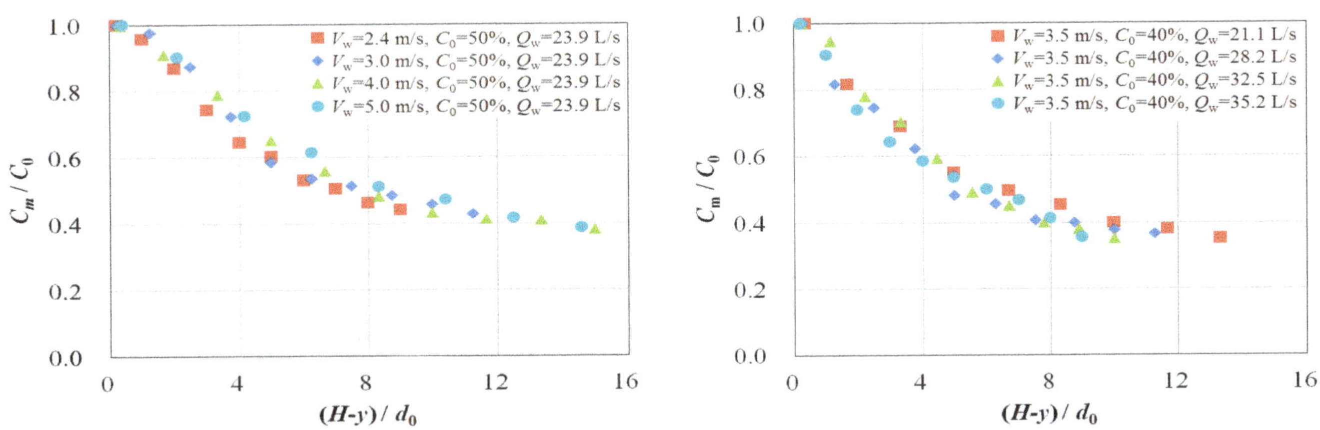

(**a**) Air concentration distribution for different V_w (**b**) Air concentration distribution for different Q_w

Figure 4. Influence of V_W and Q_W on axial air concentration distribution.

By theoretically analyzing the concentration flux of the aerated jet based on the self-similarity of the velocity and air concentration distribution, the power-law relation was found between the axial air concentration and the flow distance in the free jet region [29]. The relationship between the axial air concentration C_m and the distance from the nozzle $x' = H - y$, normalized by the initial air concentration C_0 and the jet flow thickness d_0 in the plunge pool, respectively, is written as:

$$\frac{C_m}{C_0} = k_0 \times \left(\frac{x'}{d_0}\right)^{-0.5} = k_0 \times \sqrt{\frac{d_0}{H-y}} \quad (1)$$

where k_0 is a coefficient, suggested as approximate 1.2 in this study.

The calculated results of the axial air concentration distribution in the free jet region using Equation (1) was compared with the experimental data, as shown in Figure 5a. Independent of the initial jet velocity and the clear water flow rate, the axial air concentration sharply decreased to about 50% of C_0 in a short distance ($\frac{H-y}{d_0} \approx 5$) from the nozzle exit. Due to the influence of the pressure gradient near the plunge floor, further from the nozzle exit, the air concentration distribution was more dispersed in the free jet region. It can be derived from the experimental results that the air concentration distribution along lateral cross-sections in the free jet region conformed to the Gaussian distribution law, as shown in Figure 5b. In the free jet region, no significant influence of different initial flow conditions was observed on the cross-sectional air concentration distribution, which tended to be self-similar. Downstream of the free jet region, the cross-sectional air concentration distribution was affected by the swirling jet in the plunge pool.

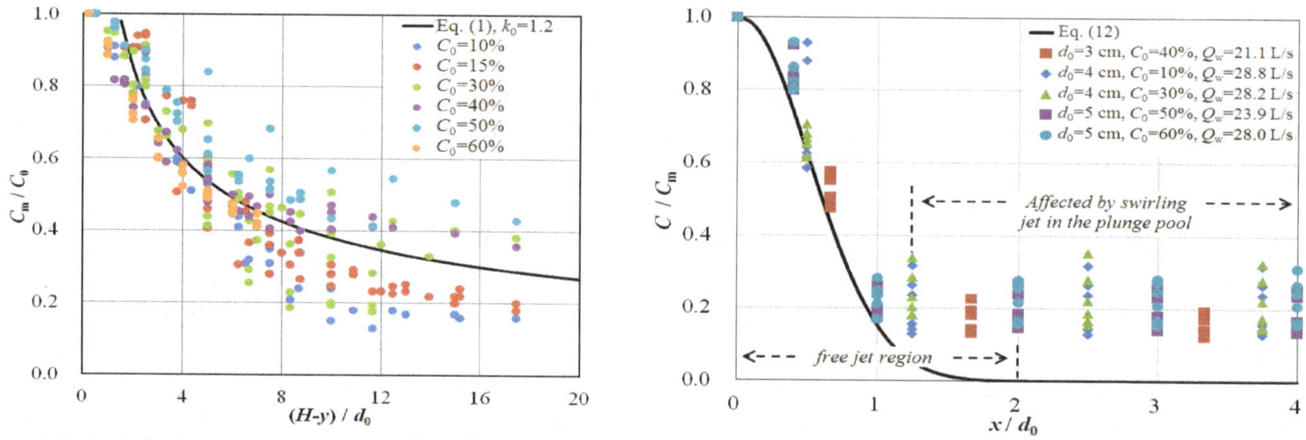

(**a**) Axial air concentration distribution (**b**) Cross-sectional air concentration distribution

Figure 5. Air concentration distribution in the free jet region.

For a two-dimensional constant jet along the x' direction in an x'–0–y' coordinate system, the continuity equation of the air concentration diffusion is:

$$\frac{\partial(CU)}{\partial x'} + \frac{\partial(CV)}{\partial y'} = \frac{\partial}{\partial y'}\left(D_y \frac{\partial C}{\partial y'}\right) \tag{2}$$

where C is the air concentration, U is the velocity in the x' direction, V is the velocity in the y' direction, and D_y is the turbulent diffusion coefficient.

The longitudinal velocity U can be approximated using the average velocity $\overline{u}$. The transverse velocity can be negligible, i.e., $V \approx 0$.

According to Prandtl's hypothesis of free turbulence theory [34], D_y can be approximated as:

$$D_y = k\overline{u}b \tag{3}$$

where k is the coefficient, and b is the thickness of the jet, which is assumed to expand linearly and can be written as:

$$b = mx' = \tan\theta \times x' \tag{4}$$

where θ is the diffusion angle.

Previous research suggests that with respect to the conventional diffusion angle (about eight degrees), the diffusion angle of aerated water increases due to the influence of bubbles [10]. Twelve degrees is taken as the diffusion angel in this study.

Due to hardly any influence of initial flow conditions on both the axial attenuation and the lateral distribution of air concentration in the free jet region, it is assumed that the air concentration distribution in the free jet region is self-similar, resembling that in the velocity distribution, which can be written as:

$$\frac{C}{C_m} = f_1(\eta), \quad \eta = \frac{y'}{b} \tag{5}$$

Combining Equations (2), (3) and (5), $\frac{C}{C_m}$ can be expressed as:

$$f_1 = \frac{C}{C_m} = A \times \exp\left[-\frac{m}{2k} \times \eta^2\right] \tag{6}$$

where A is the integral constant and determined by boundary conditions. When $\eta = \frac{y'}{b} = 0, C = C_m$, $A = 1$, Equation (6) can be rewritten as:

$$\frac{C}{C_m} = \exp\left[-\frac{m}{2k} \times \left(\frac{y'}{b}\right)^2\right] \tag{7}$$

The concentration half-width is defined as:

$$b_{1/2} = y'|_{c=0.5C_m} \tag{8}$$

It is assumed that the concentration half-width extends linearly along the path, thus:

$$b_{1/2} = \frac{1}{2}d_0 + \tan\theta \times x' \tag{9}$$

Combining Equations (7) and (8):

$$\ln\left(\frac{1}{2}\right) = -\frac{m}{2k} \times \left(\frac{b_{\frac{1}{2}}}{b}\right)^2 = -0.693 \tag{10}$$

Considering:

$$-\frac{m}{2k} \times \left(\frac{y'}{b}\right)^2 = -\frac{m}{2k} \times \left(\frac{b_{\frac{1}{2}}}{b}\right)^2 \times \left(\frac{y'}{b_{\frac{1}{2}}}\right)^2 = -0.693 \times \left(\frac{y'}{b_{1/2}}\right)^2 \tag{11}$$

Combining Equations (7) and (11), the air concentration distribution on the flow cross-section can be calculated by:

$$\frac{C}{C_m} = \exp\left[-0.693 \times \left(\frac{y'}{b_{1/2}}\right)^2\right] \tag{12}$$

where C is the transverse air concentration diffusion, C_m is the corresponding air concentration along the jet axis, and y' is the transverse distance from the jet axis.

The calculated result of Equation (12) was compared with the experimental results, as shown in Figure 5b. In the free jet region, the calculated results of the lateral diffusion were essentially consistent with the experimental data. Further away from the axis, due to the influence of the swirling jet, the experimental value of the air concentration was greater than the calculated result obtained using Equation (13). Therefore, the next section will focus on the establishment of an analysis model for air concentration diffusion in the plunge pool.

In the x–0–y coordinate system in this study, the air concentration distribution in the free jet region can be calculated by:

$$\begin{aligned} C &= C_0 \times 1.2\sqrt{\frac{d_0}{x'}} \times \exp[-0.693\left(\frac{y'}{b_{\frac{1}{2}}}\right)^2] \\ &= C_0 \times 1.2\sqrt{\frac{d_0}{H-y}} \times \exp[-0.693\left(\frac{x}{b_{1/2}}\right)^2] \end{aligned} \tag{13}$$

3.2. Air Concentration Distribution in the Swirling Region of the Plunge Pool

For one side of the plunge pool, the experimental photographs of the submerged jet motion and air concentration distribution are shown in Figure 6. After being plunged into the still water, the aerated jet moves along the vertical axis, and the aerated wall jet forms after a 90-degree deflection of the jet flow near the flume floor. While the wall jet moves downstream, it is gradually developed toward the free surface under the combined effects of turbulence and buoyancy. Due to the existence of the downstream board, part of the aerated flow moves upstream along the water surface and forms a swirling region in the plunge pool. At the downstream end of the swirling region, the direction of the aerated flow movement is almost vertically upward. While moving with the water flow, the air bubbles expand continuously and laterally into the swirling region under the effects of turbulence.

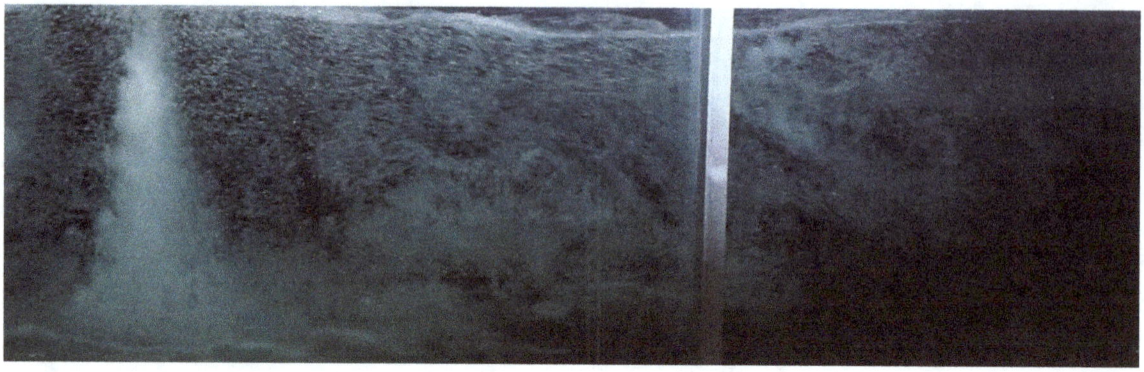

(**a**) d_0 = 2 cm, C_0 = 15%, Q_w = 16.1 L/s

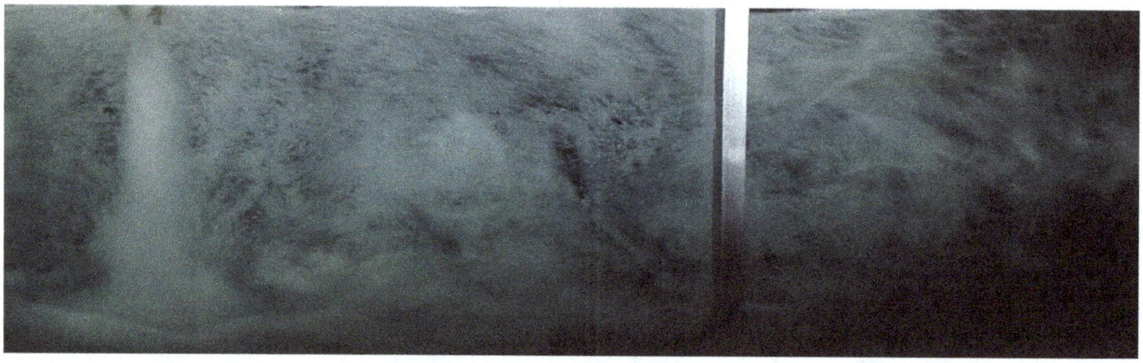

(**b**) d_0 = 3 cm, C_0 = 30%, Q_w = 21.1 L/s

(**c**) d_0 = 2 cm, C_0 = 60%, Q_w = 14.5 L/s

Figure 6. Air concentration distribution in the plunge pool.

According to Melo [13], the flow patterns in the plunge pool impinged by vertical jets are composed of three distinct regions, as shown in Figure 7a. In the free jet region, the vertical jet develops by shearing with the surrounding water body without the influence of the plunge floor. The impact of the jet against the floor builds up the pressure gradient and deflects the jet parallel to the plunge floor into the wall jet region. Due to the limited area of the plunge pool, the horizontal wall jet develops upwards and flows into the swirling region.

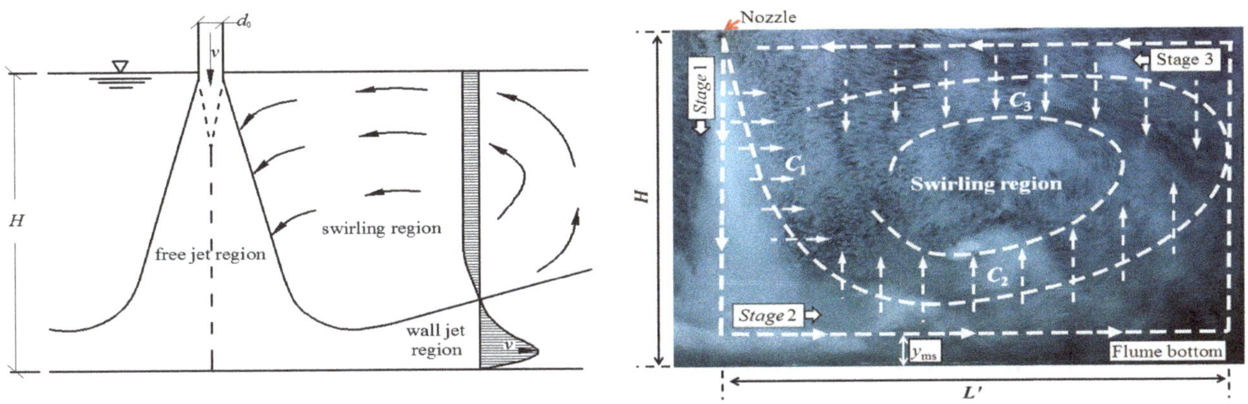

(**a**) Sketch map of diffusing jet in a pool (Melo [13]) (**b**) Stages of air concentration diffusion

Figure 7. Generalized calculation of air concentration in the swirling region.

When calculating the concentration field in the swirling region, the motion and diffusion of the air bubbles in the plunge pool are generalized into the following three stages, as shown in Figure 7b:

Stage 1: The aerated jet moves along the jet axis until it reaches the floor of the plunge pool, categorized as the free jet region in Figure 7a. The attenuation and lateral diffusion of the air concentration occurs along the path, which is consistent with the distribution law in the free jet region mentioned above. At this stage, the air concentration that diffuses laterally into the swirling region is taken as C_1;

Stage 2: The aerated flow is deflected by 90 degrees at the bottom of the plunge pool, and moves along the axis near the bottom plate, categorized as the wall jet region in Figure 7a. Regardless of the local and instantaneous air concentration loss, the initial air concentration at this stage is assumed to be the same as the air concentration at the end of the axis in the previous stage. The air concentration of the aerated jet continues to decrease along the path, until the jet reaches the end section of the swirling region. At this stage, the air concentration that diffuses vertically into the swirling region is taken as C_2;

Stage 3: The aerated flow moves upstream along the free surface until it reaches the original jet axis at stage 1, categorized as the swirling region in Figure 7a. The initial air concentration value in this stage is obtained by approximating the axis of the aerated jet in the swirling region to a rectangle with a length of L' and a width of H. The air concentration that diffuses vertically into the swirling region is taken as C_3.

In the plunge pool, excluding the jet axis, the air concentration at a certain point can be considered as the superposition of C_1, C_2, and C_3. When calculating the axial air concentration attenuation and the lateral diffusion in the three stages mentioned using Equation (13), separately, the jet flow distance x' and the transverse distance from the axis y' can be expressed as:

$$\text{stage 1}: \; x'_1 = H - y, \quad y'_1 = x \tag{14}$$

$$\text{stage 2}: \; x'_2 = H + x, \quad y'_2 = y \tag{15}$$

$$\text{stage 3}: \; x'_3 = 2H + 2L' - x, \quad y'_3 = H - y \tag{16}$$

In order to locate the end section in stage 2 of the air concentration diffusion, the swirling region length L' is defined as the distance between the jet axis and the vertical cross-section, where the mean air concentration equals 5% of C_0.

$$L' = x|_{c_{mean}=0.05c_0} \tag{17}$$

where C_{mean} is the mean air concentration.

Under current experimental conditions, the relationship between the swirling region length and the initial water flow velocity is shown in Figure 8, which can be correlated with the linear function as:

$$\frac{L\prime}{d_0} = 8 \times \frac{V_w^2}{gH} + 11, \quad 0.6 < \frac{V_w^2}{gH} < 6 \tag{18}$$

where V_w is the water velocity, and g is the acceleration of gravity. The correlation coefficient is 0.80. It should be noted that the length of the swirling region was deliberately defined using the statistical mean air concentration of the vertical cross-section. The swirling region was actually determined by the flow field. In the downstream end of the plunge pool, the decrease in flow velocity led to changes in the following behaviors of the air bubbles. This resulted in the deviation of measurements in the experiment.

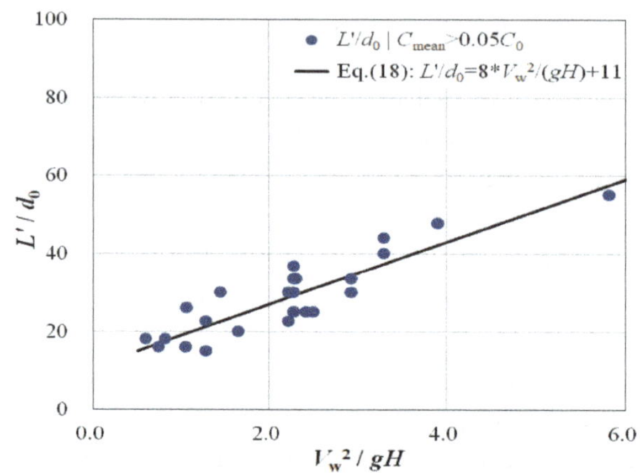

Figure 8. Distribution of swirling region length L'.

In stage 2 of the air concentration diffusion, the position of the maximum air concentration values along the vertical cross-sections in the wall jet can be estimated as:

$$y_{ms} = \frac{1}{3} \times b \tag{19}$$

where b is the thickness of the jet.

According to Beltaos [35], the relation between the maximum velocity of the wall jet u_{m1} and the initial velocity u_0 is:

$$\frac{u_{m1}}{u_0}\sqrt{\frac{H}{d_0}} = 2.77\left\{1 - \exp\left[-38.5\left(\frac{x}{H}\right)^2\right]\right\}^{\frac{1}{2}} \tag{20}$$

Considering the symmetrical aerated flow on both sides of the plunge pool, and the relation:

$$\frac{1}{2}u_{m1} \times b = \frac{1}{2}u_0 \times d_0 \tag{21}$$

Combining Equations (19)–(21), y_{ms} can be expressed as:

$$y_{ms} = \frac{1}{3} \times \frac{\sqrt{H \times d_0}}{2.77\left\{1 - \exp\left[-38.5(x/H)^2\right]\right\}^{1/2}} \tag{22}$$

The air concentration can be linearly solved on the vertical cross-sections between the axis of the wall jet and the plunge pool floor, where the air concentration value is zero.

According to the above analysis, at any point (x, y) in the swirling region of the plunge pool, the calculation of the lateral air concentration diffusion C_1 from the jet axis can be written as:

$$C_1 = C_0 \times 1.2\sqrt{\frac{d_0}{H-y}} \times \exp\left[-0.693\left(\frac{x}{b_{1/2}}\right)^2\right] \tag{23}$$

The calculation of the vertical air concentration diffusion C_2 from near the plunge pool floor can be written as:

$$C_2 = C_0 \times 1.2\sqrt{\frac{d_0}{H+x}} \times \exp\left[-0.693\left(\frac{y-y_{ms}}{b_{1/2}}\right)^2\right] \tag{24}$$

The calculation of the vertical air concentration diffusion C_3 from the free surface can be written as:

$$C_3 = C_0 \times 1.2\sqrt{\frac{d_0}{2H+2L'-x}} \times \exp\left[-0.693\left(\frac{H-y}{b_{1/2}}\right)^2\right] \tag{25}$$

Thus, the air concentration value at the point (x, y) can be obtained by:

$$C\Big|_{(x,y)} = C_1 + C_2 + C_3 \tag{26}$$

with 12 degrees of diffusion angles used in calculating C_1 and C_2, and zero degrees in calculating C_3.

Combing Equations (23)–(26), it is possible to calculate the air concentration at any point in the swirling region of the plunge pool.

3.3. Comparison of Calculated and Experimental Results

The calculated and experimental results of the air concentration distribution along the vertical cross-sections in the plunge pool are compared in Figure 9. When the initial air concentration was small, such as $C_0 < 40\%$, the calculated results agreed well with the experimental measurements. When the initial air concentration was large, such as $C_0 > 40\%$, the calculated results could essentially reflect the variation of the air concentration distribution in the region close to the jet axis, such as $x/d_0 < 15$–20. In the downstream region, on the other hand, such as $x/d_0 > 15$–20, the calculated results were somewhat different from the experimental measurements.

The reason for the deviation is that the bubble motion and diffusion process in the hypothetical analysis model is different from that in the actual jet flow. The influence of the bubble floating process by buoyancy on the actual jet flow was not considered, leading to a decrease in the adaptability of the idealized hypothetical analysis in this paper. However, the absolute value of the air concentration in the downstream region of the plunge pool was relatively small. For example, for $d_0 = 2$ cm, $C_0 = 60\%$, $Q_w = 14.5$ L/s, the absolute value of the air concentration was essentially less than 0.05 in the downstream region, such as $x/d_0 > 10$. Consequently, the present analysis model can essentially predict the air concentration distribution of aerated jet flows in the swirling region of the plunge pool.

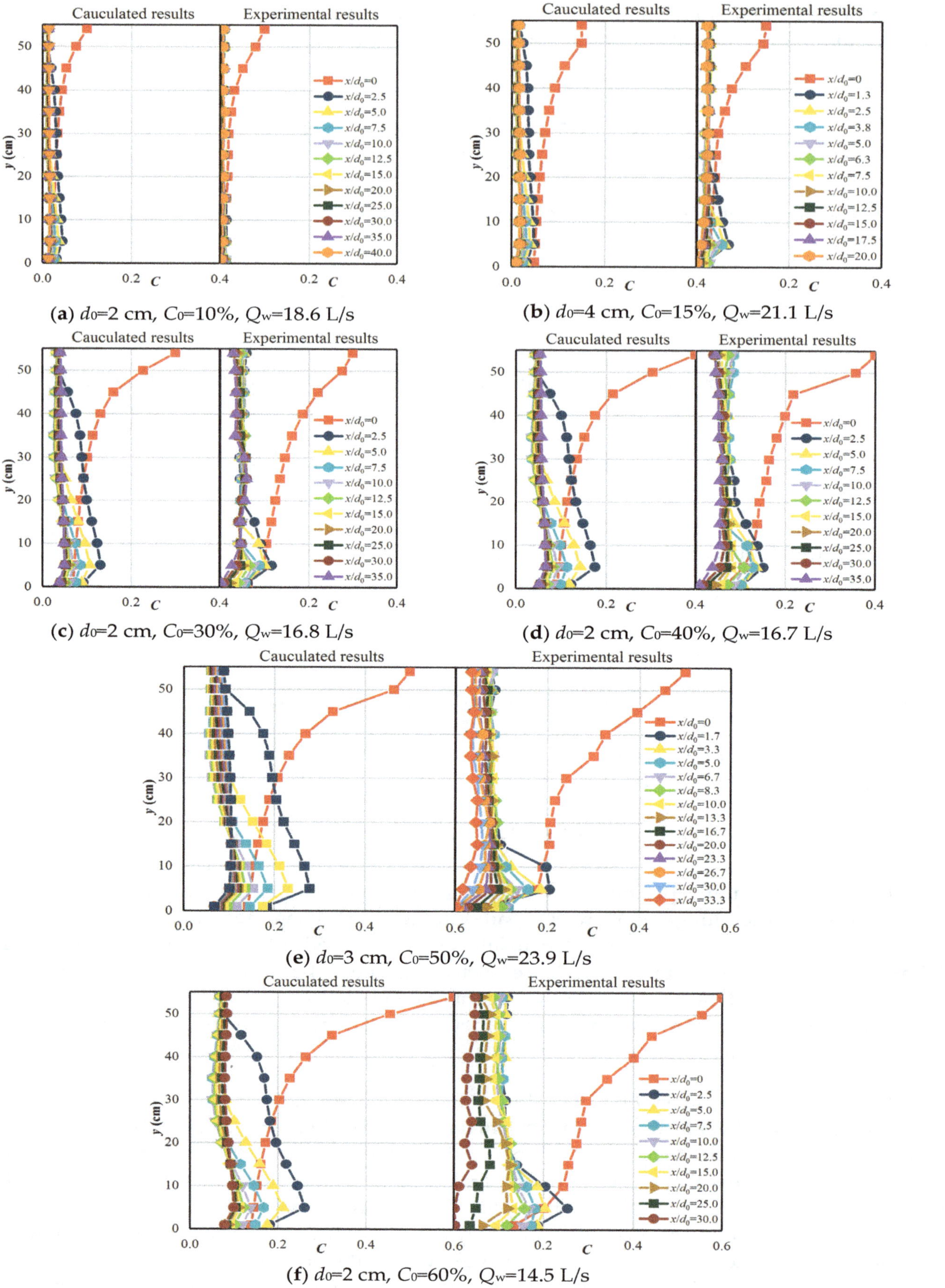

(a) d_0=2 cm, C_0=10%, Q_w=18.6 L/s

(b) d_0=4 cm, C_0=15%, Q_w=21.1 L/s

(c) d_0=2 cm, C_0=30%, Q_w=16.8 L/s

(d) d_0=2 cm, C_0=40%, Q_w=16.7 L/s

(e) d_0=3 cm, C_0=50%, Q_w=23.9 L/s

(f) d_0=2 cm, C_0=60%, Q_w=14.5 L/s

Figure 9. Comparison of calculated and experimental results.

4. Conclusions

In this paper, a series of experimental studies on the air concentration distribution of vertical submerged aerated jet flows in a plunge pool were carried out. Based on the distribution law of the air concentration attenuation along the axial and lateral directions, an analysis model of the air concentration distribution in the plunge pool was proposed. The conclusions were as follows.

In the free jet region, the axial air concentration attenuation followed a power-law distribution, and the air concentration distribution along the lateral cross-sections conformed to the Gaussian distribution law. Both the axial attenuation and the lateral distribution of air concentration in the free jet region were almost unchanged for different initial water velocities, air concentrations, and flow rate conditions. Based on the observation and analysis of the flow patterns in the swirling region in the plunge pool, the air concentration was assumed to be a superposition of lateral diffusion in three stages of the jet flow. The position of the maximum air concentration values along the vertical cross-sections in the wall jet and the length of the swirling region, as well as their relationship with the initial flow conditions, were introduced. A set of formulas was proposed for the prediction of the air concentration distribution in the plunge pool. The present analysis model, the results of which have been compared with experimental data, can essentially predict the air concentration distribution of aerated jet flows in the swirling region of a plunge pool.

In water conservancy projects, the jet flows are aerated, and they form an air concentration distribution inside the plunge pool, which has a significant effect on the downstream energy dissipation and the total dissolved gas. Meanwhile, the increase in aeration results in changes to the mass and heat transfer process at the air–water interface. This research improves our understanding of the air dissolution process downstream practical applications, and provides a new model to predict the air concentration distribution in plunge pools caused by aerated jets. The results of this study reveal the complexity of the air concentration distribution in a plunge pool caused by aerated jets. In order to further improve the prediction accuracy for the air concentration distribution in plunge pools, coupling analyses of bubble floating by buoyancy and jet motion in the air diffusion process should be the focus of future studies.

Author Contributions: The paper is the product of the joint efforts of the authors who worked together on experimental model tests. W.X. conceived the experiments, having a scientific background in applied hydraulics. C.C. and W.W. conducted the experimental investigations and analyzed the data under the supervision of W.X. C.C. wrote the manuscript with the co-authors.

References

1. Beltaos, S.; Rajaratnam, N. Impinging Circular Turbulent Jets. *J. Hydraul. Div.* **1974**, *100*, 1313–1328.
2. Beltaos, S.; Rajaratnam, N. Impingement of Axisymmetric Developing Jets. *J. Hydraul. Res.* **2010**, *15*, 311–326. [CrossRef]
3. Cooper, D.; Jackson, D.C.; Launder, B.E.; Liao, G.X. Impinging jet studies for turbulence model assessment—I. Flow-field experiments. *Int. J. Heat Mass Transfer* **1993**, *36*, 2675–2684. [CrossRef]
4. Baydar, E.; Ozmen, Y. An experimental investigation on flow structures of confined and unconfined impinging air jets. *J. Heat Mass Transfer* **2006**, *42*, 338–346. [CrossRef]
5. Schmocker, L.; Pfister, M.; Hager, W.H.; Minor, H.E. Aeration characteristics of ski jump jets. *J. Hydral. Eng.* **2008**, *98*, 90–97. [CrossRef]
6. Ervine, D.A.; Falvey, H.T. Behaviour of turbulent water jets in the atmosphere and in plunge pools. *Proc. Inst. Civil Eng.* **1987**, *83*, 295–314. [CrossRef]
7. Ervine, D.A.; Falvey, H.T.; Withers, W. Pressure fluctuations on plunge pool floors. *J. Hydraul. Res.* **1997**, *35*, 257–279. [CrossRef]

8. Castillo, L.G.; Carrillo, J.M.; Blázquez, A. Plunge pool dynamic pressures: A temporal analysis in the nappe flow case. *J. Hydraul. Res.* **2015**, *53*, 101–118. [CrossRef]
9. Chanson, H.; Aoki, S.; Hoque, A. Physical modelling and similitude of air bubble entrainment at vertical circular plunging jets. *Chem. Eng. Sci.* **2004**, *59*, 747–758. [CrossRef]
10. Wei, W.R.; Deng, J.; Liu, B. Influence of aeration and initial water thickness on axial velocity attenuation of jet flows. *J. Zhejiang Univ-Sci. A (Appl Phys & Eng).* **2013**, *14*, 362–370. [CrossRef]
11. Wei, W.R.; Deng, J.; Liu, B.; Zhang, Y.L.; Qian, X.Y.; Liu, X.J. Experimental investigation on the effects of aeration on the velocity attenuation of jet flow. *J. Sichuan Uni.* **2011**, *43*, 29–33. [CrossRef]
12. Luo, M.; Guo, Y.K. Study of diffusion of confined two-dimensional aerated jets under water surface. *J. Hydral. Eng.* **1992**, *7*, 29–34. [CrossRef]
13. Melo, J. Reduction of plunge pool floor dynamic pressure due to jet air entrainment. In Proceedings of the International Workshop on Rock Scour due to Falling High-Velocity Jet, Lausanne, Switzerland, 25–28 September 2002; Schleiss, A.J., Bollaert, E.F., Eds.; Swets and Zei tlinger: Lisse, The Netherlands, 2002; pp. 125–136.
14. Deng, J.; Xu, W.L.; Liu, S.J.; Wang, W. Influence of water jet aeration on pressure in scour pool and plunge pool. *Adv. Water Sci.* **2009**, *20*, 373–378. [CrossRef]
15. Xu, W.; Deng, J.; Qu, J.; Liu, S.; Wang, W. Experimental investigation on influence of aeration on plane jet scour. *J. Hydral. Eng.* **2004**, *130*, 160–164. [CrossRef]
16. Duarte, R.; Schleiss, A.J.; Pinheiro, A. Influence of jet aeration on pressures around a block embedded in a plunge pool bottom. *Environ. Fluid Mech.* **2015**, *15*, 673–693. [CrossRef]
17. Duarte, R.; Pinheiro, A.; Schleiss, A.J. An Enhanced Physically Based Scour Model for Considering Jet Air Entrainment. *Eng.* **2016**, *2*, 294–301. [CrossRef]
18. Sarkar, A.; Dey, S. Review on Local Scour Due to Jets. *Int. J. Sediment. Res.* **2004**, *19*, 53–81.
19. Pagliara, S.; Hager, W.; Minor, H.-E. Hydraulics of Plane Plunge Pool Scour. *J. Hydral. Eng.* **2006**, *132*, 450–461. [CrossRef]
20. Bollaert, E.; Schleiss, A. Scour of rock due to the impact of plunging high velocity jets Part II: Experimental results of dynamic pressures at pool bottoms and in one- and two-dimensional closed end rock joints. *J. Hydraul. Res.* **2003**, *41*, 465–480. [CrossRef]
21. Bin, A.K. Gas entrainment by plunging liquid jets. *Chem. Eng. Sci* **1993**, *48*, 136. [CrossRef]
22. Chanson, H. References. In *Air Bubble Entrainment in Free-Surface Turbulent Shear Flows*; Chanson, H., Ed.; Academic Press: San Diego, CA, USA, 1996; pp. 239–261. ISBN 978-0-12-168110-4.
23. Harby, K.; Chiva, S.; Muñoz-Cobo, J.L. An experimental study on bubble entrainment and flow characteristics of vertical plunging water jets. *Exp. Therm. Fluid Sci.* **2014**, *57*, 207–220. [CrossRef]
24. Brattberg, T.; Chanson, H. Air entrapment and air bubble dispersion at two-dimensional plunging water jets. *Chem. Eng. Sci* **1998**, *53*, 4113–4127. [CrossRef]
25. Kiger, K.T.; Duncan, J.H. Air-Entrainment Mechanisms in Plunging Jets and Breaking Waves. *Annu. Rev. Fluid Mech.* **2011**, *44*, 563–596. [CrossRef]
26. Miwa, S.; Moribe, T.; Tsutstumi, K.; Hibiki, T. Experimental Investigation of Air Entrainment by Vertical Plunging Liquid Jet. *Chem. Eng. Sci* **2018**, *181*. [CrossRef]
27. Brouilliot, D.; Lubin, P. Numerical simulations of air entrainment in a plunging jet of liquid. *J. Fluids Struct.* **2013**, *43*, 428–440. [CrossRef]
28. Samanta, A.; Vinuesa, R.; Lashgari, I.; Schlatter, P.; Brandt, L. Enhanced secondary motion of the turbulent flow through a porous square duct. *J. Fluid Mech.* **2015**, *784*, 681–693. [CrossRef]
29. Dong, Z.Y.; Wu, C.G. Characteristics of concentration diffusion of submerged aerated jets. *J. Chengdu Univ. Sci. Technol.* **1994**, *4*, 31–38.
30. Chanson, H. Air entrainment in two-dimensional turbulent shear flows with partially developed inflow conditions. *Int. J. Multiphase Flow* **1995**, *21*, 1107–1121. [CrossRef]
31. Khaled, J.H. Liquid Jet Impingement on a Free Liquid Surface: PIV Study of the Turbulent Bubbly Two-Phase Flow. In Proceedings of the ASME 2010 Joint Us-European Fluids Engineering Summer Meeting Collocated with International Conference on Nanochannels, Microchannels, and Minichannels ASME 2010 Joint Us-European Fluids Engineering Summer Meeting: Volume 1, Symposia—Parts A, B, and C, Montreal, QC, Canada, 1–5 August 2010; pp. 2877–2885.

32. Vinuesa, R.; Schlatter, P.; Nagib, H.M. Role of data uncertainties in identifying the logarithmic region of turbulent boundary layers. *Exp. Fluids* **2014**, *55*, 1751. [CrossRef]
33. Albertson, M.L.; Dai, Y.B.; Jensen, R.A.; Rouse, H. Diffusion of Submerged Jets. *Trans. Amer. Soc. Civil Eng.* **1950**, *115*, 639–664. [CrossRef]
34. Prandtl, L. Bemerkungen zur Theorie der freien Turbulenz. *Z.A.M.M.* **1942**, *22*, 241–243. [CrossRef]
35. Beltaos, S.; Rajaratnam, N. Plane Turbulent Impinging Jet. *J. Hydraul. Res.* **1973**, *11*, 29–59. [CrossRef]

6

Numerical Investigation on the Hydraulic Properties of the Skimming Flow over Pooled Stepped Spillway

Shicheng Li and Jianmin Zhang *

State Key Laboratory of Hydraulics and Mountain River Engineering, Sichuan University, Chengdu 610065, China; tyutscu@163.com

* Correspondence: zhangjianmin@scu.edu.cn

Abstract: Pooled stepped spillway is known for high aeration efficiency and energy dissipation, but the understanding for the effects of pool weir configuration on the flow properties and energy loss is relatively limited, so RNG $k-\varepsilon$ εturbulence model with VOF method was employed to simulate the hydraulic characteristics of the stepped spillways with four types of pool weirs. The calculated results suggested the flow in the stepped spillway with staggered configuration of' two-sided pooled and central pooled steps (TP-CP) was highly three dimensional and created more flow instabilities and vortex structures, leading to 1.5 times higher energy dissipation rate than the fully pooled configuration (FP-FP). In FP-FP configuration, the stepped spillway with fully pooled and two-sided pooled steps (FP-TP) and the spillway with fully pooled and central pooled steps (FP-CP), the pressure on the horizontal step surfaces presented U-shaped variation, and TP-CP showed the greatest pressure fluctuation. For FP-TP and FP-CP, the vortex development in the transverse direction presented the opposite phenomenon, and the maximum vortex intensity in TP-CP occurred at Z/W = 0.25, while FP-FP illustrated no significant change in the transverse direction. The overlaying flow velocity distribution in the spanwise direction demonstrated no obvious difference among FP-FP, FP-TP, and FP-CP, while the velocity in TP-CP increased from the axial plane to the sidewalls, but the maximum velocity for all cases were approximately the same.

Keywords: numerical simulation; pooled stepped spillway; pool weir; flow property

---

1. Introduction

Stepped spillway, an important water-transferring hydraulic infrastructure, has long been used in water related projects owing to its high efficiency of energy dissipation and air entrainment, making it possible to release massive and high speed flood flows while reducing and preventing dangerous scour and cavitation. Additionally, economic benefits will be enjoyed due to the reduction in the construction of the downstream energy dissipation basin [1,2]. Due to the advantages of stepped spillway, a growing number of this particular hydraulic structure is being applied to engineering projects.

For the past decades, stepped spillway has been studied globally by the means of laboratory experiments and numerical simulations [3–7], providing significant insights into the complicated air-water flow properties, momentum exchange process, and energy loss mechanism. For example, model tests performed by Felder and Chanson [8] to investigate the influence of uniform and non-uniform step height on the hydraulic performance illustrated that there was no significant change between the two configurations regarding the residual head. Channel slope also plays a vital role in the air-water flow properties, such as velocity profile and turbulence intensity reported by Felder and Chanson [9]. The time-averaged pressure on the step faces was experimentally measured by Zhang et al. [10], drawing the conclusions that the pressure distributions on the horizontal step surfaces exhibited S-shaped variations with the maximum and minimum values occurring at the downstream end and upstream start, respectively. Physical models were established by Zhang and Chanson [11] to examine the effect of step edge and cavity geometry on the air entrainment and energy loss performance in stepped chutes. Apart from experimental studies, thanks to the advances in the computational performance of computers, numerical methods have also been successfully used to study the detailed flow properties in complex contexts [12–15]. Baylar et al. [16] numerically examined the energy loss and aeration efficiency of the stepped chutes and found that both of them in nappe flow regime were greater than that of the skimming flow regime. Numerical simulations by researchers [17–19] demonstrated its high performance in investigating the detailed flow field and pressure distribution over stepped spillway.

In recent years, an innovative design has been made to stepped spillway to improve the energy dissipation performance. Unlike the conventional stepped spillway whose horizontal step faces are flat, this novel kind of stepped spillway, pooled stepped spillway, is equipped with pool weirs at the step edges. A few of the researches have been carried out to investigate the air-water structures in this unique type of stepped spillway. For instance, the differences concerning the air entrainment processes of flat, pooled, and a combination of flat and pooled stepped spillway have been clarified by André [20] and Kökpinar [21]. Thorwarth [22] reported self-induced instabilities over pooled stepped spillways with channel slopes of 8.9° and 14.6°. The detailed air-water flow properties over several kinds of stepped pooled spillway were also physically studied by Felder [7] and Felder and Chanson [9].

While the existing researches have gained understanding into this particular type of stepped spillway, little has been known about the impact of pool weir configurations on the flow properties. Therefore, in order to expand the limited insights into the effect of pool weir's geometrical properties on the flow characteristics, numerical investigations of pooled stepped spillways with different pool weir configurations were carried out.

2. Numerical Simulation

In the present study, ANSYS 16.0 (ANSYS®, Canonsburg, PA, USA) was utilized to investigate the flow characteristics over stepped spillway with different pool weir configurations. The numerical model consists of 4 parts: the upstream water tank, the broad-crested weir with an upstream rounded corner (radius = 0.08 m), the stepped spillway and the downstream channel. The length, L_{crest}, and width, W, of the broad-crested weir are 1.01 m and 0.52 m, respectively. The spillway has 10 steps with vertical step height, h, and horizontal step length, l, of 0.1 m and 0.2 m, respectively. The channel slope is 26.6° and constant in each configuration. Moreover, the height, d, and the thickness, l_w, of pool weir are, respectively, 0.09 m and 0.015 m. The schematics of the spillways are shown in Figure 1.

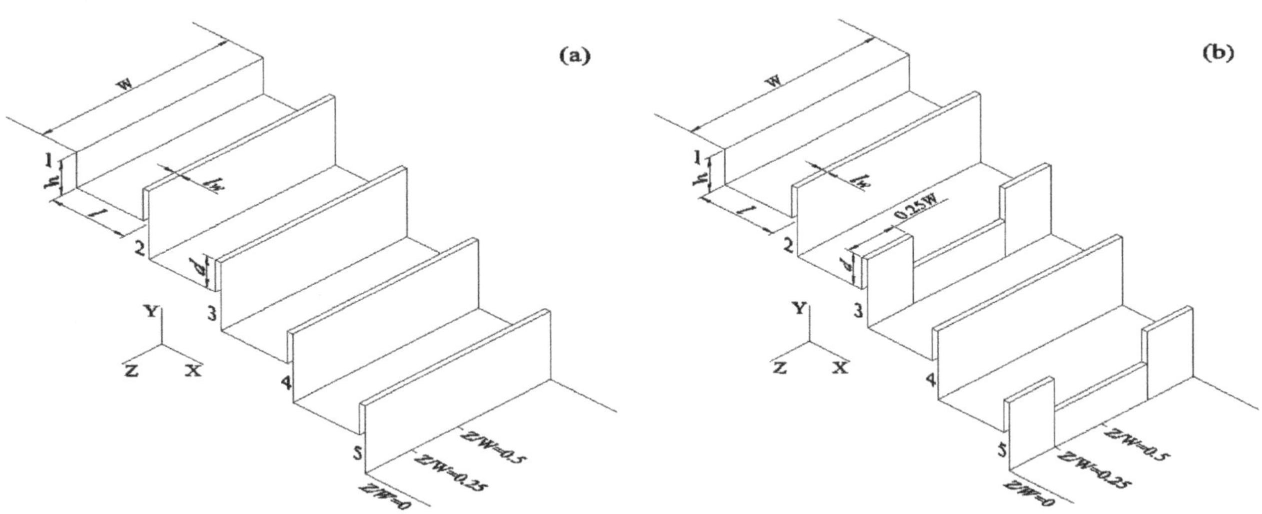

Figure 1. *Cont.*

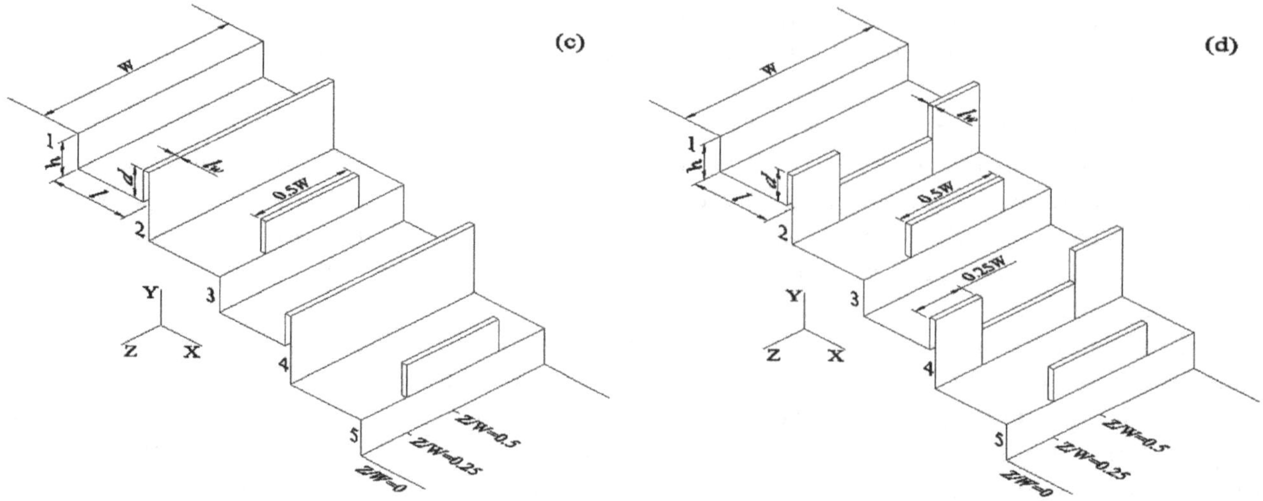

Figure 1. Sketch of pooled stepped spillway configurations. (**a**) Fully pooled steps (FP-FP); (**b**) fully pooled and two-sided pooled steps (FP-TP); (**c**) fully pooled and central pooled steps (FP-CP); and (**d**) two-sided pooled and central pooled steps (TP-CP).

2.1. Volume of Fluid Method

Developed by Hirt and Nichols [23], the volume of fluid (VOF) model is believed to be a highly effective method to determine the position of the interface of two or more immiscible flows [19,24–26]. For the purpose of tracking the air-water interface, the continuity equation for the volume fraction of water is adopted in the following format:

$$\frac{\partial \alpha_w}{\partial t} + u_i \frac{\partial \alpha_w}{\partial x_i} = 0 \tag{1}$$

where α_w is the volume fraction of water, u_i and x_i are the velocity components and coordinates ($i = 1, 2, 3$), respectively. Hence, by solving the continuity equation, the free surface of water can be determined. In addition, it is clear that the sum of their volume fractions, α_w and α_a, respectively, is unity in every computational cell, which can be described as:

$$\alpha_w + \alpha_a = 1 \tag{2}$$

Obviously, the maximum and minimum value for the volume fraction of water, α_a, is 1 and 0, indicating the given cell is full of water or air, respectively, between which there must be a mixture of them containing the interface.

As for the density, ρ, and molecular viscosity, μ, they can be given as

$$\rho = \alpha_w \rho_w + (1 - \alpha_w)\rho_a \tag{3}$$

$$\mu = \alpha_w \mu_w + (1 - \alpha_w)\mu_a \tag{4}$$

where ρ_w and ρ_a represents the density of water and air, respectively, while μ_w and μ_a are the molecular viscosity of water and air, respectively. Therefore the values of ρ and μ can be known by iterating the Equations (3) and (4).

2.2. Turbulence Model

Proposed by Yakhot and Orszag [27], the Renormalization Group (RNG) $k - \varepsilon$ closure model has been employed to simulate flow over stepped spillways in many studies [28–30], yielding reliable results compared with laboratory experiments. The continuity equation, momentum equation, and transport equations of turbulent kinetic energy, and the turbulent dissipation rate, used in the model, are shown below.

Continuity equation:

$$\frac{\partial \rho}{\partial t} + \frac{\partial \rho u_i}{\partial x_i} = 0 \tag{5}$$

Momentum equation:

$$\frac{\partial \rho u_i}{\partial t} + \frac{\partial}{\partial x_j}(\rho u_i u_j) = -\frac{\partial p}{\partial x_i} + \frac{\partial}{\partial x_j}\left[(\mu + \mu_t)\left(\frac{\partial u_i}{\partial x_j} + \frac{\partial u_j}{\partial x_i}\right)\right] \tag{6}$$

k equation:

$$\frac{\partial(\rho k)}{\partial t} + \frac{\partial(\rho_i k)}{\partial x_i} = \frac{\partial}{\partial x_i}\left[\left(\mu + \frac{\mu_t}{\sigma_k}\right)\frac{\partial k}{\partial x_i}\right] + G_k - \rho\varepsilon \tag{7}$$

ε equation:

$$\frac{\partial(\rho\varepsilon)}{\partial t} + \frac{\partial(\rho u_i \varepsilon)}{\partial x_i} = \frac{\partial}{\partial x_i}\left[\left(\mu + \frac{\mu_t}{\sigma_\varepsilon}\right)\frac{\partial \varepsilon}{\partial x_i}\right] + C_{1\varepsilon}\rho\frac{\varepsilon}{k}G_k - C_{2\varepsilon}\rho\frac{\varepsilon^2}{k} \tag{8}$$

where ρ and μ respectively represent the average density of the volume fraction and the molecular viscosity. p is the modified pressure, and μ_t is the turbulent viscosity, which can be calculated from the following equations consisting of k and ε. $\mu_t = C_\mu \frac{k^2}{\varepsilon}$, $C_{1\varepsilon} = 1.42 - \frac{\eta\left(1-\frac{\eta}{\eta_0}\right)}{1+\beta\eta^3}$, $\eta = \frac{Sk}{\varepsilon}$, $S = \sqrt{2S_{ij}S_{ij}}$. The constants in the above expressions are given in the Table 1.

Table 1. Constants of governing equations.

η_0	438
β	0.012
C_μ	0.085
$C_{2\mu}$	1.68
σ_k	0.7179
σ_ε	0.7179

2.3. Boundary Conditions

Combined with VOF method, RNG $k-\varepsilon$ turbulence model was applied to numerically study the flow characteristics over stepped spillway with different pool weirs and the entire simulation domain was meshed with structured grids, as shown in Figure 2. The boundary conditions for the mathematical models were set as follows:

(a) Inflow boundary: Velocity of the flow calculated from the discharge, which is 0.113 m^3/s in all cases, was given as the inlet boundary;

(b) Outflow boundary: Pressure outlet boundary was selected at the outlet and there was no normal gradient for all variables;

(c) Free surface: Pressure inlet was employed and its value was the standard atmospheric pressure; and

(d) Wall boundary: No-slip velocity boundary and wall function were chosen for the wall surfaces and near-wall regions, respectively.

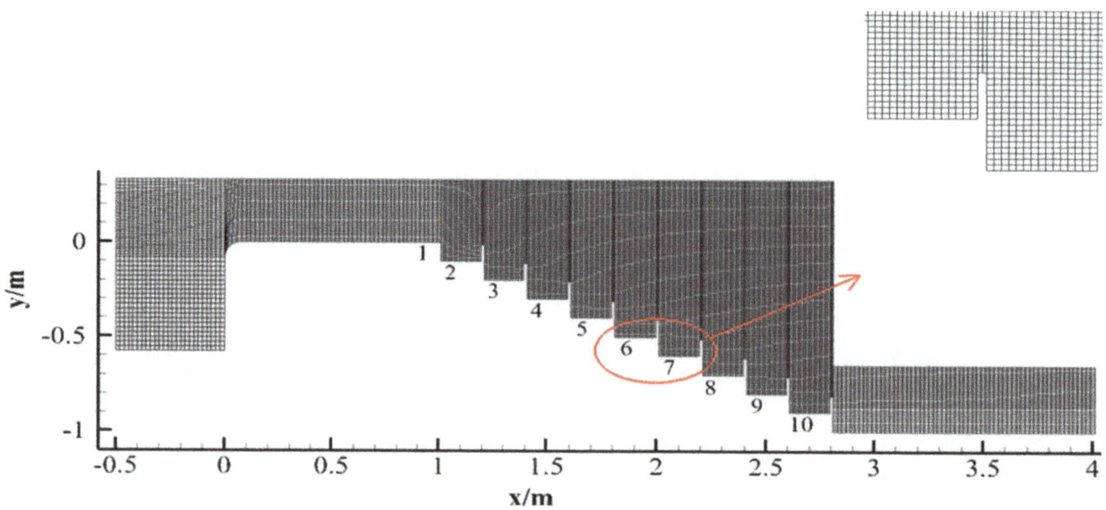

Figure 2. Meshing pattern in the computational domain.

2.4. Grid Testing and Model Verification

Before applying the numerical model to further analysis, it is necessary to validate its reliability. Therefore, the accuracy of the calculation results were tested by utilizing grid convergence index (GCI) [31] with grid numbers of approximately 0.31 million, 0.55 million and 0.76 million. Figure 3 presents the impact of the grid sizes on the uncertainty of the computational velocity distribution at the axial plane. The discretization uncertainties at most locations were quite small. The maximum GCI values for the velocity profiles over step 6 and step 7 were 7.7% and 5.6%, corresponding to ±0.19 m/s and ±0.15 m/s, respectively. Considering the simulation efficiency and accuracy, the grid number set as 0.55 million is reasonable.

In addition, experimental data obtained from Felder et al. [32] were used to verify the numerical results by comparing the velocity distribution over the pool at steps 8, 9, and 10, as shown in Figure 4. The schematic of the physical model used for validation was the same as the FP-FP configuration in the present article, except the pool weir height. In the research of Felder et al. [32], the pool weir height was 0.031 m. It can be seen from Figure 4 that, at most locations, the calculated results of velocity distribution were fairly consistent with that of laboratory tests. The maximums of the relative error for steps 8, 9 and 10 were 14.8%, 7.1% and 6.5%, respectively, indicating that the numerical simulations produced reliable and acceptable results.

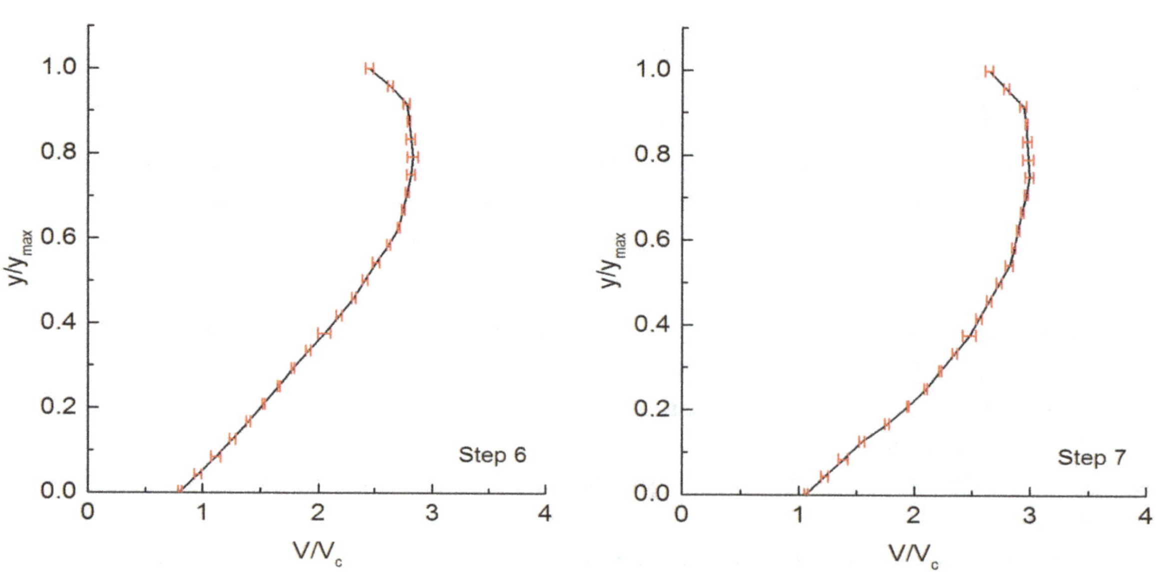

Figure 3. Fine-grid solution with discretization error bars computed using the GCI index.

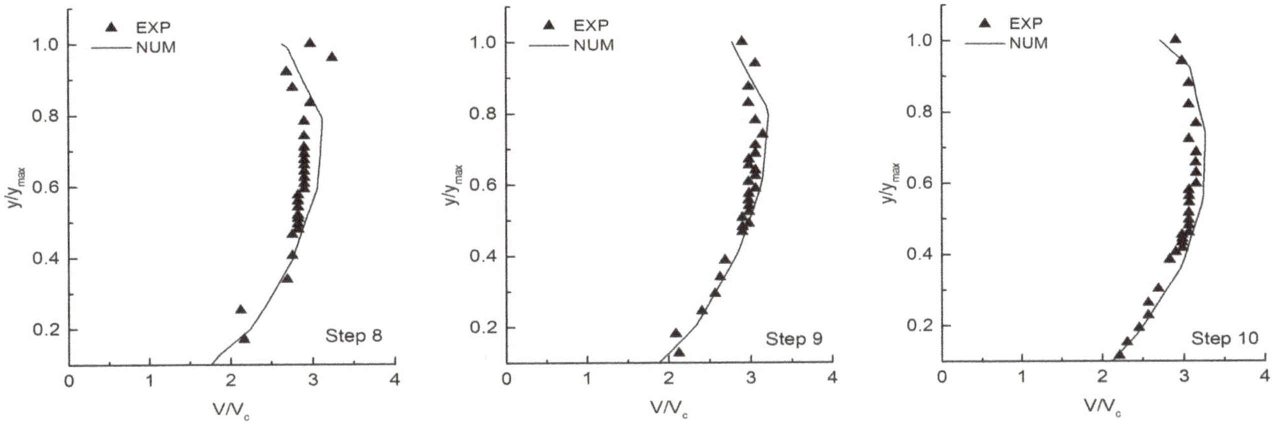

Figure 4. Comparison of numerical and experimental data. Notes: V_c: critical flow velocity; y represents the distance normal to the pseudo-bottom formed by the pool edges with $y = 0$ at the pool edges.

3. Results and Analysis

3.1. Flow Pattern

After ensuring the accuracy of the calculation, the effects of pool weir geometrical parameters on the hydraulic behaviors were numerically examined. Figure 5 illustrates the flow pattern at the axial plane in the calculated cases. It is noticeable that the free water surface of FP-FP configuration was parallel to the pseudo-bottom, and showed the least instability among the simulated configurations. FP-TP and FP-CP showed the similar free surface profile, but slight flow turbulences caused by the partially pooled steps can be observed. An obvious feature of the flow in TP-CP was the highly three-dimensional flow motion in the spanwise direction, so TP-CP highlighted strong fluctuations in free water surface.

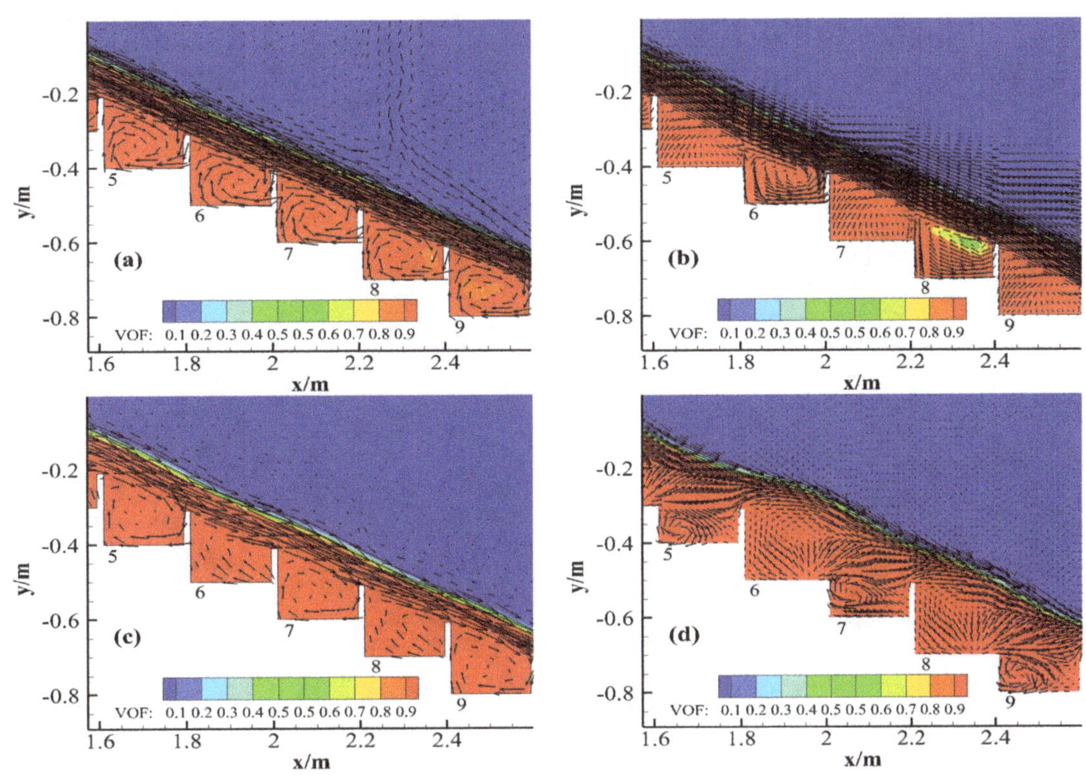

Figure 5. Flow pattern at the axial plane in different configurations. (**a**) FP-FP; (**b**) FP-TP; (**c**) FP-CP; and (**d**) TP-CP.

Figure 6 highlights the streamlines on the representative steps in each configuration. As can be seen, like the traditional flat stepped spillway, the flow surface over the pool weir was approximately parallel to the pseudo-bottom formed by the pool edges, although TP-CP presented some instabilities caused by the staggered configurations of central pooled and two-sided pooled steps. For FP-FP configuration, vortexes with the similar size and position were formed in every pool and there was no significant change in the transverse direction, which was different from other configurations, that is, the flow pattern in FP-TP, FP-CP, and TP-CP was three-dimensional. In FP-TP, the scale of the vortex on the two-sided pooled steps increased from the axial plane to the sidewalls, while the scale of the vortex on the fully pooled steps decreased from the axial plane to the sidewalls. FP-CP presented the opposite phenomenon compared with FP-TP. In TP-CP configuration, the maximum vortex intensity occurred at Z/W = 0.25 and no eddy can be found on the central pooled steps at the sidewalls and on the two-sided pooled steps at the axial plane.

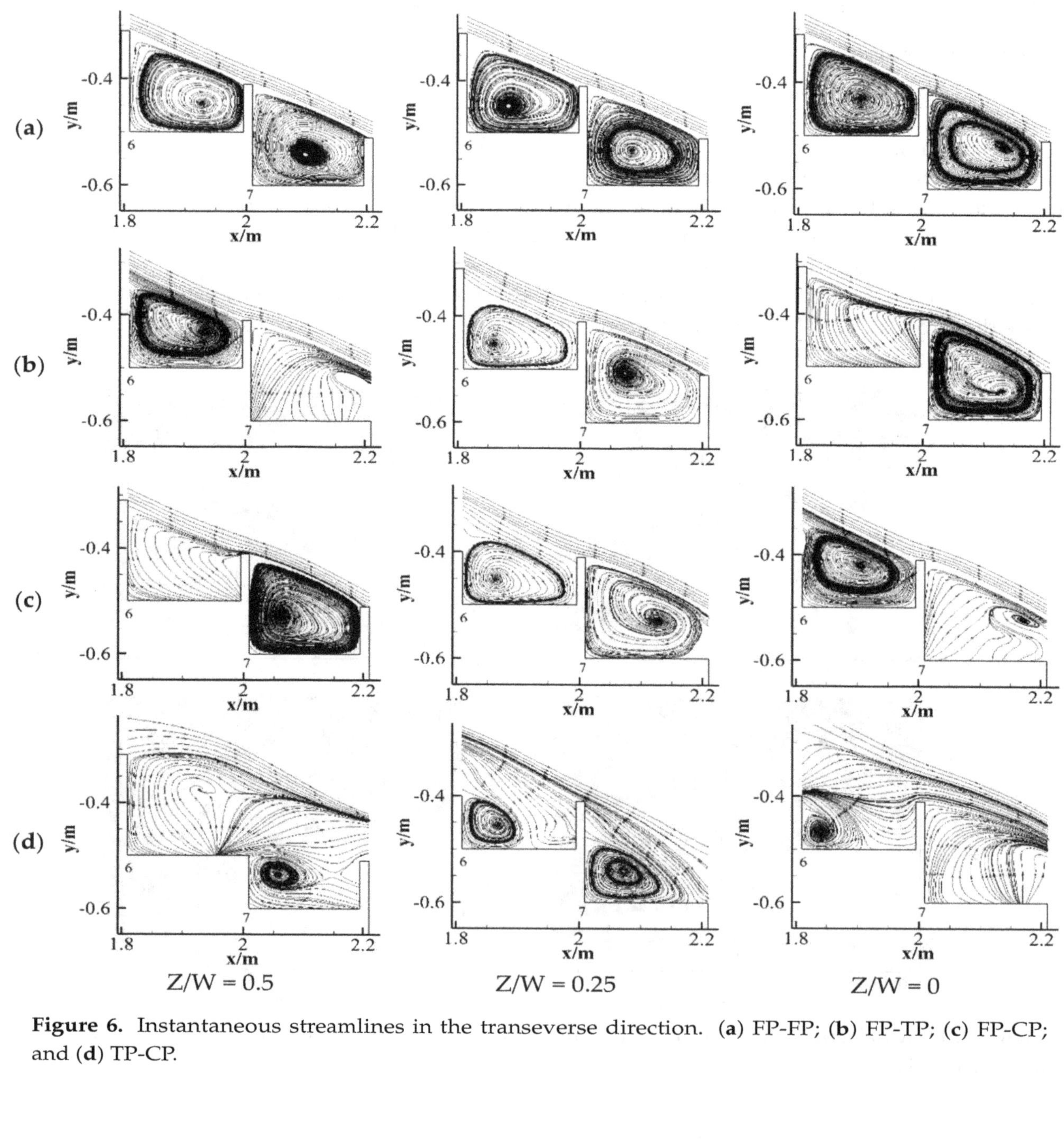

Figure 6. Instantaneous streamlines in the transeverse direction. (**a**) FP-FP; (**b**) FP-TP; (**c**) FP-CP; and (**d**) TP-CP.

For the identification of turbulence structures, Q criterion [33] is introduced to examine the effects of pool weir geometry on the vortex flow. The Q criterion is defined as follows:

$$Q = \frac{1}{2}\left(\|\Omega_{ij}\|^2 - \|S_{ij}\|^2\right) \tag{9}$$

where

$$\Omega_{ij} = \frac{1}{2}\left(\frac{\partial u_i}{\partial x_j} - \frac{\partial u_j}{\partial x_i}\right) \tag{10}$$

represents the rate of rotation tensor corresponding to the pure rotational motion, and

$$S_{ij} = \frac{1}{2}\left(\frac{\partial u_i}{\partial x_j} + \frac{\partial u_j}{\partial x_i}\right) \tag{11}$$

represents the rate of strain tensor corresponding to the irrotational motion.

The parameter Q is an effective approach to reveal the main motion pattern of fluid element. It can been seen from Equation (9) that the rotation motion will dominate the region where the value of Q is positive, while the deformation motion will dominate when Q is negative. However, it should be noted that the vortex structures cannot be accurately represented in the region with $Q = 0$, due to experimental error and viscous diffusion.

Figure 7 demonstrates the iso-surface of $Q = 100$. In FP-FP configuration, the number of lumps of the Q iso-surface distributing on each step was approximately the same and there was no apparent change in the transverse direction. The distribution of lumps in FP-TP and FP-CP exhibited the similar pattern. The major difference was that small lumps can be found near the sidewalls on the two-sided pooled steps in FP-TP, while small lumps were located near the axial plane in FP-CP, meaning turbulent structures existed at the two locations. It is noticeable that distribution of lumps in TP-CP configuration was much denser than that of other cases, indicating greater vortex and more turbulent structures.

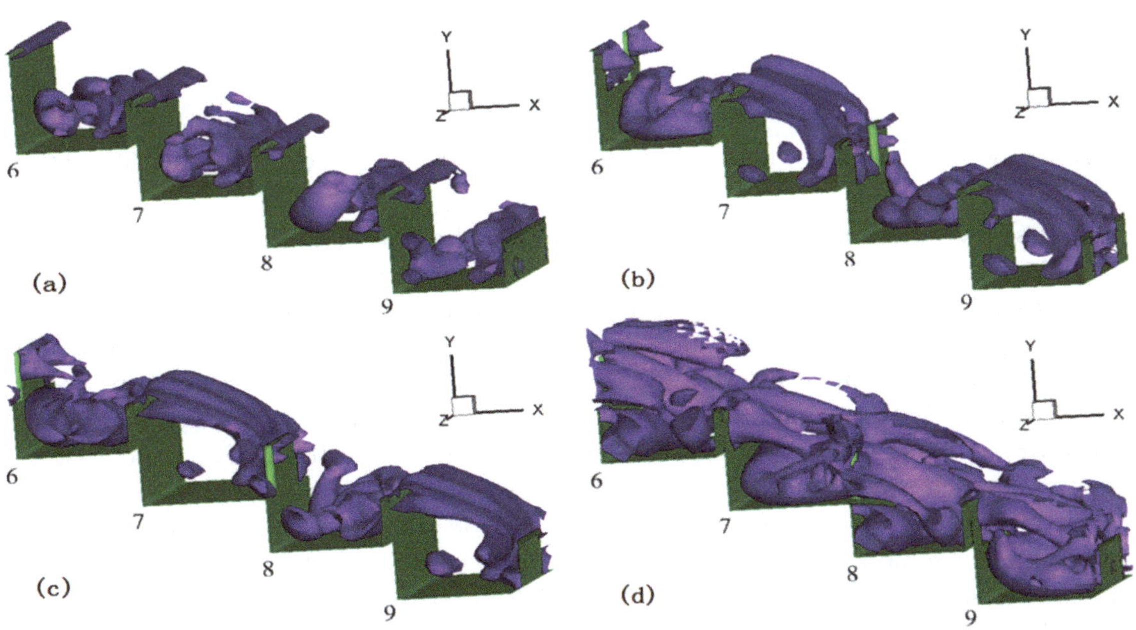

Figure 7. Distributions of lumps of the iso-surface of $Q = 100$. (**a**) FP-FP; (**b**) FP-TP; (**c**) FP-CP; and (**d**) TP-CP.

3.2. Velocity Distribution

Figure 8 illustrates the velocity vectors at the axial plane. It implies that, in all configurations, the skimming flow transferring from the upstream to the downstream over the steps was partly trapped in the recirculating zones under the pseudo-bottom. However, it is evident that stable vortices can only be found on some steps at the axial plane in FP-TP, FP-CP, and TP-CP, which is consistent with the analysis from flow pattern. Furthermore, the velocity magnitude of the overlaying flow was significantly greater than that of the vortex.

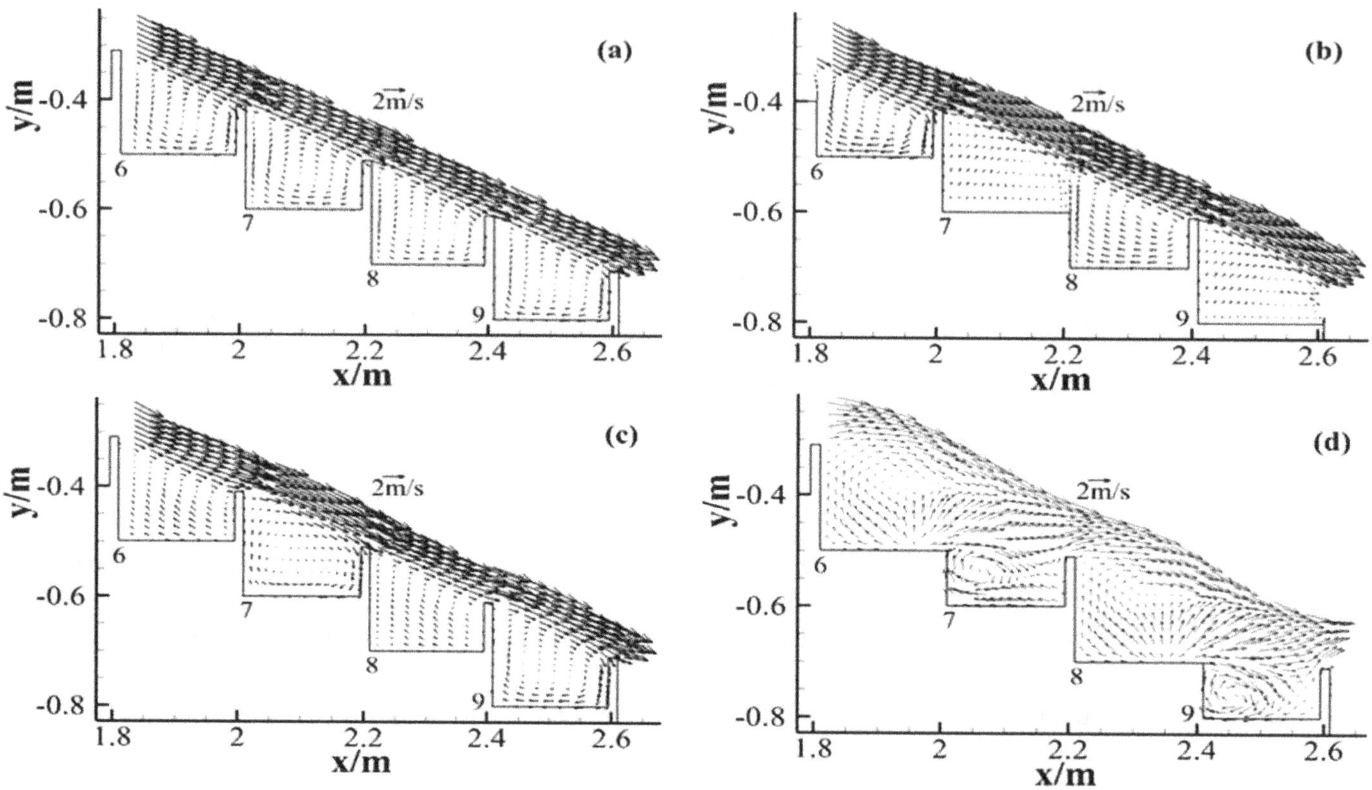

Figure 8. Velocity vectors at the axial plane. (**a**) FP-FP; (**b**) FP-TP; (**c**) FP-CP; and (**d**) TP-CP.

Figure 9 displays the velocity distributions above the center of the horizontal step surfaces. In FP-FP, the velocity showed the same distribution in the transverse direction; the velocity distributions on each step presented the similar pattern, and no significant change between different steps can be observed for $y/y_{\max} < 0.8$, but in the flow surface region, the velocity above the downstream step surfaces was greater than that of the upstream ones; the minimum and maximum velocity above each step occurred at the same position: $y/y_{\max} = 0.3$ and $y/y_{\max} = 0.9$, respectively. In FP-TP configuration, at the axial plane, the velocity distributions above the fully pooled and two-sided pooled steps were obviously different in the bottom region, with the former greater than the latter; However, at Z/W = 0.25, there was no major difference in the velocity distribution above different steps except the free surface region. In FP-CP, at Z/W = 0.5, the velocity at $0.5 < y/y_{\max} < 0.8$ for different steps indicated the same distribution; different from Z/W = 0.5, the velocity over different steps at Z/W = 0.25 exhibited similar distributions. In TP-CP, the velocity distributions presented a totally different pattern from other configurations; the velocity demonstrated the different distribution on two-sided pooled steps and central pooled steps, and velocity fluctuated greatly in the spanwise direction, proving its highly three dimensional characteristics.

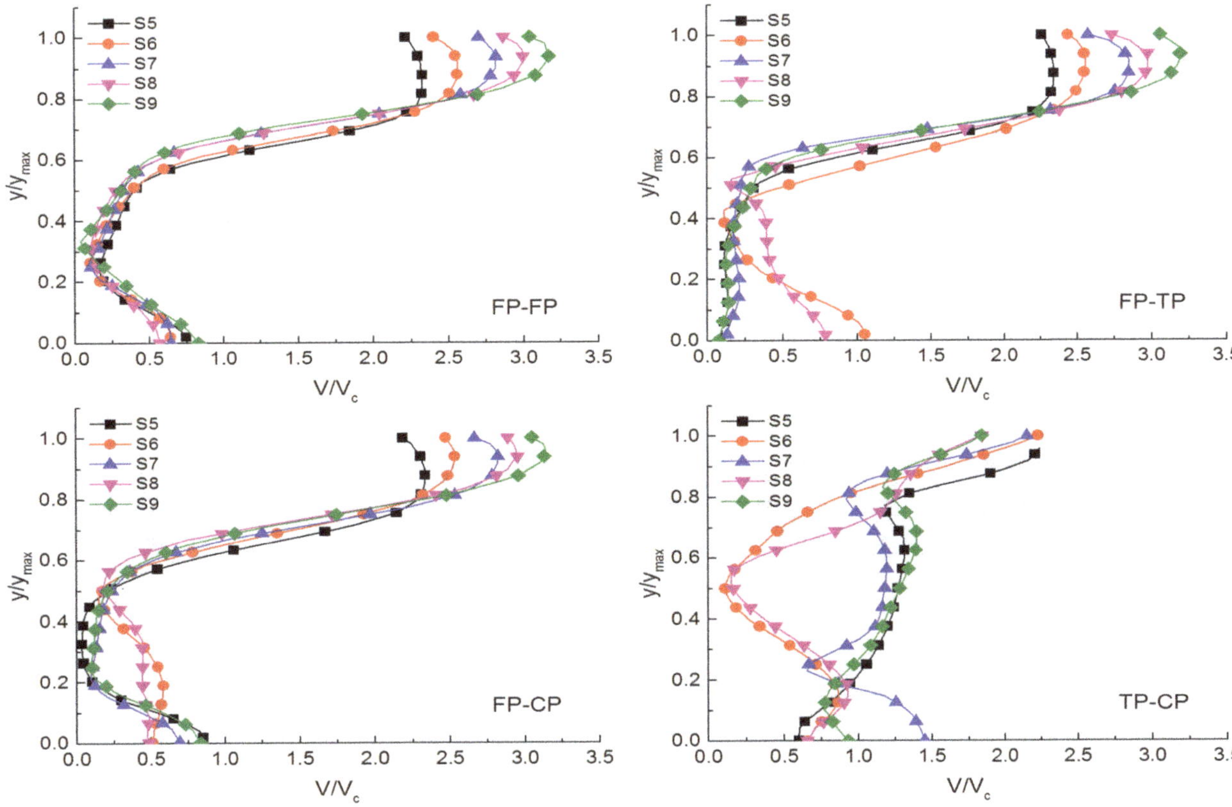

(a) Z/W = 0.5

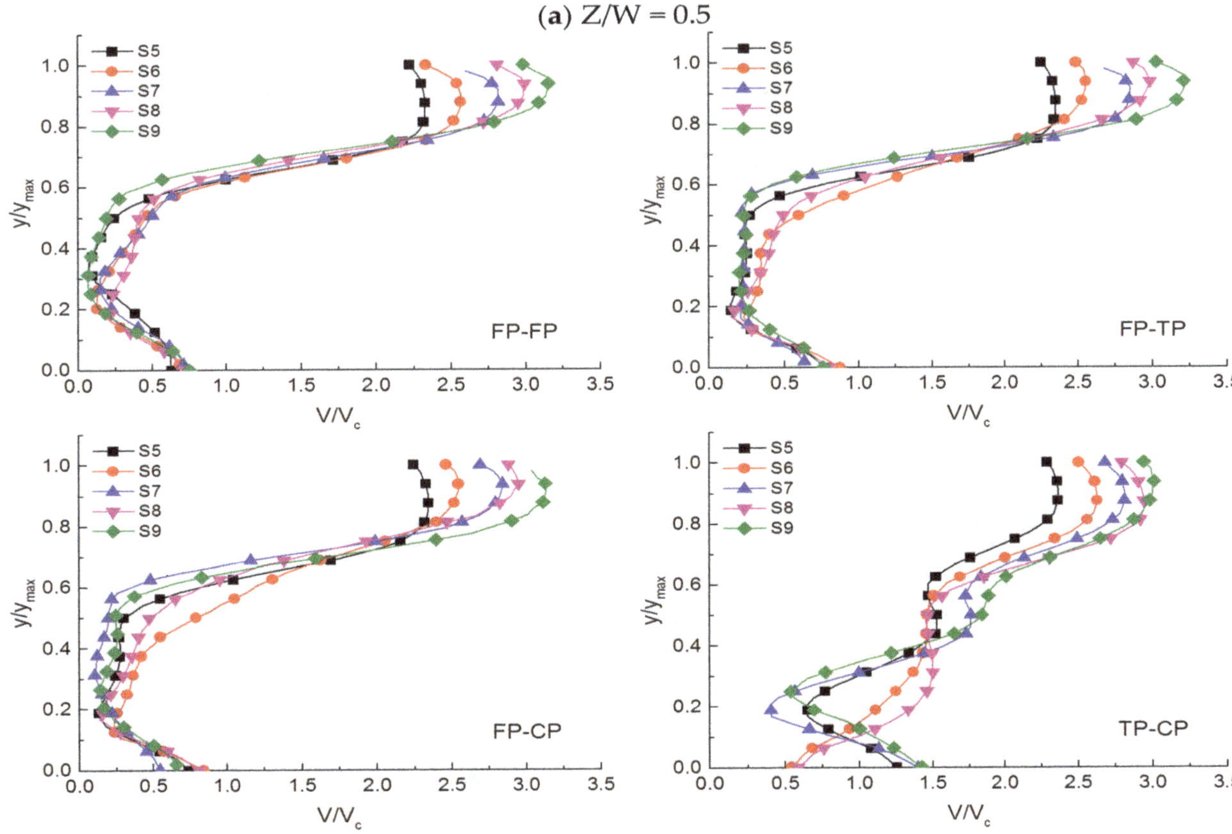

(b) Z/W = 0.25

Figure 9. Velocity distributions above the step surfaces in each configuration. (**a**) Z/W = 0.5; and (**b**) Z/W = 0.25.

In order to better compare the velocity field in the transverse direction, velocity contour maps at different cross-sections are shown in Figure 10. It can be noticed that, in all cases, the velocity distribution along the transverse direction was symmetrical and the overlaying flow over the pool weir showed the maximum velocity magnitude, while the minimum velocity value occurred near the bottom, but both the minimum and maximum indicated no apparent change in different configurations; In FP-FP, FP-TP and FP-CP, the velocity of the mainstream illustrated no significant difference along the spanwise direction, but the velocity in TP-CP fluctuated greatly with its value increasing from the axial plane to the sidewalls.

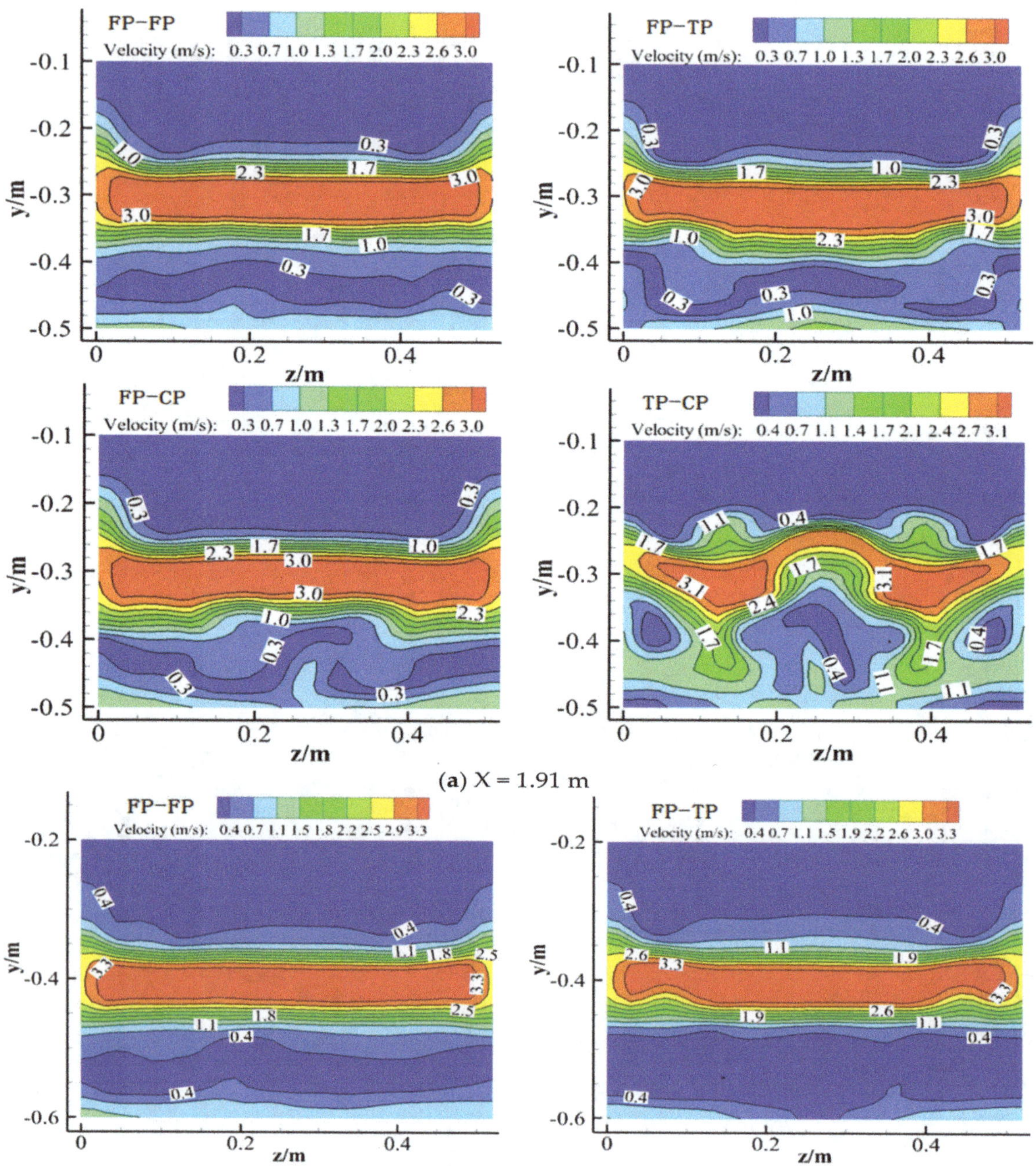

Figure 10. *Cont.*

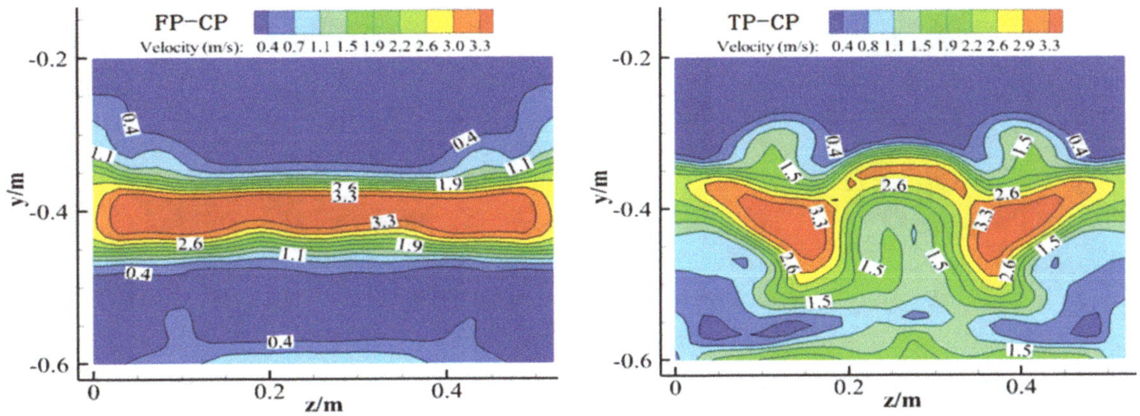

(**b**) X = 2.11 m

Figure 10. Velocity countour maps at the cross-sections. (**a**) X = 1.91 m; and (**b**) X = 2.11 m.

3.3. Pressure Distribution

Figure 11 highlights the pressure distributions on the horizontal step surfaces of two representative steps (steps 6 and 7). In FP-FP configuration, the minimum and maximum pressure on different horizontal step surfaces showed no significant change, as well as the locations where the minimum and maximum pressure occurred. The pressure distribution patterns on the fully pooled steps in FP-TP and on the central pooled steps in FP-CP were similar with the extreme pressure occurring at the axial plane near the downstream end of the step face and the minimum pressure occurring in the middle the step surface; the pressure distributions on the two-sided pooled steps in FP-TP and on the fully pooled steps in FP-CP demonstrated the similar patterns with the extreme pressure occurring at the sidewalls near the downstream end of the step face. As for TP-CP configuration, the extreme pressures on the two-sided pooled and central pooled steps occurred near the sidewalls and axial plane, respectively.

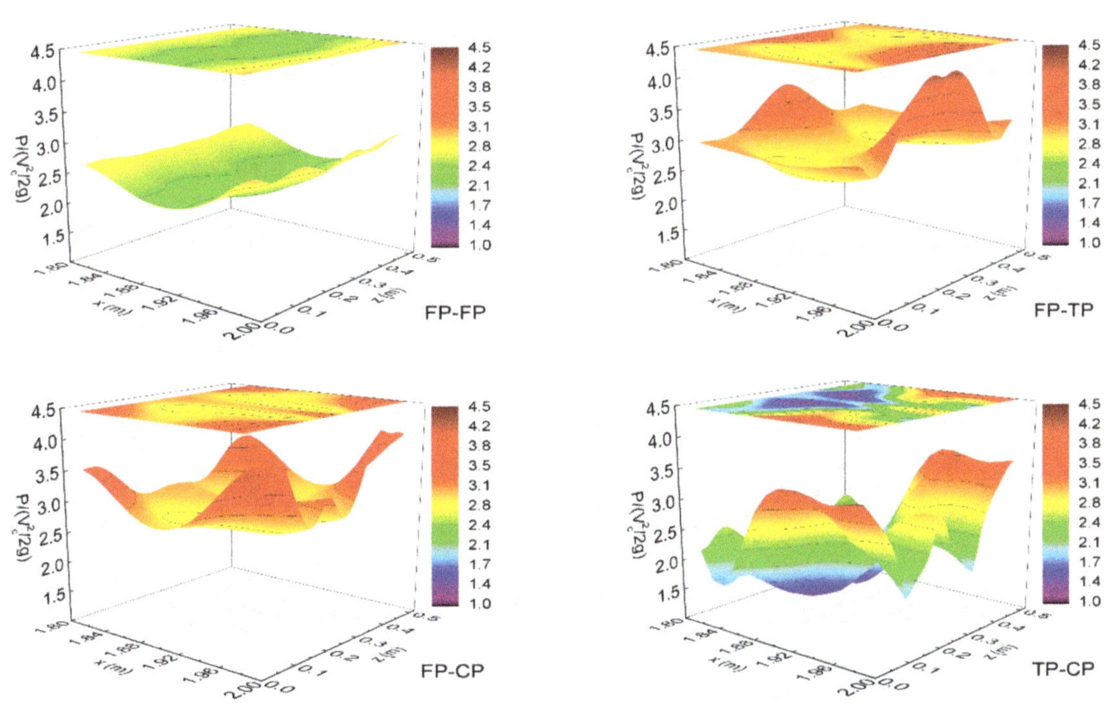

(**a**) Step 6

Figure 11. *Cont.*

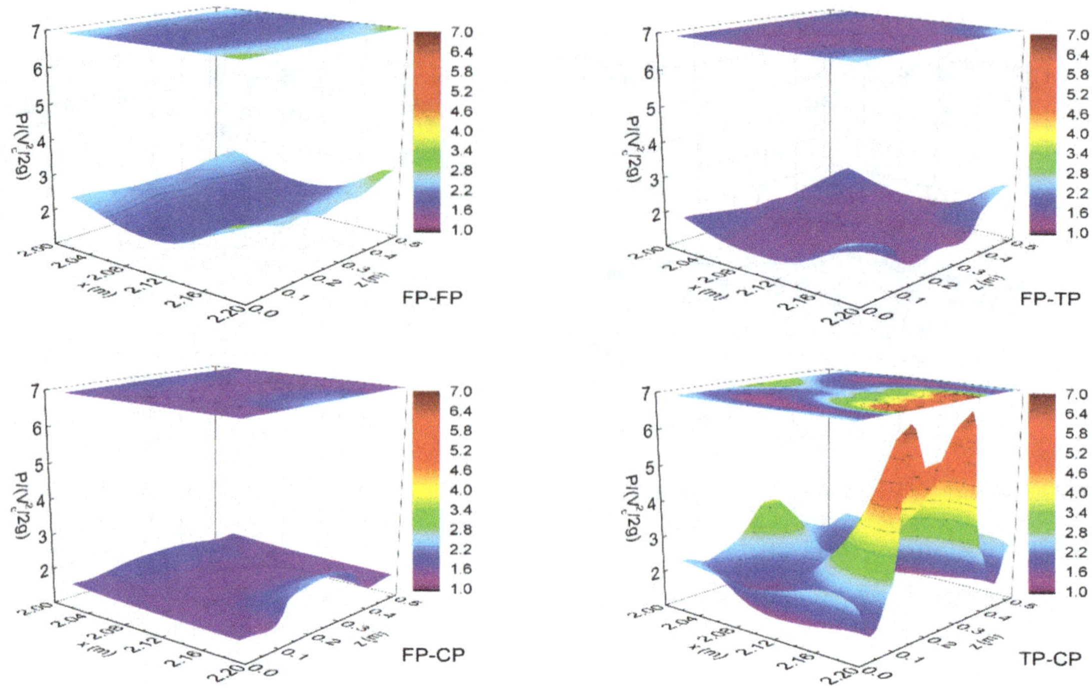

(**b**) Step 7

Figure 11. Pressure distributions on the horizontal step surfaces. (**a**) Step 6; (**b**) Step 7.

Figure 12 illustrates the pressure distributions on the vertical step surfaces of two representative steps (steps 6 and 7). As can been seen, in FP-FP, FP-TP and FP-CP, the pressure distributions on the vertical step surfaces highlighted the same pattern: the value of pressure decreased with the increasing in water depth, but the pressure for TP-CP presented some variations in the vertical direction. The maximum pressures on odd-number step surfaces in different configurations were in the following order: FP-CP > TP-CP > FP-TP > FP-FP, while for the even-number steps, it was TP-CP > FP-FP > FP-TP > FP-CP.

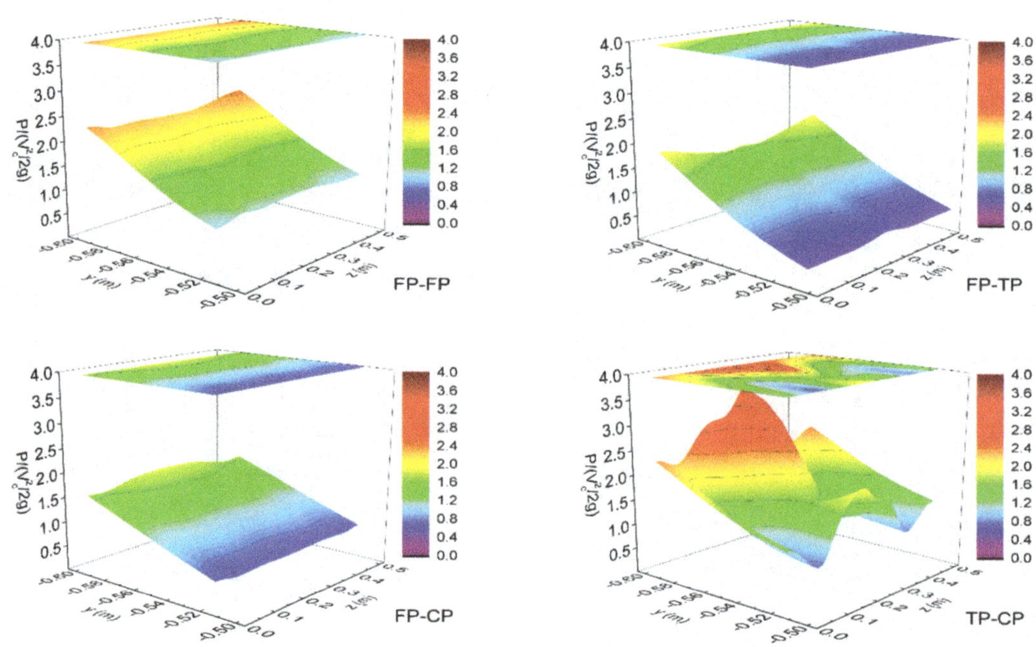

(**a**) Step 6

Figure 12. *Cont.*

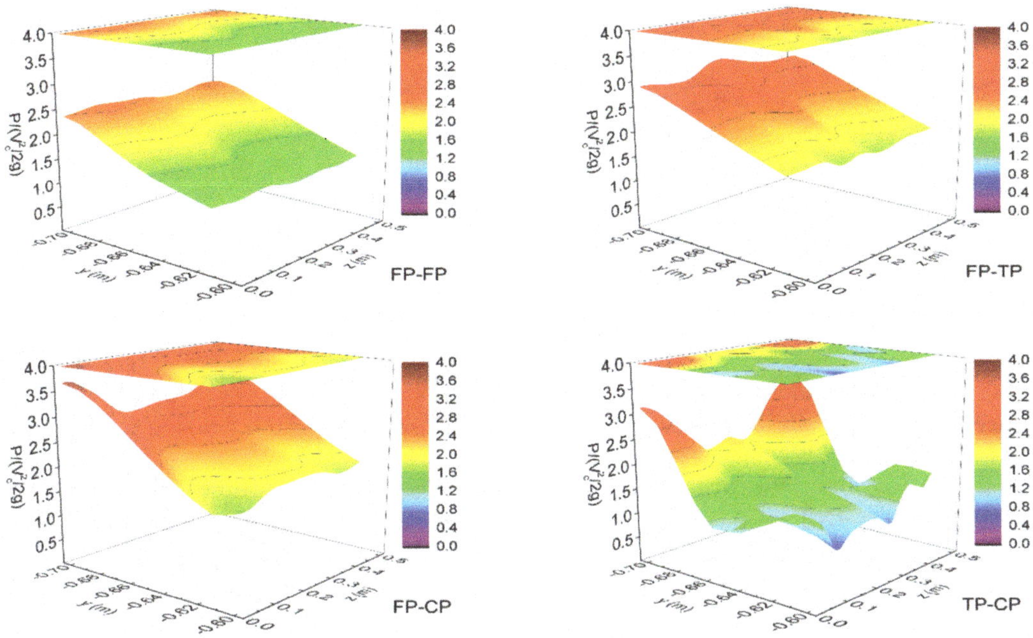

(**b**) Step 7

Figure 12. Pressure distributions on the vertical step surfaces. (**a**) Step 6; and (**b**) Step 7.

Figure 13 presents the pressure distributions on the horizontal step surfaces in the transverse direction. In FP-FP, FP-TP and TP-CP, the pressure on the upstream step surfaces was greater than that of the downstream ones, but it was quite chaotic and showed no obvious trend in TP-CP. The pressure distributions on horizontal step surfaces of FP-FP, FP-TP, and TP-CP showed U-shaped variation with the maximum pressures occurring at the downstream end of the step surface. In addition, the greatest fluctuation in pressure in the spanwise and streamwise direction can be found in TP-CP, indicating its flow complexities and highly three-dimensional characteristics.

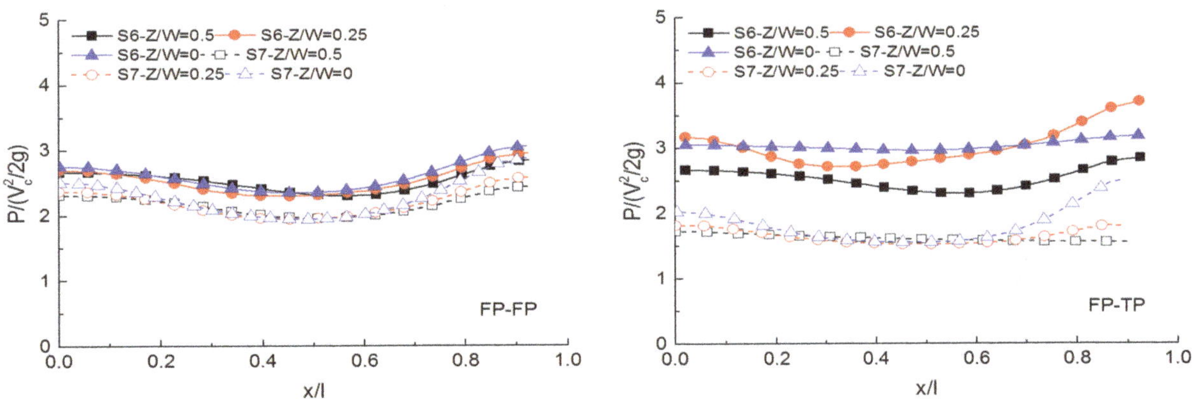

Figure 13. *Cont.*

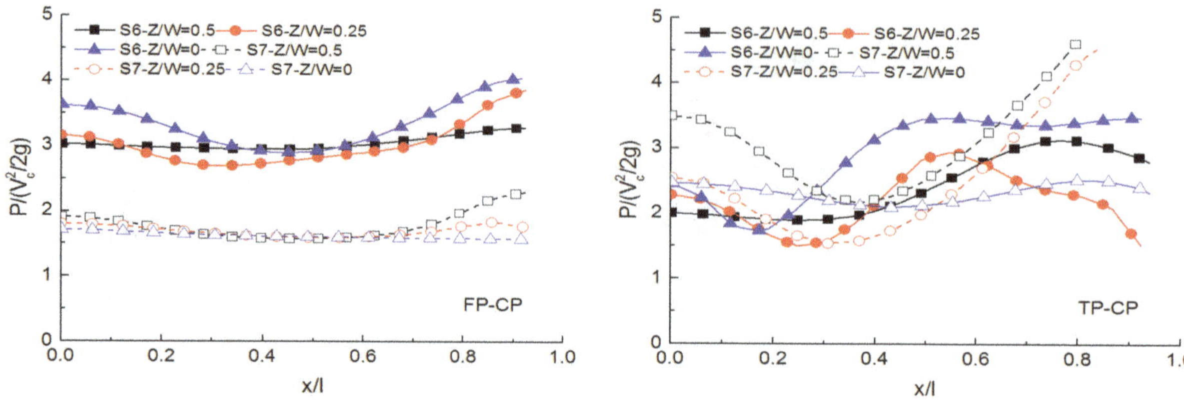

Figure 13. Pressure distributions on horizontal step surfaces in the transverse direction.

Figure 14 presents the pressure distributions on the vertical step surfaces in the transverse direction. In FP-FP, FP-TP, and FP-CP, the pressure distributions on the vertical step surfaces displayed no significant change in the transverse direction, but great pressure fluctuation can be seen in TP-CP. In FP-TP and FP-CP, the pressure on the vertical step surfaces of the downstream steps was moderately greater than that of the upstream ones, while FP-FP showed no evident difference.

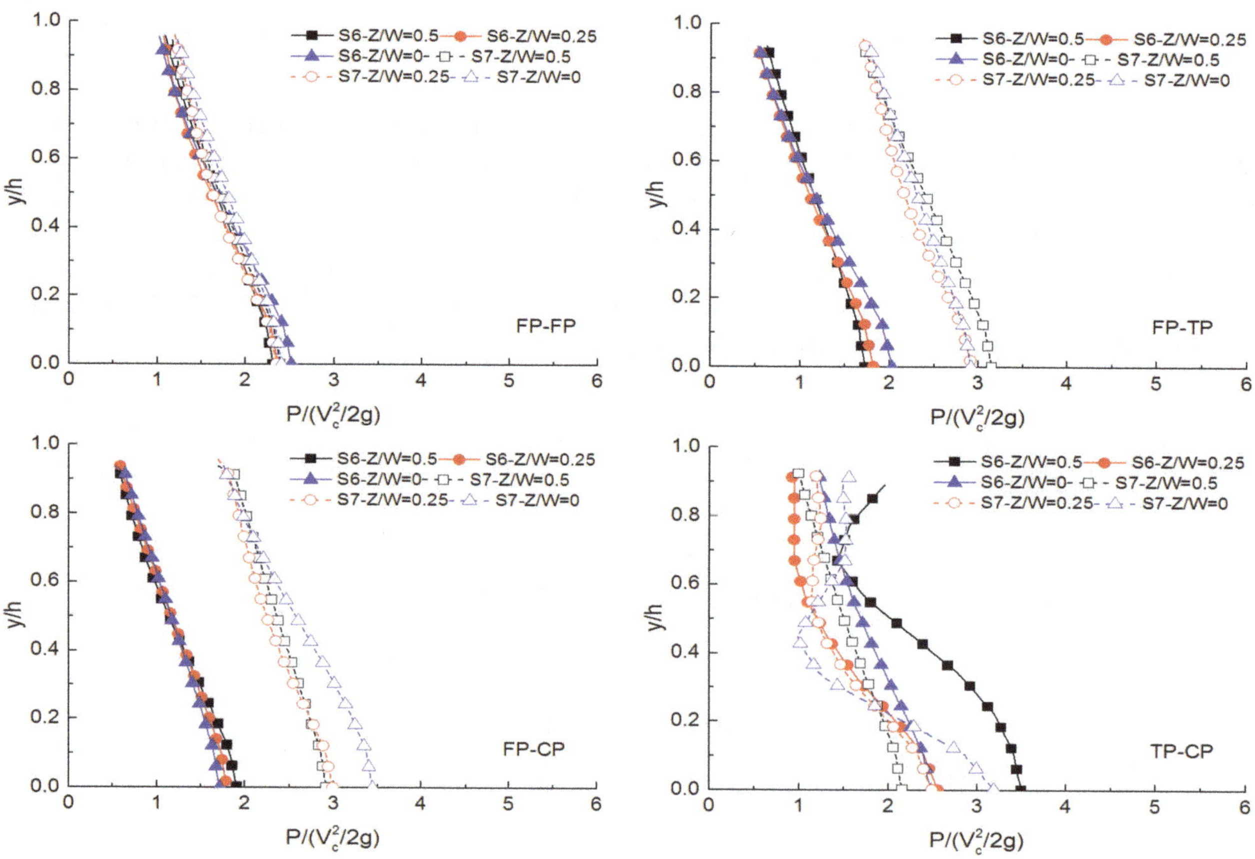

Figure 14. Pressure distributions on vertical step surfaces in the transverse direction.

3.4. Residual Head and Energy Dissipation

For the conservation of the downstream hydraulic structures, it is critical to quantify the residual head at the downstream of the spillway and the relative energy dissipation rate over the stepped spillway. The residual head, H_{res}, at the last pool weir can be estimated through the following equation:

$$H_{res} = d_e \times \cos\theta + \frac{U_w^2}{2 \times g} + d \tag{12}$$

where d_e is the equivalent clear water depth, θ represents the channel slope, U_w and d indicate the flow velocity and the pool weir height, respectively.

Together with the previous studies in Table 2, the dimensionless residual head, H_{res}/d_c, is shown in Figure 15. The flow discharge of the studies in Table 2 was the same and the flow regime was skimming flow. The residual energy in FP-FP configuration between the present work and the study of Morovati et al. [29] showed little difference, indicating the accuracy of the simulation. For configuration 1~3, the residual head showed no significant change for a given pool height, but the residual data increased with the decrease in pool height. In addition, the residual energy for the spillway with the combination of two-sided pooled steps and central pooled steps (TP-CP) decreased significantly compared with the spillway with only two-sided pooled steps or central pooled steps.

Table 2. Summary of the pooled stepped spillways in previous and present studies.

Reference	Step Geometry
Morovati et al. [29]	Fully pooled steps: h = 0.1 m, l = 0.2 m, d = 0.09 m In-line configuration (pooled and flat steps in-line): h = 0.1 m, l = 0.2 m, d = 0.09 m Staggered configuration (pooled and flat steps staggered): h = 0.1 m, l = 0.2 m, d = 0.09 m Two-sided pooled steps: h = 0.1 m, l = 0.2 m, d = 0.09 m Central-pooled steps: h = 0.1 m, l = 0.2 m, d = 0.09 m
Felder et al. [32]	Fully pooled steps: h = 0.1 m, l = 0.2 m, d = 0.031 m In-line configuration (pooled and flat steps in-line): h = 0.1 m, l = 0.2 m, d = 0.031 m Staggered configuration (pooled and flat steps staggered): h = 0.1 m, l = 0.2 m, d = 0.031 m
Present study	FP-FP: h = 0.1 m, l = 0.2 m, d = 0.09 m FP-TP: h = 0.1 m, l = 0.2 m, d = 0.09 m FP-CP: h = 0.1 m, l = 0.2 m, d = 0.09 m TP-CP: h = 0.1 m, l = 0.2 m, d = 0.09 m

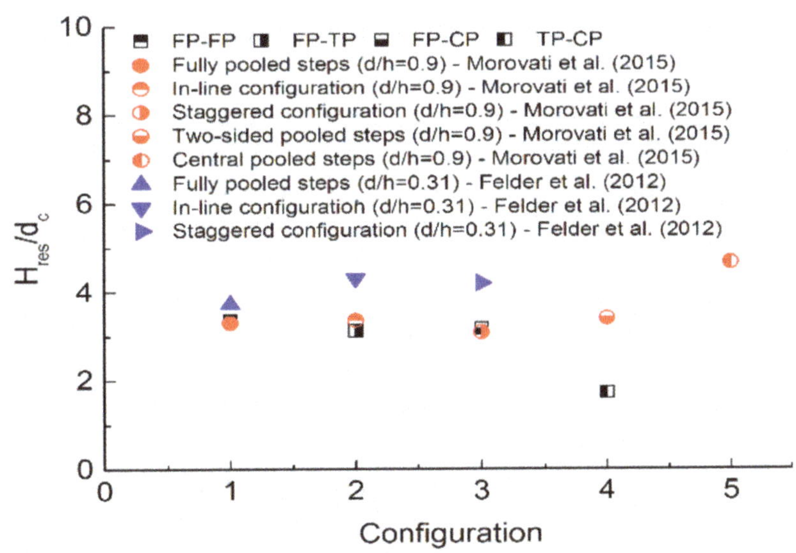

Figure 15. Residual energy at the downstream end of the spillway.

The energy dissipation rate, $\Delta H/H_{\max}$, is an important indicator to evaluate the relative energy loss from the start to the end of the stepped spillway. The energy dissipation rate is given as:

$$\frac{\Delta H}{H_{\max}} = \frac{H_{\max} - H_{res}}{H_{\max}} = \frac{H_{dam} + 1.5d_c - H_{res}}{H_{dam} + 1.5d_c} \tag{13}$$

where H_{dam} represents the dam height. The rate of energy dissipation is presented in Figure 16. It can be seen that TP-CP showed the best energy dissipation performance compared with other configurations, and the energy dissipation rate for TP-CP was almost 1.5 times larger than that of FP-FP, but no obvious change can be observed in FP-FP, FP-TP and FP-CP. That is because there were more turbulent structures in TP-CP caused by the concave-convex pool weirs, leading to greater momentum change between the overlaying flow and the vortex flow, thus creating more energy loss.

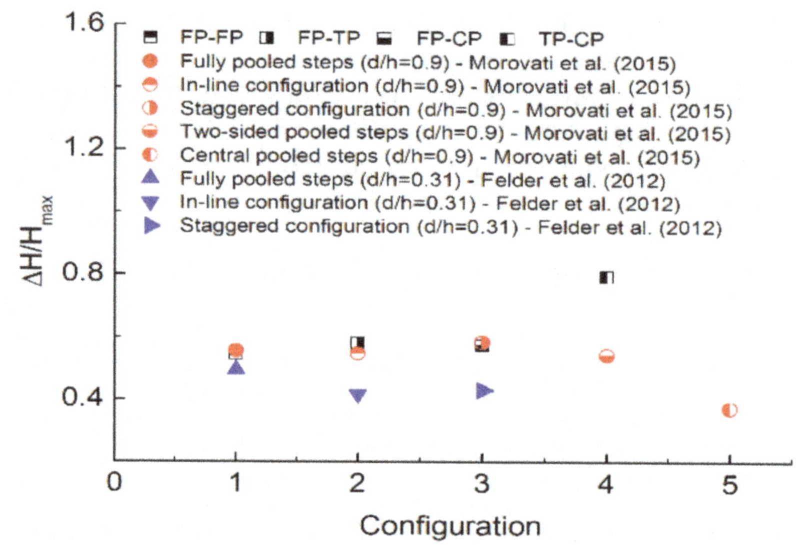

Figure 16. Energy dissipation rate.

4. Conclusions

In the present study, four types of pooled stepped spillway with different pool weirs were numerically investigated by using RNG $k-\varepsilon$ turbulence model with VOF method. The numerical results agreed well with the experimental data from Felder et al. [32] and the following conclusions can be drawn:

1. The flow pattern in FP-FP presented no significant change in the transverse direction and was similar on different steps. But changing half of the fully pooled steps into partially pooled steps (two-sided pooled or central pooled) can result in three dimensional flow motion. In FP-TP, the maximum vortex intensity on the fully pooled steps and partially pooled steps occurred at the axial plane and sidewalls, respectively, while the opposite phenomenon can be seen in FP-CP. When replacing all the fully pooled steps by staggered configuration of two-sided pooled and central pooled steps (TP-CP), the flow was characterized by highly three dimensional motion, and the vortex development in the transverse direction showed a unique pattern with the maximum intensity of vortex occurring at $Z/W = 0.25$. Besides, TP-CP created more flow instabilities and turbulent structures.
2. The velocity distributions over the steps in FP-FP highlighted the similar pattern with the minimum and maximum values occurring at $y/y_{\max} = 0.3$ and 0.9, respectively. In FP-TP, FP-CP and TP-CP, the velocity distributions on the odd-number steps was different from that of the even-number steps, but the maximum velocity in all configurations indicated no difference. In the transverse direction, the velocity distribution in FP-FP showed the smallest variation, while TP-CP presented the greatest change.

3. The pressure on the horizontal step surfaces of FP-FP, FP-TP and FP-CP showed U-shaped variations with the maximum pressures occurring at the downstream end of the step surface, and TP-CP highlighted the greatest pressure fluctuation in both streamwise and transverse direction. The pressure distribution patterns on the fully pooled steps in FP-TP and on the central pooled steps in FP-CP were similar; the pressure distributions on the two-sided pooled steps in FP-TP and on the fully pooled steps in FP-CP demonstrated the similar patterns. For the vertical steps, the maximum pressures on odd-number step surfaces were in the following order: FP-CP > TP-CP > FP-TP > FP-FP, while for the even-number steps, it was TP-CP > FP-FP > FP-TP > FP-CP. There was no negative pressure on the step surfaces.
4. From the largest to the smallest, the energy dissipation rates were in the following order: TP-CP > FP-CP > FP-TP > FP-FP. Specifically, the energy loss for FP-TP and FP-CP was quite close, but slightly higher than FP-FP, and the energy dissipation rate for TP-CP was 1.5 times larger than FP-FP. Thus, changing half of the fully pooled steps into central pooled or two-sided pooled steps presented no obvious effect on the energy dissipation ratio, while shifting all the fully pooled steps into the combination of central pooled and two-sided pooled steps can significantly improve the energy dissipation performance.

Author Contributions: Conceptualization, S.L.; Methodology, S.L.; Software, S.L.; Formal Analysis, S.L.; Investigation, S.L.; Writing-Original Draft Preparation, S.L.; Writing-Review & Editing, J.Z.; Supervision, J.Z.; Project administration, J.Z.; Funding acquisition, J.Z.

References

1. Boes, R.M.; Hager, W.H. Hydraulic design of stepped spillways. *J. Hydraul. Eng.* **2002**, *129*, 671–679. [CrossRef]
2. Chanson, H. Hydraulic Design of stepped spillways and downstream energy dissipators. *Dam Eng.* **2001**, *11*, 205–242.
3. Chanson, H.; Luke, T. Hydraulics of stepped chutes: The transition flow. *J. Hydraul. Res.* **2004**, *42*, 43–54. [CrossRef]
4. Chanson, H.; Toombes, L. Experimental investigations of air entrainment in transition and skimming flows down a stepped chute. *Can. J. Civ. Eng.* **2002**, *29*, 145–156. [CrossRef]
5. Felder, S.; Chanson, H. Turbulence, dynamic similarity and scale effects in high-velocity free-surface flows above a stepped chute. *Exp. Fluids* **2009**, *47*, 1–18. [CrossRef]
6. Wang, J.; Fu, L.; Xu, H.; Jin, Y. Numerical study on flow over stepped spillway using Lagrangian method. In Proceedings of the International Conference on Energy Engineering and Environmental Protection, Sanya, China, 20–22 November 2017.
7. Felder, S. Air-Water Flow Properties on Stepped Spillways for Embankment Dams: Aeration, Energy Dissipation and Turbulence on Uniform, Non-Uniform and Pooled Stepped Chutes. Ph.D. Thesis, University of Queensland, Brisbane, Australia, 2013.
8. Felder, S.; Chanson, H. Energy dissipation down a stepped spillway with non-uniform step heights. *J. Hydraul. Eng.* **2011**, *137*, 1543–1548. [CrossRef]
9. Felder, S.; Chanson, H. Aeration, flow instabilities, and residual energy on pooled stepped spillways of embankment dams. *J. Irrig. Drain. Eng.* **2013**, *139*, 880–887. [CrossRef]
10. Zhang, J.M.; Chen, J.G.; Wang, Y.R. Experimental study on time-averaged pressures in stepped spillway. *J. Hydraul. Res.* **2012**, *50*, 236–240. [CrossRef]
11. Zhang, G.; Chanson, H. Effects of step and cavity shapes on aeration and energy dissipation performances of stepped chutes. *J. Hydraul. Eng.* **2018**, *144*, 04018060. [CrossRef]
12. Peng, Y.; Zhou, J.G.; Burrows, R. Modelling the free surface flow in rectangular shallow basins by lattice Boltzmann method. *J. Hydraul. Eng.* **2011**, *137*, 1680–1685. [CrossRef]
13. Peng, Y.; Zhou, J.G.; Burrows, R. Modelling solute transport in shallow water with the lattice Boltzmann method. *Comput. Fluids* **2011**, *50*, 181–188. [CrossRef]
14. Peng, Y.; Mao, Y.F.; Wang, B.; Xie, B. Study on C-S and P-R EOS in pseudo-potential lattice Boltzmann model

for two-phase flows. *Int. J. Mod. Phys. C* **2017**, *28*, 1750120. [CrossRef]
15. Peng, Y.; Wang, B.; Mao, Y.F. Study on force schemes in pseudopotential lattice Boltzmann model for two-phase flows. *Math. Probl. Eng.* **2018**, 6496379. [CrossRef]
16. Baylar, A.; Unsal, M.; Ozkan, F. The effect of flow patterns and energy dissipation over stepped chutes on aeration efficiency. *KSCE J. Civ. Eng.* **2011**, *15*, 1329–1334. [CrossRef]
17. Dong; Zhi-yong; Joseph; Hun-wei. Numerical simulation of skimming flow over mild stepped channel. *J. Hydrodyn.* **2006**, *18*, 367–371. [CrossRef]
18. Zhenwei, M.U.; Zhang, Z.; Zhao, T. Numerical simulation of 3-D flow field of spillway based on VOF method. *Procedia Eng.* **2012**, *28*, 808–812. [CrossRef]
19. Tabbara, M.; Chatila, J.; Awwad, R. Computational simulation of flow over stepped spillways. *Comput. Struct.* **2005**, *83*, 2215–2224. [CrossRef]
20. André, S. High Velocity Aerated Flows on Stepped Chutes with Macro-Roughness Elements. Ph.D. Thesis, EPFL, Lausanne, Switzerland, 2004.
21. Kökpinar, M.A. Flow over a stepped chute with and without macro-roughness elements. *Can. J. Civ. Eng.* **2004**, *31*, 880–891. [CrossRef]
22. Thorwarth, J. Hydraulics of Pooled Stepped Spillways—Self-Induced Unsteady Flow and Energy Dissipation. Ph.D. Thesis, University of Aachen, Aachen, Germany, 2008.
23. Hirt, C.W.; Nichols, B.D. Volume of fluid (VOF) method for the dynamics of free boundaries. *J. Comput. Phys.* **1981**, *39*, 201–225. [CrossRef]
24. Li, X.; Wang, Y.R.; Zhang, J.M. Numerical simulation of an offset jet in bounded pool with deflection wall. *Math. Probl. Eng.* **2017**, *2017*, 1–11. [CrossRef]
25. Zhang, J.M.; Chen, J.G.; Xu, W.L.; Wang, Y.R. Three-dimensional numerical simulation of aerated flows downstream sudden fall aerator expansion-in a tunnel. *J. Hydrodyn.* **2011**, *23*, 71–80. [CrossRef]
26. Chen, Q.; Dai, G.; Liu, H. Volume of Fluid Model for Turbulence Numerical Simulation of Stepped Spillway Overflow. *J. Hydraul. Eng.* **2002**, *128*, 68–688. [CrossRef]
27. Yakhot, V.; Orszag, S.A. Renormalization group analysis of turbulence. I. Basic theory. *J. Sci. Comput.* **1986**, *1*, 3–51. [CrossRef]
28. Bai, Z.L.; Zhang, J.M. Comparison of different turbulence models for numerical simulation of pressure distribution in V-shaped stepped spillway. *Math. Probl. Eng.* **2017**, *2017*, 1–9. [CrossRef]
29. Morovati, K.; Eghbalzadeh, A.; Javan, M. Numerical investigation of the configuration of the pools on the flow pattern passing over pooled stepped spillway in skimming flow regime. *Acta Mech.* **2015**, *227*, 1–14. [CrossRef]
30. Carvalho, R.F.; Rui, M. Stepped spillway with hydraulic jumps: Application of a numerical model to a scale model of a conceptual prototype. *J. Hydraul. Eng.* **2014**, *135*, 615–619. [CrossRef]
31. Celik, I.B.; Ghia, U.; Roache, P.J.; Freitas, C.J. Procedure of estimation and reporting of uncertainty due to discretization in CFD applications. *J. Fluids Eng.* **2008**, *130*, 078001–078004. [CrossRef]
32. Felder, S.; Guenther, P.; Hubert, C. *Air-Water Flow Properties and Energy Dissipation on Stepped Spillways: A Physical Study of Several Pooled Stepped Configurations*; The University of Queensland: Brisbane, Australia, 2012; ISBN 9781742720555.
33. Hunt, J.C.R.; Wray, A.A.; Moin, P. Eddies, streams and convergence zones in turbulent flows. In *Proceedings of the Summer Program 1988*; Stanford University: Stanford, CA, USA; 1988.

7

Experimental Study on the Impact Characteristics of Cavitation Bubble Collapse on a Wall

Jing Luo, Weilin Xu *, Jun Deng, Yanwei Zhai and Qi Zhang

State Key Laboratory of Hydraulics and Mountain River Engineering, Sichuan University, Chengdu 610065, China; luojing@scu.edu.cn (J.L.); djhao2002@scu.edu.cn (J.D.); zaiyanwei0422@163.com (Y.Z.); scu_zq@163.com (Q.Z.)

* Correspondence: xuwl@scu.edu.cn.

Abstract: As a hydrodynamic phenomenon, cavitation is a main concern in many industries such as water conservancy, the chemical industry and medical care. There are many studies on the generation, development and collapse of cavitation bubbles, but there are few studies on the variation of the cyclic impact strength on walls from the collapse of cavitation bubbles. In this paper, a high-speed dynamic acquisition and analysis system and a pressure measuring system are combined to study the impact of a cavitation bubble generated near a wall for various distances between the cavitation bubble and the wall. The results show that (1) with the discriminating criteria of the impact pressure borne by the wall, the critical conditions for the generation of a micro-jet in the collapse process of the cavitation bubbles are obtained, and therefore collapses of cavitation bubbles near the wall are mainly divided into primary impact area collapses, secondary impact area collapses and slow release area collapses; (2) it can be seen from the impact strength of the cavitation bubble collapse on the wall surface that the impact of cavitation bubbles on the wall surface during the first collapse decreases as γ (the dimensionless distance between the cavitation bubble and the wall) increases, but the impact of the second collapse on the wall surface increases first and then decreases sharply. When γ is less than 1.33, the impact on the wall surface is mainly from the first collapse. When γ is between 1.33 and 2.37, the impact on the wall surface is mainly from the second collapse. These conclusions have potential theoretical value for the utilization or prevention and control technologies for cavitation erosion.

Keywords: cavitation bubble; high-speed photography; impact pressure; micro-jet

---

1. Introduction

When the collapse of cavitation bubbles occurs near a wall, the wall will bear the cyclic impact from the collapse of the cavitation bubble. The characteristic of the interaction between the cavitation bubble and the wall has become the focus of many technical fields, such as water conservancy, shipbuilding, the chemical industry and many other industries.

Theoretical study about cavitation bubbles can be tracked back to the last century [1–3]. The study of the cavitation field mainly focuses on the damage and destruction of a solid wall surface upon cavitation bubble collapse [4,5].

Shaw et al. [6] used a laser to generate a cavitation bubble and combined this with Schlieren photography to conduct a study on the impact performance of the impact wave and micro-jet generated during the first collapse of a cavitation bubble on a wall. Adopting high-speed photographic equipment with a camera speed of up to 100 million frames per second to study the evolution properties of cavitation bubbles around a wall, Lindau and Lauterborn [7] discovered the annular form during the collapse of a cavitation bubble, the formation of a contrajet and the shock wave. Liu et al. [8] used the method of laser-inducing a cavitation bubble to conduct a study on the impacts of a micro-jet caused by laser-induced cavitation bubble collapse on a metallic copper surface, and found that the liquid-jet was the main damage mechanism in cavitation erosion. Xu et al. [9] studied the cavitation bubble dynamic characteristics on an aluminum surface by using the technology of a sensitive fiber-optic sensor based on the optical beam deflection principle. Ren et al. [10] used a hydrophone to conduct a study on the strength mechanism of impact from cavitation collapse within the range of the relative bubble–wall dimensionless distance γ ($\gamma = h/R_{max}$, R_{max} is the maximal radius of the cavitation bubble, h is the distance between the center of the cavitation bubble and the wall) which is between 0.5 and 2.5. The above studies on the collapse of cavitation bubbles either focus on the first collapse or adopt the indirect method of hydrophone measurement. In addition, from the research on the impact of the micro-jet or high-velocity droplet on a wall with the use of high-speed photography, only qualitative rules can be obtained. However, measuring the microscopical area of the wall under the impact of the micro-jet or high-velocity droplet with the pressure test system provides a new quantitative method for solving millimeter-sized impact-related issues [11].

In recent years, some new research directions on cavitation have been proposed, such as extracorporeal shock-wave lithotripsy [12] and ultrasonic cleaning [13,14]. In addition, studies on the interaction between cavitation bubbles and cavitation bubbles, cavitation bubbles and air bubbles, cavitation bubbles and particles, and cavitation bubbles and ice blocks have also been reported. Fong et al. [15] adopted the underwater discharge technique to induce multiple cavitation bubbles so as to study the coupling effect between two cavitation bubbles generated at different instants. Pain et al. [16] studied the jet flow in an air bubble induced by a nearby cavitation bubble and discovered that the speed of such a jet flow can be up to 250 m/s. Luo et al. [17,18] used the method of high-voltage discharge technology to generate a cavitation bubble to study the interaction between a cavitation bubble and an air bubble, and the interaction between two cavitation bubbles and a rigid wall. Goh et al. [19] studied the interactions between cavitation bubbles horizontally placed below the underwater slab and air bubbles attached to it and discovered that the ratio between the oscillation time of the air bubbles and the oscillation time of the cavitation bubbles is the important parameter impacting the micro-jet caused by cavitation bubble collapse. The interaction between a cavitation bubble and a particle was researched by adopting a high-voltage discharge technology to induce a cavitation bubble [20] and low-voltage underwater discharge technology to induce a cavitation bubble [21]. The interaction between ice and cavitation bubbles was researched by Cui et al. [22], and the direction of the micro-jet and the propagation of shock waves were captured.

Due to the very small volume and short life cycle of cavitation bubbles during the evolution, great difficulties have been imposed on related studies. Therefore, numerical simulation studies have been performed by various researchers. Coupled boundary element and finite element methods have been performed on bubble–structure interactions [23–25]. The boundary element method is used for research on the toroidal configuration and jet impact characteristics in the later stage of the collapse of cavitation bubbles in the inviscid and irrotational field [26–28] numerically simulated the late rebound phenomena of a cavitation bubble by the method of vortex lines. Instead of using a vortex surface, Klaseboer et al. [29] simulated the ring stage of a cavitation bubble with a vortex ring, and profoundly understood the dynamic characteristics of a cavitation bubble and the interaction with walls for different distances between the bubble and the wall, which was in agreement with the experimental results.

For the past few decades, many researchers have tried to find out the impulses associated with cavitation bubble collapse. In most existing studies, these results are only qualitative due to the non-uniform spatial sensitivity of the transducer. Therefore, in this paper, cavitation bubbles are induced by adopting a low voltage spark-discharge system, the cavitation bubble evolution and corresponding wall pressure change processes are measured simultaneously by the piezoresistive pressure sensor. The impact strength of the first collapse and rebound–regeneration collapse of the cavitation bubble on the wall surface is mainly analyzed with the high-speed photographic image and pressure testing data of the wall surface, and the influence of the dimensionless distance of the bubble wall on the impact strength of the wall surface is revealed.

2. Experimental Setups

2.1. Cavitation Bubble Induction Device

In order to get the impact characteristics of the cavitation bubble collapse on the wall, the experimental setup requires a cavitation bubble induction device, a high-speed dynamic collection and analysis system and a transient pressure testing system, as shown in Figure 1.

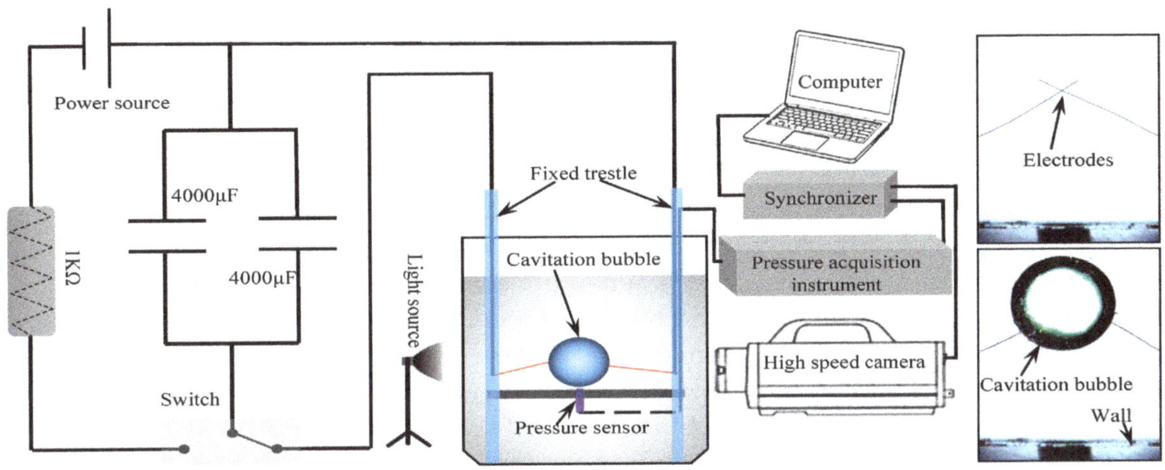

Figure 1. Schematic diagram of experimental facilities.

There are three ways to induce cavitation bubbles in quiescent water: by focusing a laser, by a high-frequency ultrasound wave or by an underwater discharge. The laser technique to induce a cavitation bubble has many advantages, such as its higher accuracy in controlling the focal spot, but the focusing method of the laser is rather complex and costly [30]. In the current work, the method of low-voltage underwater discharge was adopted. The method can accurately control the position and size of the cavitation bubble. To minimize possible interference, the diameter of the electrodes used is 0.15 mm, and the bubbles created have a maximum radius of 30 times the diameter of the electrodes. The copper electrodes (length 50 mm) are fixed in the trestles. The trestles are placed in a water tank (40 cm × 30 cm × 40 cm), which is filled with twice deionized water (at a constant temperature of 22 °C). To create the bubble, two capacitors are fully charged to the specified voltage (the adjustable range is 0 V~100 V) through a 1 KΩ resistor, then switched to the discharge circuit, and the capacitor is discharged through the pair of crossed electrodes.

The life cycle of cavitation bubbles is very short; in order to study the impact process of the collapse of the cavitation bubble on the wall surface, a high-speed dynamic collection and analysis system and a transient pressure testing system must be adopted for synchronized recording. The high-speed dynamic collection and analysis system is composed of a high-speed camera, lens assembly, lighting equipment, etc., and the transient pressure testing system is composed of the resistance-type pressure sensor and the data acquisition instrument. These two systems are respectively connected to a computer through a synchronous trigger device. The Fastcam SA-Z high-speed camera (maximum acquisition rate: 1,000,000 fps, Photron Inc., Tokyo, Japan,) is used. The radius is of millimeter-size when the cavitation bubble generated after the discharge between electrodes in the water body evolves to its maximum volume. Thus, in order to obtain the clear outline of the cavitation bubbles, a micro-lens and LED lamp (150 W) must be adopted in the acquisition system and synchronized with the high-speed photography system. A framing rate of 150,000 frames per second is used to capture the bubble dynamics in this paper, with a shutter speed of 5.06 μs. The width of the frame is 256 pixels.

2.2. Measurement of Cavitation Bubbles During the Evolution Process and the Impact Process

Considering the size of the cavitation bubbles (the radius of cavitation bubbles produced from the system in the experiment ranges from 5.00 mm to 12.00 mm) and the time for the impact of the collapse of the bubble on the wall surface being very short, the size of the sensor in the transient pressure testing system must be small enough, and the sampling frequency and accuracy must be high enough; in addition, by observing the high-speed photographic images, the time for the impact of the micro-jet on the wall due to the collapse of the cavitation bubbles induced by the experimental system lasted about 100 μs. The minimum rise time was at least 10 μs. The pressure sensor used in the experimental system has a high-frequency bandwidth (200 kHz). Therefore, with the existing sensor technology, the piezoresistive pressure sensor used in this paper can meet the requirements for the measurement of the impact of the collapse of cavitation bubbles on the wall surface. The radius of the sensing area of the sensor adopted in the test is 1.50 mm (the maximum measuring range is 40 MPa). The sampling frequency is 2 MHz, and the accuracy is 0.5% of the maximum measuring range.

The method of image gray processing is used to obtain the maximum radius of the cavitation bubble. In this paper, a circle radius with the same number of the pixels is regarded as the equivalent radius R_{eq} of the cavitation bubble (R_{max} is R_{eq} when the cavitation bubble radius reaches maximum), and the centroid of the cavitation bubble in the high-speed photographic image is regarded as the center of the cavitation bubble. The dimensionless distance γ is the ratio of the distance h between the centers of the cavitation bubble and pressure sensor and the maximum value of the maximal equivalent radius R_{max} of the cavitation bubble. Figure 2 shows the data processing process: therein, the scatter diagram shows the relationship between the first cycle of the cavitation bubble time and the corresponding R_{max}, and it can be seen that this data processing method provides good reproducibility.

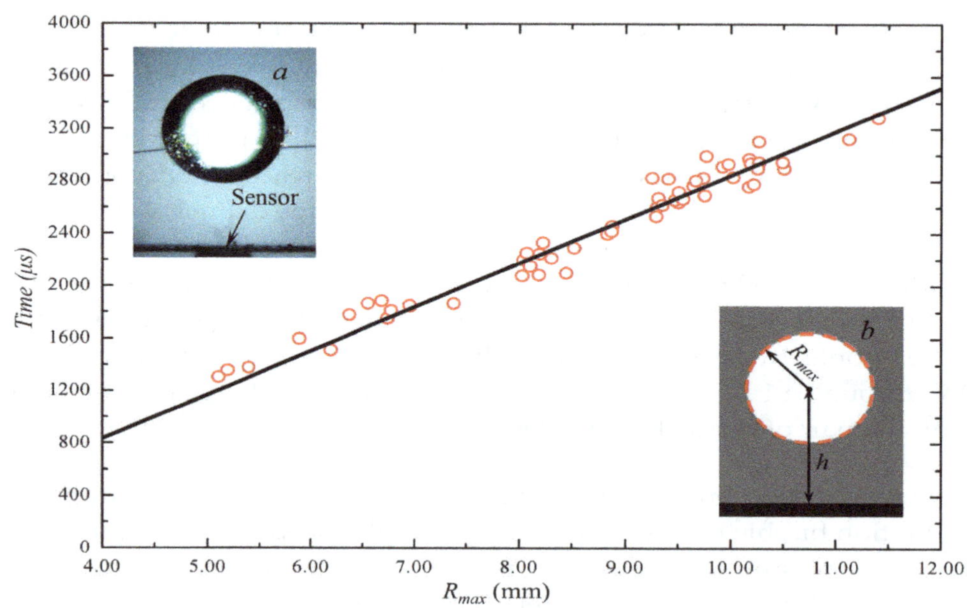

Figure 2. Processing method of characteristic parameters. (**a**) Cavitation bubble photograph; (**b**) characteristic parameters.

3. Results and Discussion

3.1. Influence of Distance Between the Cavitation Bubble and Wall on the Development of Micro-Jetting

A selected set of high-speed photographic images of the cavitation bubbles in the collapse stage for different γ (dimensionless distances between the cavitation bubble and the wall) are shown in Figure 3. Therein, the wall surface is located at the bottom of the image. In the three sets of tests, the dimensionless distances γs are 2.91, 1.91 and 0.83, respectively, the maximum values R_{max} of the equivalent radius R_{eq} are 9.40 mm, 9.31 mm and 9.47 mm respectively, and the distances h from the center of the cavitation bubbles to the wall's surface when the cavitation bubbles expand to their maximum radii are 27.35 mm, 17.76 mm and 7.86 mm, respectively.

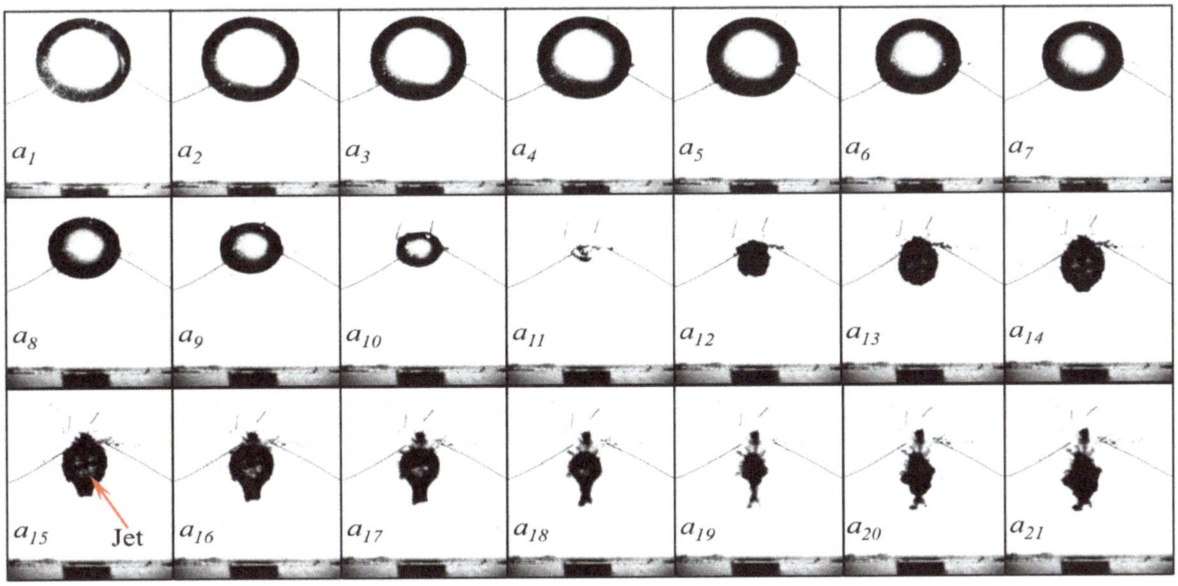

(**a**) $R = 9.40$ mm; $\gamma = 2.91$

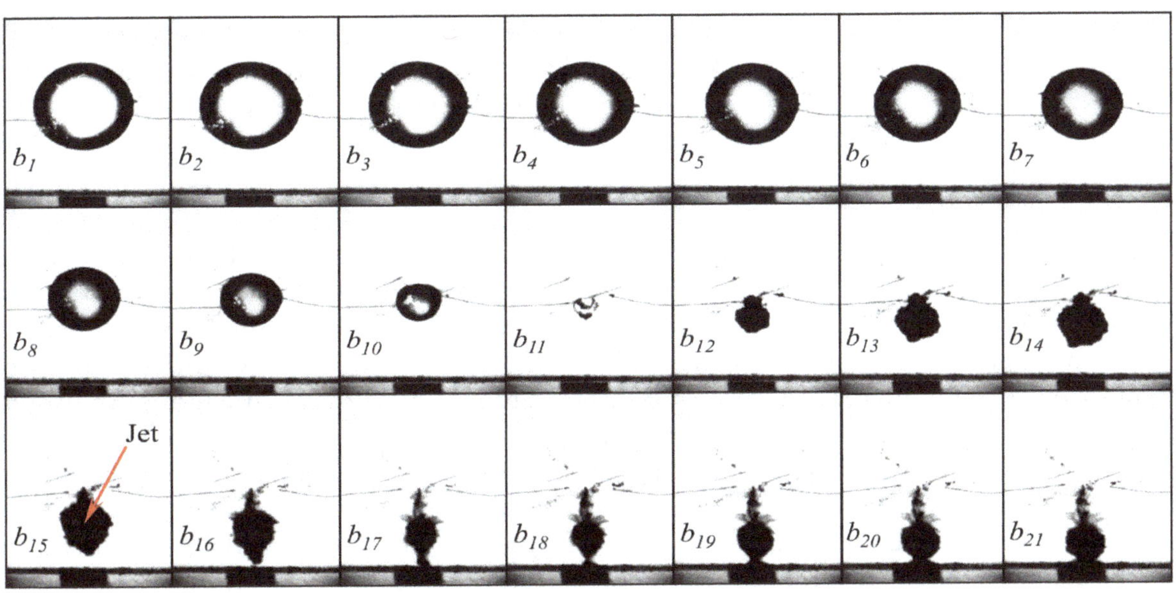

(**b**) $R_{max} = 9.31$ mm; $\gamma = 1.91$

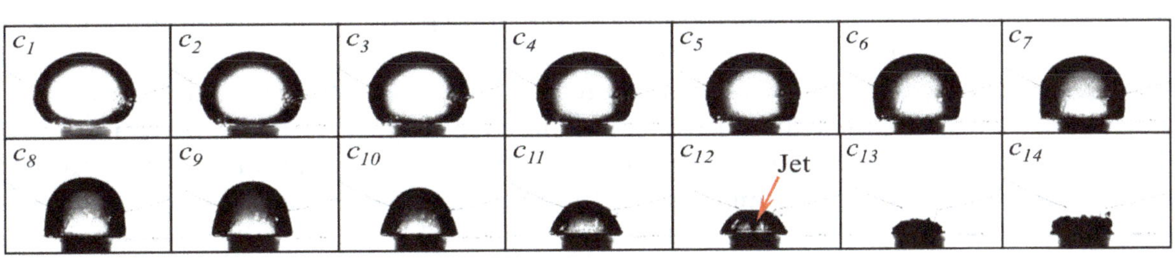

(**c**) $R_{max} = 9.47$ mm; $\gamma = 0.83$

Figure 3. High-speed photographic images of the cavitation bubbles in the contraction collapse stage under the conditions of different distances γ (frame-rate: 150,000 fps; exposure time: 5.06 μs; frame width: 32.00 mm).

When the cavitation bubble expands to the maximum radius in the testing, as shown in Figure 3 (a_1,b_1,c_1), the cavitation bubbles begin the contraction stage through an action of the external water body. With the gradual contraction of the cavitation bubble, due to the existence of the wall under the cavitation bubble, the surrounding water begins to fill the surrounding space released by the cavitation bubbles in their contraction process, and the upper and lower surfaces of the cavitation bubble begin to form asymmetrical shapes gradually. The wall surface under the cavitation bubble blocks the filling of water, resulting in a slow contraction speed of the lower surface of the cavitation bubble, while the upper surface belongs to the unbounded domain and contracts quickly. Thus, this asymmetric contraction gradually begins to form from the upper surface (as shown in a_9, b_9 and c_{11} in Figure 3). The development of the asymmetric collapse of the surface has been accompanied by the shrinkage of cavitation bubbles to the minimum volume (as shown in a_{11}, b_{11} and c_{13} in Figure 3). The cavitation bubbles will rebound–regenerate after they shrink to the minimum volume, and as shown in Figure 3, the surface of the rebound cavitation bubble is not very smooth (as shown in a_{12} and b_{12} in Figure 3), but just as in the first cycle, the cavitation bubble will again undergo the expansion–contraction process. In the second expansion–contraction process, the surface is not smooth enough, but this non-smoothness is not enough to change the collapse direction of cavitation bubbles. In Figure 3b, the rebound cavitation bubbles move quickly toward the wall surface in the expansion–collapse process and impact onto the wall in the second collapse.

From the above three tests, the following can be concluded: when the dimensionless distance γ is small, the micro-jet directly impacting the wall surface can be formed during the first asymmetric contraction of the cavitation bubble; as the dimensionless distance γ has been increasing continuously, the cavitation bubbles will move to the wall surface quickly and will eventually impact on the wall surface. With the further increase of γ, cavitation bubbles do not impact on the wall surface through two or three times of contraction. Therefore, the collapse of cavitation bubbles near the wall surface can be divided into main impact area collapse, secondary impact area collapse and slow release area collapse as per the parameters of the distance between the cavitation bubble and wall.

3.2. *Impact of Cavitation Bubble Collapse on Wall*

The collapse position of cavitation bubbles near the wall surface can be obtained directly through high-speed photography for the development of the micro-jet under the above different γ conditions, but the impact strength of cavitation bubbles on the wall surface during the first collapse, whether the impact can reach the wall surface during the second collapse, and the impact strength of the second collapse on the wall surface cannot be ascertained. With these questions, the high-speed dynamic collection and analysis system and transient pressure testing system are combined in this part, obtaining the impact process of the first and second cavitation bubble collapses on the wall surface (that is, the main impact area and secondary impact area). Figure 4 shows the impact process of the two groups of cavitation bubble collapses on the wall surface when the dimensionless distances γ are 0.91 and 1.79 respectively. The wall surface is located at the bottom of the image, and the pressure sensor is placed inside of the wall surface.

The maximum radius of the cavitation bubble in Figure 4a is 9.98 mm, and when the cavitation bubble reaches the maximum volume, the distance from the center to the wall surface is 9.08 mm. When the electric spark discharges into the water during the induction of the cavitation bubble, the pressure value appears to fluctuate slightly, and the pressure returns to a normal condition after discharge. The subsequently generated cavitation bubbles will begin the expansion–contraction–collapse–rebound stage. In the cavitation bubble development process, the pressure on the wall surface is stable; when the cavitation bubble contracts to the minimum volume (the first expansion-contraction cycle is 2.91 ms), the pressure on the wall surface sharply increases, abd from the contents of the previous section it can be seen that because of the short dimensionless distance γ, the collapse generated due to asymmetrical cavitation bubble contraction directly forms the micro-jet impacting the wall surface (shown as the image in Figure 4a), and the maximum peak value of impact pressure of the first cavitation bubble collapse on the wall surface is 19.37 MPa. After that, the peak pressure gradually decreases when the cavitation bubble is in the rebound–regeneration stage, the pressure on the wall surface is relatively stable for the whole rebound regeneration stage, and the pressure borne on the wall surface greatly increases again when the cavitation bubble contracts to the minimum volume again. The time period of the whole regeneration rebound is 1.23 ms; the wall surface bears the collapse impact of the rebound cavitation bubble again, and by this time, the maximum pressure on the wall surface is 6.50 MPa. It can be seen from the whole impact process of the cavitation bubbles on the wall surfaces in the condition of $\gamma = 0.91$ that the strength of the two impacts of cavitation bubbles on the wall surfaces will gradually reduce from 19.37 MPa in the first impact to 6.50 MPa in the second impact.

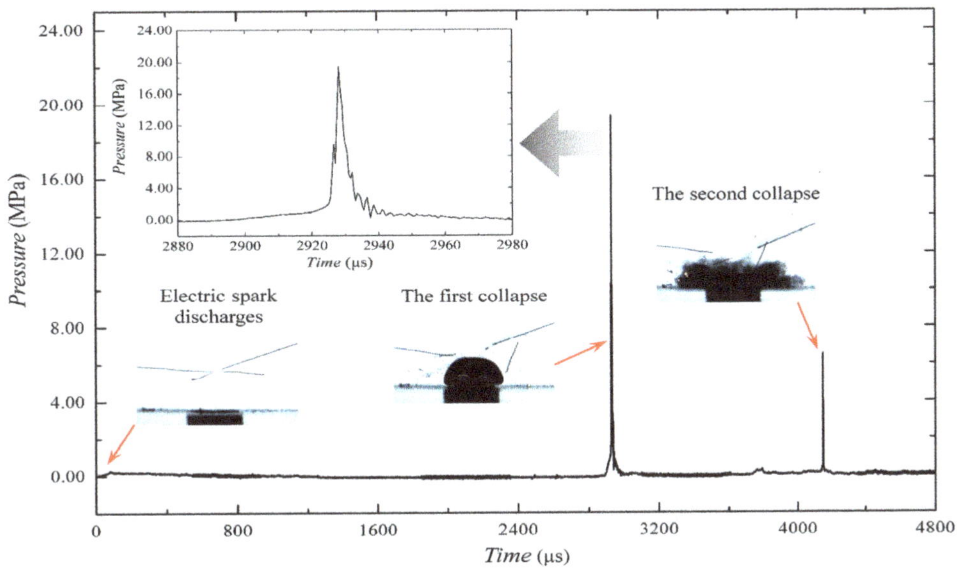

(**a**) R_{max} = 9.98 mm; γ = 0.91

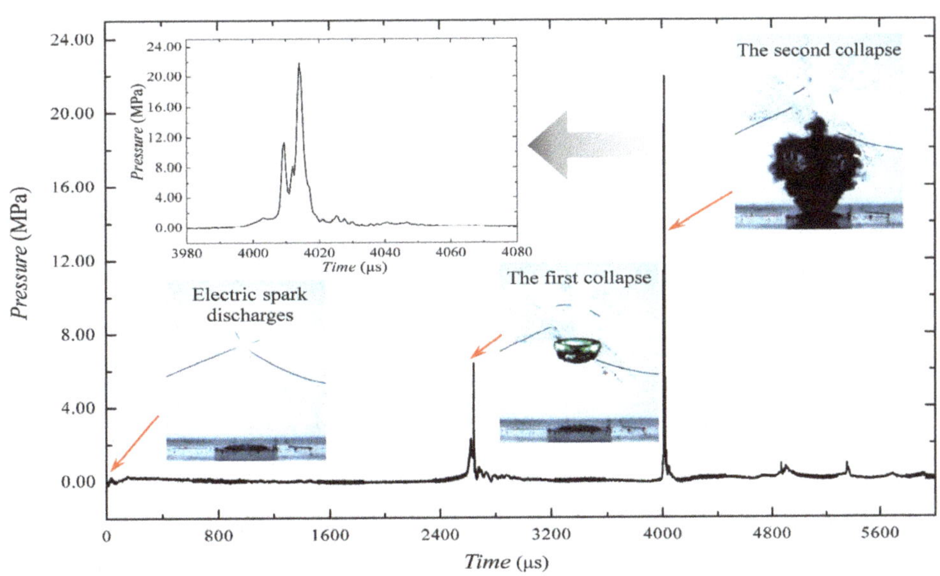

(**b**) R_{max} = 11.02 mm; γ = 1.79

Figure 4. The impulsive process generated by cavitation bubble collapse for (**a**) $\gamma = 0.91$; (**b**) $\gamma = 1.79$.

The maximum radius of the cavitation bubble in Figure 4b is 11.02 mm, and the distance from the center to the wall surface is 19.73 mm when the cavitation bubble reaches the maximum radius.

The first peak and the second peak are the impact on the wall surface during the first collapse and rebound collapse of the cavitation bubble, with the peak pressure values of 6.38 MPa (2.64 ms) and 21.89 MPa (4.01 ms), respectively. The minimum distance to the wall surface when the cavitation bubble shrinks to the minimum volume for the first time is 13.81 mm, and at this moment, the dimensionless distance from the center of the cavitation bubble to the wall surface is 5.06. The cavitation bubble will quickly move to the wall surface in the rebound evolution process; when the cavitation bubble contracts to the minimum volume once more, the cavitation bubble is close to the wall surface, which is as shown in Figure 4b.

It can be clearly seen from Figure 4 that when the dimensionless distance γ is small, the impact of the first collapse of the cavitation bubble on the wall surface is greater than the impact during rebound collapse; when γ can meet a certain condition and the cavitation bubble completes the impact of the first collapse on the wall surface, the rebound cavitation bubble will quickly move to the wall surface, and the impact pressure greater than that of the first collapse will act on the wall surface once more.

Shaw et al. [6] used laser-induced cavitation bubbles and a pressure (voltage) transducer to obtain the waveform of the impact of the first collapse of cavitation bubbles on the wall when γ is between 0.56 and 1.5. The waveform of the impact of the first collapse of cavitation bubbles on the wall reported in the literature is very similar to that in Figure 4a. The similarities between the two are as follows: (1) there are three peaks in the pressure rise. Except the maximum peak, the other two are distributed on both sides of the maximum pressure value; (2) the time interval between the two peaks before the pressure in the experiment increases to the maximum value is 8 μs, and the time interval during the pressure drop is 3 μs; (3) the time of the pressure rise is slightly greater than that of the pressure drop. The differences between the two are as follows: (1) the time span between several peaks is different. The time span in the experiment in this paper is greater than the corresponding value in the literature; (2) the time span of the pressure rise and that of the pressure drop differ. The time of the pressure rise in this experiment is about 13 μs, and that of the pressure drop about 6 μs, which are both greater than that in the literature. The above phenomenon is mainly caused by the influence of the size of the cavitation bubbles. The maximum radius of the corresponding cavitation bubble is 1.17 mm when γ is 0.89 in the literature, and 9.98 mm when γ is 0.91 in this paper.

The impact of the jet on the wall will cause a water hammer effect. The water hammer pressure keeps a linear relation with the jet velocity and is calculated as follows [3]:

$$P_{wh} = \frac{\rho_w c_w \rho_s c_s}{\rho_w c_w + \rho_s c_s} v_t \tag{1}$$

where ρ_w and c_w are the density and sound velocity of the jet medium, respectively; ρ_s and c_s are the density and sound velocity of the solid wall, respectively; and v_t is the velocity of the jet as it impacts the wall. The wall used in this paper is made of plexiglass, with a density and sound velocity of 2700 kg/m^3 and 2692 m/s, respectively. The density and sound velocity of the jet medium are 1000 kg/m^3 and 1435 m/s, respectively.

Figure 5 is a high-speed photographic image of the collapse of cavitation bubbles in Figure 4b, where v is the micro-jet velocity. The frame-rate used in the experiment was 150,000 frames per second, and the pictures in Figure 5 were acquired every 15 frames based on the experimental rate. The micro-jet was formed during the first collapse of the cavitation bubble when the velocity of the micro-jet was approximately 74.96 m/s. When the cavitation bubble entered the rebound stage, a micro-jet was again formed with impact on the wall. The micro-jet developed gradually and eventually imposed an impact on the wall, with its speed decreasing to approximately 13.00 m/s when arriving at the wall. According to the formula for the water hammer pressure, the impact pressure of the micro-jet formed from the rebound cavitation bubbles against the wall is estimated to be 21.36 MPa, and the pressure peak measured by the pressure sensor is 21.89 MPa, both of which are very close. It can be seen that the impact of the micro-jet formed from the rebound cavitation bubbles against the wall mainly comes from the impact of the micro-jet. For the impact of the micro-jet from the first collapse on the wall when the distance between the bubble and the wall is small, due to the limited speed of the high-speed camera used in this experiment, no clear shock wave can be obtained. Therefore, it is currently impossible to tell whether the main reason for the pressure peak of the impact of the cavitation bubble on the wall from the first collapse is the shock wave or the micro-jet.

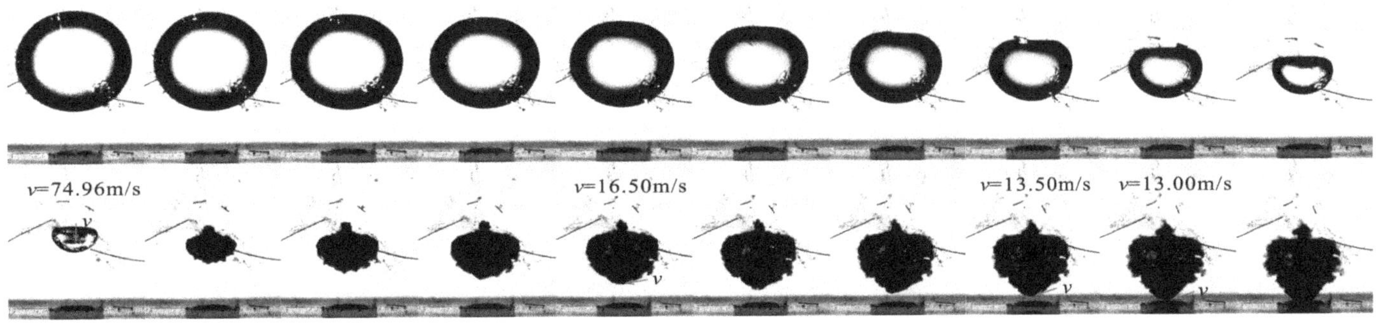

Figure 5. Process of the impact of micro-jet of rebound cavitation bubbles on a wall (frame-rate: 150,000 fps; exposure time: 5.06 μs; R_{max} = 11.02 mm, γ = 1.79).

3.3. Influence of the Distance Between the Cavitation Bubble and Wall on the Impact Strength of the Wall Surface

In the whole process of cavitation bubble evolution, when the dimensionless distance γ meets a certain condition, the second collapse impact strength on the wall surface is greater than the first collapse impact strength. Thus, in this section, the dimensionless distance γ and the impact strength on the wall surface are systematically studied to gain the relationship of the peak impact pressure on the wall surface at cavitation bubble collapse stages under different γ conditions.

In order to obtain the evolution process and impact process of the cavitation bubble with different dimensionless distances γ on the wall surface, the distance from the discharge electrode to the wall surface (with pressure sensor buried inside) will be slightly adjusted in the testing, and the characteristic radius R_{max} of the cavitation bubble will be changed under each h condition. For the peak value of the impact strength of the cavitation bubble on the wall surface, the maximum pressure of the first impact and second impact during the impact of the cavitation bubble on the wall surface is selected in this section to reflect the impact strength of the cavitation bubble on the wall surface.

Figure 6 shows the relationship between the dimensionless distance γ and the wall peak pressure; therein, the horizontal axis indicates the dimensionless distance γ and the vertical axis indicates the corresponding pressure peaks on the wall surface during the first and second collapses of the cavitation bubble. It can be seen from Figure 6 that for the first collapse of the cavitation bubble, the relationship between the dimensionless distance γ and pressure peak on the wall surface shows an exponential-type distribution overall: with the gradual decrease of γ, the maximum value of the impact on the wall surface rapidly increases. Then, the peak value will increase first and then rapidly decrease as a whole during the impact of the second collapse of the cavitation bubble on the wall surface. When the distance between the cavitation bubble and the wall increases, the peak pressure on the wall can be divided into the following three parts: when γ is less than 1.33, the impact of the first collapse of the cavitation bubble on the wall surface is greater than the impact of the second collapse on the wall surface; when γ between the cavitation bubble and wall is greater than 1.33 and less than 2.37, the impact strength on the wall surface caused by the second collapse of the cavitation bubble will be greater than the impact strength of the first collapse on the wall surface; when γ is greater than 2.37, the impact of the first collapse of the cavitation bubble on the wall surface is greater than the impact of the second collapse on the wall surface, and the maximum of the first impact peak is less than 4 MPa.

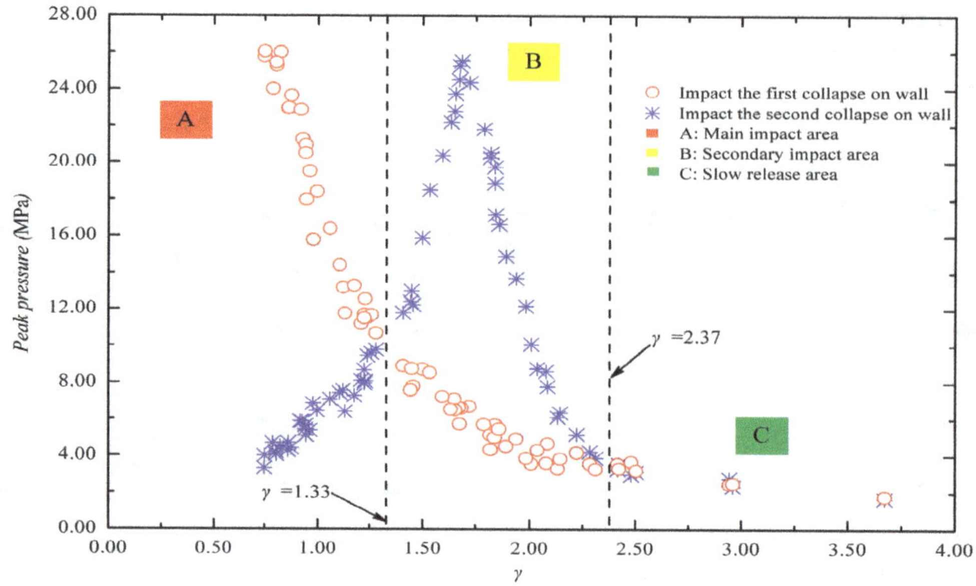

Figure 6. The relationship between the dimensionless distance (γ) and the wall peak pressure.

It can be seen from the above analysis that for the collapse of cavitation bubbles near the wall surface, when the dimensionless distance γ is less than 2.37, the impact of the first and second collapses of the cavitation bubble on the wall surface is huge; when γ is greater than 1.33 and less than 2.37, the impact on the wall surface caused due to the second collapse of the cavitation bubble is greater than the impact on the wall surface caused due to the first collapse. The reason why the above impact characteristics are seen can be analyzed from the following two aspects: on the one hand, when the bubble is very close to the wall surface, during the first bubble-asymmetric shrinkage, a micro-jet with a direct impact on the wall is formed. When the cavitation bubble is a little further away from the wall surface, the impact of the wall on the first shrinkage form of the cavitation bubble is weakened, and a micro-jet is formed during the shrinkage of the cavitation bubble. The micro-jet towards the wall surface drives the cavitation bubble, and finally, during the second shrinkage, the micro-jet is formed. This micro-jet can impose an impact on the wall. Although the energy of the cavitation bubble itself is attenuated after the first cycle of evolution, the impact of the micro-jet from the second shrinkage on the wall is still considerably large. As the distance between the cavitation bubble and the wall surface further increases, the bubble during the second shrinkage obviously moves toward the wall surface, but the resulting micro-jet from the second collapse is not sufficient to impose an impact on the wall, and then the bubble enters the next expansion and contraction stage. On the other hand, the cavitation bubble has a certain energy after the birth. The energy is dissipated due to the viscosity of the water body and other factors after the first and second expansion. After the two-fold expansion and shrinkage, the energy is significantly reduced. As a result, less obvious movement toward the wall is observed and the impact on the wall is reduced greatly in the subsequent evolution of the cavitation bubble.

Philipp and Lauterborn [3] studied the deterioration of the wall under different bubble–wall conditions when the cavitation bubble's R_{max} = 1.45 mm with the laser-induced cavitation technology. There were over 100 groups of samples. By analyzing the deterioration and bubble–wall distance γ, it is found that when $\gamma \geq 2.2$, no deterioration was found on the solid wall; when $2.2 \geq \gamma \geq 1.5$, the area of deterioration increased with the decrease of γ. By comparing the deterioration in the literature and the impact of the cavitation bubble on the wall obtained in this paper, it can be found that the critical values are basically consistent. Since a low-pressure discharge-induced cavitation bubble is used here, it is more difficult to achieve cavitation under the condition of smaller values of γ. Furthermore, the cavitation bubble was larger with the low-voltage underwater discharge technology used in this paper, and the size in the literature is smaller, which may cause the slight difference in the critical value between the paper and the literature. This difference, however, does not affect the zoning characteristics of the impact of the cavitation bubble on the wall.

4. Conclusions

Through the combination of the high-speed dynamic collection and analysis system and transient pressure testing system, the impact characteristics of the collapse of the low-voltage spark-discharge cavitation bubble on the wall surface are studied in this paper, drawing the following important conclusions:

(1) The dimensionless distance between the cavitation bubble and wall (γ) has an important influence on the formation of the micro-jet in the late evolution stage of the cavitation bubble. When the dimensionless distance γ is less than 1.33, a micro-jet directly impacting the wall surface can be formed during the first contraction of the cavitation bubble; subsequently, the collapse of the main impact area will occur. When γ is within 1.33~2.37, the second collapse of the cavitation bubble after rebound will cause a greater impact on the wall surface, then the collapse of the secondary impact area will occur. When γ is greater than 2.37, the influence of wall on the development of a micro-jet is not obvious, that is, collapse in the slow release area occurs.

(2) Concerning the peak value of the impact pressure of the cavitation bubble collapse on the wall surface, the impact of cavitation bubbles on the wall surface during the main impact gradually decreases with the increase of the distance between the cavitation bubble and the wall, and the impact of the cavitation bubble on the wall surface during the secondary impact increases first and then decreases. More specifically, when the dimensionless distance γ is less than 1.33, the impact on the wall surface due to cavitation bubble collapse is mainly from the first collapse, and with increasing γ, the impact of the second collapse on the wall surface increases. When γ is within 1.33~2.37, the impact of the cavitation bubble on the wall surface during the first collapse will be less than the impact strength of the re-collapse after rebound on the wall surface with increase of γ. When γ is greater than 2.37, the impact of the first and second collapses of the cavitation bubble on the wall surface will decrease considerably.

The above conclusions have potential theoretical value for the anti-erosion design of hydraulic engineering, the erosion resistance of blades in hydraulic machinery and the better exertion of the cavitation effect in ultrasonic cleaning technology.

Author Contributions: Conceptualization, W.X. and J.L.; Methodology, J.D. and J.L.; Software, J.L.; Validation, W.X. and J.D.; Formal Analysis, J.L.; Investigation, J.L.; Resources, J.L.; Data Curation, Y.Z. and Q.Z.; Writing-Original Draft Preparation, J.L.; Writing-Review & Editing, J.D.; Visualization, J.L.; Supervision, W.X.; Project Administration, W.X.; Funding Acquisition, J.D.

References

1. Shima, A.; Takayama, K.; Tomita, Y.; Ohsawa, N. Mechanism of impact pressure generation from spark-generated bubble collapse near a wall. *Aiaa J.* **1983**, *21*, 55–59. [CrossRef]
2. Tomita, Y.; Shima, A. Mechanisms of impulsive pressure generation and damage pit formation by bubble collapse. *J. Fluid Mech.* **1986**, *169*, 535–564. [CrossRef]
3. Philipp, A.; Lauterborn, W. Cavitation erosion by single laser-produced bubbles. *J. Fluid Mech.* **1998**, *361*, 75–116. [CrossRef]
4. Mitelea, I.; Bordeaşu, I.; Pelle, M.; Crăciunescu, C. Ultrasonic cavitation erosion of nodular cast iron with ferrite–pearlite microstructure. *Ultrason. Sonochem.* **2015**, *23*, 385–390. [CrossRef] [PubMed]
5. Dong, F.; Li, X.; Zhang, L.; Ma, L.; Li, R. Cavitation erosion mechanism of titanium alloy radiation rods in aluminum melt. *Ultrason. Sonochem.* **2016**, *31*, 150–156. [CrossRef] [PubMed]

6. Shaw, S.J.; Schiffers, W.P.; Emmony, D.C. Experimental observations of the stress experienced by a solid surface when a laser-created bubble oscillates in its vicinity. *J. Acoust. Soc. Am.* **2001**, *110*, 1822. [CrossRef] [PubMed]
7. Lindau, O.; Lauterborn, W. Cinematographic observation of the collapse and rebound of a laser-produced cavitation bubble near a wall. *J. Fluid Mech.* **2003**, *479*, 327–348. [CrossRef]
8. Liu, X.M.; He, J.; Lu, J.; Ni, X.W. Effect of surface tension on a liquid-jet produced by the collapse of a laser-induced bubble against a rigid boundary. *Opt. Laser Technol.* **2009**, *41*, 21–24. [CrossRef]
9. Xu, R.; Zhao, R.; Cui, Y.; Lu, J.; Ni, X. The collapse and rebound of gas-vapor cavity on metal surface. *Opt. Int. J. Light Electron. Opt.* **2009**, *120*, 115–120. [CrossRef]
10. Ren, X.D.; He, H.; Tong, Y.Q.; Ren, Y.P.; Yuan, S.Q.; Liu, R.; Zou, C.Y.; Wu, K.; Sui, S.; Wang, D.S. Experimental investigation on dynamic characteristics and strengthening mechanism of laser-induced cavitation bubbles. *Ultrason. Sonochem.* **2016**, *32*, 218–223. [CrossRef] [PubMed]
11. Rochester, M.C.; Brunton, J.H. Surface pressure distribution during drop impingement. In Proceedings of the Fourth International Conference on Rain Erosion and Related Phenomena, Neues Schloss, Meersburg, Germany, 8–10 May 1974; pp. 371–393.
12. Klaseboer, E.; Fong, S.W.; Turangan, C.K.; Khoo, B.C.; Szeri, A.J.; Calvisi, M.L.; Sankin, G.N.; Zhong, P. Interaction of lithotripter shockwaves with single inertial cavitation bubbles. *J. Fluid Mech.* **2007**, *593*, 33–56. [CrossRef] [PubMed]
13. Lauterborn, S.; Urban, W. Ultrasonic cleaning of submerged membranes for drinking water applications. *J. Acoust. Soc. Am.* **2008**, *123*, 3291. [CrossRef]
14. Verhaagen, B.; Fernández, R.D. Measuring cavitation and its cleaning effect. *Ultrason. Sonochem.* **2016**, *29*, 619–628. [CrossRef] [PubMed]
15. Fong, S.W.; Adhikari, D.; Klaseboer, E.; Khoo, B.C. Interactions of multiple spark-generated bubbles with phase differences. *Exp. Fluids* **2009**, *46*, 705–724. [CrossRef]
16. Pain, A.; Terence Goh, B.H.; Klaseboer, E.; Ohl, S.W.; Cheong Khoo, B. Jets in quiescent bubbles caused by a nearby oscillating bubble. *J. Appl. Phys.* **2012**, *111*, 137. [CrossRef]
17. Luo, J.; Xu, W.L.; Niu, Z.P.; Luo, S.J.; Zheng, Q.W. Experimental study of the interaction between the spark-induced cavitation bubble and the air bubble. *J. Hydrodyn.* **2013**, *25*, 895–902. [CrossRef]
18. Luo, J.; Xu, W.L.; Li, R. High-speed photographic observation of collapse of two cavitation bubbles. *Sci. Chin. Technol. Sci.* **2016**, *59*, 1707–1716. [CrossRef]
19. Goh, B.H.T.; Ohl, S.W.; Klaseboer, E.; Khoo, B.C. Jet orientation of a collapsing bubble near a solid wall with an attached air bubble. *Phys. Fluids* **2014**, *26*, 221–240. [CrossRef]
20. Xu, W.L.; Zhang, Y.L.; Luo, J.; Arong; Zhang, Q.; Zhai, Y.W. The impact of particles on the collapse characteristics of cavitation bubbles. *Ocean Eng.* **2017**, *131*, 15–24. [CrossRef]
21. Bing, H.T.G.; Shi, W.G.; Ohl, S.W.; Khoo, B.C. Spark-generated bubble near an elastic sphere. *Int. J. Multiphase Flow* **2016**, *90*, 156–166. [CrossRef]
22. Cui, P.; Zhang, A.M.; Wang, S.; Khoo, B.C. Ice breaking by a collapsing bubble. *J. Fluid Mech.* **2018**, *841*, 287–309. [CrossRef]
23. Duncan, J.H.; Zhang, S. On the interaction of a collapsing cavity and a compliant wall. *J. Fluid Mech.* **1991**, *226*, 401–423. [CrossRef]
24. Chahine, G.L.; Kalumuck, K.M.; Duraiswami, R. Bubble dynamics fluid-structure interaction simulation by coupling fluid BEM and structural FEM codes. *Int. J. Multiphase Flow* **1995**, *9*, 861–883. [CrossRef]
25. Chahine, G.L.; Kalumuck, K.M.; Hsiao, C.T. Simulation of surface piercing body coupled response to underwater bubble dynamics utilizing 3dynafs, a three-dimensional bem code. *Comput. Mech.* **2003**, *32*, 319–326. [CrossRef]
26. Zhang, Y.L.; Yeo, K.S.; Khoo, B.C.; Wang, C. 3d jet impact and toroidal bubbles. *J. Comput. Phys.* **2001**, *166*, 336–360. [CrossRef]
27. Wang, C.; Khoo, B.C. An indirect boundary element method for three-dimensional explosion bubbles. *J. Comput. Phys.* **2004**, *194*, 451–480. [CrossRef]
28. Best, J.P. The formation of toroidal bubbles upon the collapse of transient cavities. *J. Fluid Mech.* **1993**, *251*, 79–107. [CrossRef]

29. Klaseboer, E.; Hung, K.C.; Wang, C.; Wang, C.W.; Khoo, B.C.; Boyce, P.; Debono, S.; Charlier, H. Experimental and numerical investigation of the dynamics of an underwater explosion bubble near a resilient/rigid structure. *J. Fluid Mech.* **2005**, *537*, 387–413. [CrossRef]
30. Goh, B.H.; Oh, Y.D.; Klaseboer, E.; Ohl, S.W.; Khoo, B.C. A low-voltage spark-discharge method for generation of consistent oscillating bubbles. *Rev. Sci. Instrum.* **2013**, *84*, 014705. [CrossRef] [PubMed]

8

Laboratory Study of Secondary Flow in an Open Channel Bend by using PIV

Ruonan Bai [1], Dejun Zhu [1], Huai Chen [2] and Danxun Li [1,*]

1 State Key Laboratory of Hydroscience and Engineering, Tsinghua University, Beijing 100084, China; bairuonan1991@163.com (R.B.); zhudejun@tsinghua.edu.cn (D.Z.)

2 State Key Laboratory of Hydrology, Water Resources and Hydraulic Engineering, Nanjing Hydraulic Research Institute, Nanjing 210029, China; chenhuai@nhri.cn

* Correspondence: lidx@tsinghua.edu.cn.

Abstract: The present paper aims to gain deeper insight into the evolution of secondary flows in open channel bend. A U-shaped open channel with long straight inflow/outflow reaches was used for experiments. Efforts were made to precisely specify flow conditions and to achieve high precision measurement of quasi-three-dimensional velocities with a multi-pass, two-dimensional PIV (Particle Image Velocimetry) method. The experimental results show that the flow begins to redistribute before entering the bend and it takes a long distance to re-establish to uniform conditions after exiting the bend. Complex secondary flow patterns were found to be present in the bend, as well as in the straight inflow and outflow reaches. A "self-breaking" (process was identified, which correlates stream-wise velocity with the intensity of flow circulation.

Keywords: open channel bend; secondary flow; velocity distribution; PIV

1. Introduction

Complex flow patterns exist in open channel bends and they play important roles in alluvial and ecological processes. Curvature-induced secondary flows, for instance, occur as an intrinsic balance between the centrifugal force and the pressure gradient due to water surface tilting. These secondary flows act to redistribute bulk velocity, alter sediment transport, shape bar-pool topography, enhance mass mixing, increase energy loss, and reduce conveyance capacity of the channel [1–10].

There have been abundant experimental studies on secondary flow in open channel bends. In particular, two important issues were repeatedly addressed: the strength of the secondary flow [11–15] and the interaction between the secondary flow pattern and the main flow [11,16–19].

Miao et al. [12] reported that, depending on the bend curvature, the normalized magnitude of the secondary flow grows linearly or un-linearly with the ratio of flow depth over bend radius. The Reynolds number was found to play an important role in shaping the secondary flow [11], and the maximum secondary flow strength occurred at the second half of the bend [13]. Roca et al. [14] observed that the maximum vorticity increases along a curved channel from the bend entry until the cross-sections from 40° to 60°. Ramamurthy et al. [15] found that bends with vanes exhibit a lower intensity of secondary flow.

Experiments on the interaction between main flow and secondary flow show that the transverse transport of main flow momentum by the secondary circulation is the principal cause of velocity redistribution, and the deflection of the maximum velocity toward the outer wall of the channel occurs at the cross-section where the secondary flow reaches its maximum strength [11,16–18,20]. Particularly, two parameters for the total additional secondary flow term have been defined and examined for their impact on velocity distributions [17].

An extensive literature review also reveals the lack of a reliable quantitative description of the secondary flow along the entire channel. The insufficiency of previous experimental studies can be attributed to three aspects.

Firstly, the limitation in instrumentation hinders precise measurement in some important flow areas. For instance, experiments with ADV/ADVP (Acoustic Doppler Velocity/Acoustic Doppler Velocity Profiler) fail to sample the near-surface and near-bed flow regions [6,13], making it difficult, if not impossible, to explore the number and position of secondary flow cells at each cross-section [17,21]. Other instruments, e.g., velocity meter PROPLER [20] and P-EMS velocimeter [18], are unable to provide three components of a velocity vector. The popularization of PIV (Particle Image Velocimetry) in fluid mechanics shows promise as being a better instrumentation for measurement of curved flows [22–26]. In particular, some novel multi-pass window deformation approaches have been used to improve the performance of PIV in measurement of various flows [27–29].

Secondly, the experimental design in most curved flow studies is insufficient for a closer observation of the evolution of the flow along the entire channel, including the bend and the straight inflow/outflow reaches. Some experiments either examined merely one cross-section [4,17,30] or used too long a distance between cross-sections to track the evolution of secondary patterns [11]. In the studies by Blanckaert [31], Vaghefi et al. [13], Abhari et al. [18], Bai et al. [11], Zeng et al. [32], and Booij [30], velocity measurements did not cover the straight inflow or outflow reaches at all.

Thirdly, the influence of the straight inflow/outflow reaches on the secondary flow receives little attention. The inflow/outflow reaches are often short, e.g., only 2 m long compared with a 3.77 m bend in the study by Bai et al. [11]. The curved flume used by Blanckaert [33] has no straight outflow reach at all. The influence of the inflow/outflow reaches still remains unclear, and as a result, the evolution of secondary flow in the bend involves much ambiguity. Some quantitative guidelines obtained by previous studies introduced too many assumptions [12].

To further investigate the evolution of secondary flow in open channel bend, a U-shaped bend with long straight inflow/outflow reaches has been designed and experiments have been conducted. In particular, the paper aims to illustrate the characteristics of the secondary flow at simple but precisely specified conditions to avoid possible ambiguity.

The paper has the following objectives:

(1) To set up a well-designed experimental system that facilitates accurate specification of flow conditions and detailed observation of the three components of flow velocity in a U-shaped open channel flume with long straight inflow/outflow reaches. The experimental design effectively eliminates ambiguity in both the inflow and the outflow reaches and enhances measurement reliability.
(2) To report and interpret detailed and high-quality data on the secondary flow, with focus on the distribution of the main flow, the topography of water surface, and the evolution of the circulation cells.
(3) To present high-quality data on the secondary flows that may be used for validation and calibration of numerical schemes.

2. Experimental Set-Up and Methods

2.1. U-Shaped Flume

Experiments were conducted to investigate flow properties in a U-shaped flume with a 180° bend. Focus was given to the evolution of secondary flows along the entire channel.

The flume, sketched in Figure 1a, was built with a glass bed and glass side walls. It had a width (B) of 0.3 m, a length along the centerline of approximately 26 m, and a centerline curvature radius (R_c) of 1.5 m. The bend, with a ratio of $R_c/B = 5$, falls into the category of medium bend [8]. The straight inflow and outflow reaches are both 10 m long, more than twice the length of the bend (center line 4.7 m). A honeycomb was placed at the flume entry to stabilize the flow and attenuate initial large-scale flow structures. A rolling shutter as a tail gate helps to control the water level. The overall stream-wise channel slope is adjustable, but it was fixed at S = 0.001 in the present study.

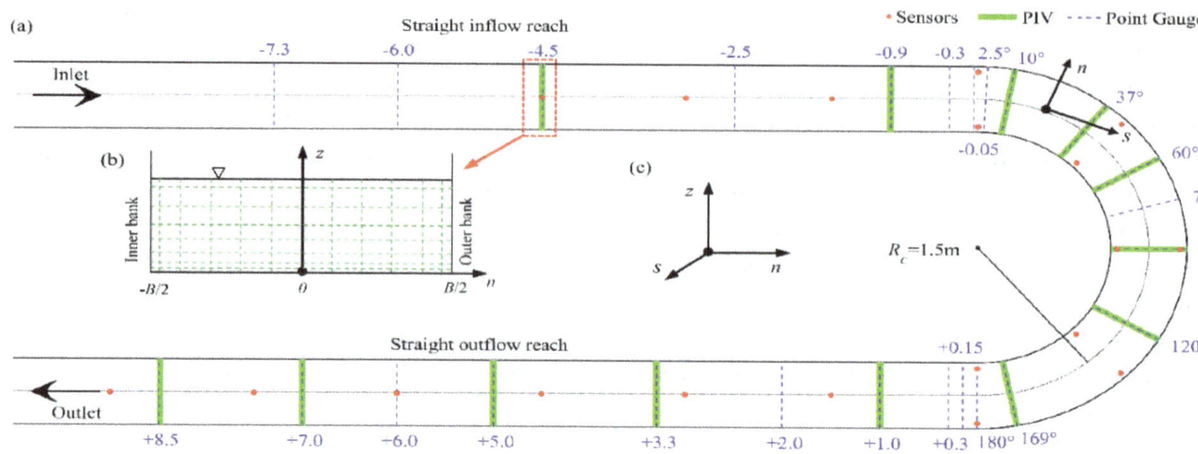

Figure 1. (**a**) Plan of the U-shaped flume with indication of positions of measurements; (**b**) Measuring grids of PIV (Particle Image Velocimetry) system (11 vertical lines and 7 horizontal lines); (**c**) the definition of the coordinate system: stream-wise (s), span-wise (n), and vertical (z).

Figure 1b shows the grids for PIV measurements. Details of the PIV technique will be given later in Figure 2.

The stream-wise, span-wise, and vertical directions are denoted by s, n, and z, respectively (Figure 1c); three components of the time-averaged velocity are U, V, and W.

Flow discharge was measured and controlled by an electromagnetic flow meter. Water surface elevations along the flume were automatically recorded with 20 ultrasonic sensors (red dots in Figure 1a) placed at 15 cross-sections. At each cross-section in the bend (0°, 45°, 90°, 135°, 180°), two sensors were used to monitor the difference in the water level along the span-wise direction. The ultrasonic sensors have a measuring range of 20–200 mm and a measuring error of ±0.2 mm. Real-time monitoring of discharge and water depth facilitates quick stabilization of the flow to steady and uniform conditions. Note that these ultrasonic sensors were used only to monitor the flow rather than to record the water level for further analysis. More precise measurements of water level were achieved with manual point gauge.

2.2. Flow

The flow in the straight inflow/outflow reaches was kept steady and uniform during the experiment. Table 1 shows the hydraulic parameters based on measurements at the cross-section −4.5, 4.5 m upstream of the bend entry. The Reynolds number and Froude number in the straight reaches were 14,917 and 0.557, respectively.

The flow in the experiment is sub-critical, thus the bend plays an important role on the secondary circulation and exerts influence on the inflow by upstream propagation. For sub-critical flows, such influence will diminish.

A water depth of H = 6 cm was used in the experiment, and the corresponding hydraulic radius was R = 4.29 cm. The straight reach of the flume is about 233 R, sufficiently long for the flow to fully develop compared with previous studies by Odgarrd and Bergs [34] and Yen [35], which recommended a straight length of 52 R and 100 R, respectively.

Table 1. Hydraulic and geometric conditions.

B (cm)	Q (L/s)	S	H (cm)	R (cm)	g (ms^{-2})	U_f (cm/s)	Re	Fr	u* (cm/s)	Re*
30	7.15	0.001	6	4.29	9.8	39.7	14817	0.557	2.05	764.4297

B is flume width, Q is flow discharge, S is bed slope, H is flow depth, R = BH/(B + 2H) is hydraulic radius, g is gravitational acceleration, $U_f = Q/(BH)$ is cross-averaged stream-wise velocity, Re = UR/ν is Reynolds number, $Fr = U/(gH)^{(1/2)}$ is Froude number, Re* = u*R/ν is shear Reynolds number, $u^* = \sqrt{gRJ}$ is shear velocity, J is hydraulic slope.

2.3. Measurements

Measurements of water levels and velocity distribution were conducted by means of point gauge and PIV, respectively.

Water surface was measured by manual point gauge with an accuracy of ±0.02 mm at 25 cross-sections (denoted by blue dot lines in Figure 1a) within the entire flume. Information on these 25 cross-sections is given in Table 2. The cross-sections in the straight reaches are denoted by their distance from bend entry (inflow, with minus sign) or from bend exit (outflow, with plus sign), and sections within the bends are denoted by their central angles in degree. For each cross-section, thirteen water depths in the span-wise direction were obtained, and their locations are n = 0, ±3, ±6, ±9, ±12, ±13, and ±14 cm, respectively. Note that the coordinate origin was set at the bottom of the mid-span plane of the channel (see Figure 1a) at each cross section.

Table 2. Cross-sections of water surface measurements.

Inflow		Bend		Outflow	
−7.3	−0.3	2.5°	90°	+0.15	+5.0
−6.0	−0.05	10°	120°	+0.3	+6.0
−4.5		37°	169°	+1.0	+7.0
−2.5		60°	180°	+2.0	+8.5
−0.9		74°		+3.3	

Thirteen locations within the width (B = 30 cm) of the channel have been measured for each cross-section (shown in Figure 1b).

A PIV system was employed to measure velocity distributions at 13 cross-sections (green lines in Figure 1a), which are located at the straight inflow reach (−4.5 m and −0.9 m), the bend (10°, 37°, 60°, 90°, 120°, and 169°), and the straight outflow reach (+1.0 m, +3.3 m, +5.0 m, +7.0 m, and +8.5 m).

At each cross-section, images in the s-z plane were recorded at eleven span-wise locations, i.e., n = 0, ±3, ±6, ±9, ±12, and ±14 (Figure 1b), to investigate velocity distributions of the stream-wise and vertical components. The image plane, which is normal to the channel bed and parallel to flume wall, is about 8 mm in the stream-wise direction and 6 cm in the vertical direction.

The laser and camera were placed vertically and horizontally, respectively, as shown in Figure 2a. To facilitate operation, a frame was designed with a gear system through which both the laser and the camera can be flexibly adjusted and precisely positioned.

Measurements at the s-n plane were achieved by exchanging the positions of the laser and the camera, as shown in Figure 2b. Images at seven layers were recorded, i.e., z/H = 0.05, 0.1, 0.2, 0.35, 0.5, 0.7, and 0.9, respectively (see Figure 1b). Thus, the stream-wise and span-wise velocity distributions at seven different vertical locations were obtained. A combination of measurements at both s-z and s-n planes provides three-component velocity data in a 11 × 7 grid.

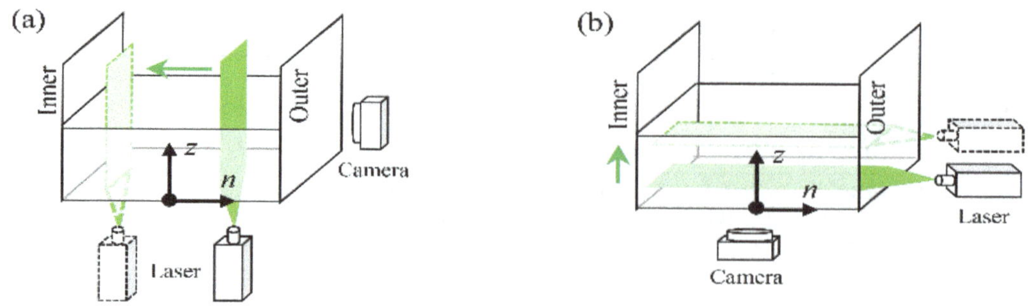

Figure 2. PIV setup with a laser and a camera: (**a**) s-z plane; (**b**) s-n plane.

The PIV system uses a continuous wave laser (8 W) at 532 nm for illumination (laser sheet 1 mm thick). The flow was seeded manually with spherical polyamide particles with a density of 1.03×10^3 kg/m^3 and a mean diameter of 5 µm. Particle images were recorded at a frame rate of 500 Hz by an 8-bit 2560 × 1920 pixel high-speed CMOS camera with a Canon EF 50 mm f/1.2 USM lens. Each measurement yields a sample of 15,000 images for analysis, corresponding to a sampling interval of 30 s. The image sizes are 96 pixels × 1120 pixels in Figure 2a, and 2192 pixel × 88 pixel in Figure 2b. Particle images were analyzed using a multi-pass, iterative multi-grid window deformation method. For grid refinement, the following sequence of the interrogation window size in terms of pixel was used: 64 × 64, 32 × 32, and 16 × 16. There is an overlap of 50% in the horizontal direction of the window. Details of the PIV algorithms can be found in Chen et al. [36] and Zhong et al. [37].

As the measurements were conducted in repeated experiments, the repeatability is essential for the validity of the measurements. This was ensured by using a sufficiently large water reservoir to minimize discharge fluctuation, carefully positioning the laser device and camera, and precisely establishing the flow with fixed depth and energy gradient. Comparison of water depth, energy slope, and velocity distribution in the vertical has indicated that the influence of environment change on measurement repeatability was negligible.

2.4. Methods for Data Analysis

2.4.1. Ensemble Average

It is worth noting that the current PIV measurements cannot be used to reconstruct the instantaneous three-dimensional velocity field (or the instantaneous vorticity field) because they refer to different instants. The quasi-3D information of the velocity field is achieved through ensemble average.

For each case, a total number of 15,000 consecutive images were captured at a frame rate of 500 Hz. Previous experience has indicated that such a sample size/interval is sufficient for statistical analysis of mean velocity [36,37]. Based on these instantaneous velocity (u, v, w) samples, the mean velocities (U, V, W) and fluctuating velocity components (u′, v′, w′) were calculated through ensemble average. Examination of the results against the log-law velocity profile and the linear Reynolds stress profile has shown that the data are adequate to evaluate overall properties of the flow.

2.4.2. Data Interpolation

Based on water level measurements at 25 cross-sections, a linear interpolation function has been applied within the bend at an interval of 4.5°.

At each PIV measurement cross-section, based on the three-dimensional velocities at a 7 × 11 grid obtained by matching the s-z plane velocity and s-n plane velocity, an interpolation for the mean velocity has been made to get a final velocity field at a 27 × 57 grid for further analysis.

2.4.3. Calculation of Secondary Flow Strength

Secondary flows at channel bends can be conveniently visualized with velocity vectors and streamlines [13,18,26,38,39]. A more detailed description, however, necessitates the introduction of a quantitative parameter, such as vorticity [4,8,10,13,21,40,41].

The stream-wise component of the instantaneous vorticity vector can be calculated as follows:

$$w_s = \frac{\partial v}{\partial z} - \frac{\partial w}{\partial n} = \frac{\partial v^*}{\partial z} - \frac{\partial w}{\partial n}, \quad (1)$$

where v and w represent instantaneous span-wise and vertical velocity respectively, as defined in Figure 1. Obviously, $v = V + v'$, where V represents averaged span-wise velocity and v' represents fluctuating velocity. However, v can be decomposed into a translator part, v_n, called instantaneous depth-averaged span-wise velocity, and a circulatory part v^*, called the span-wise component of the secondary flow as well as $v = v_n + v^*$.

The vorticity provides a good basis for the analysis of transverse circulation.

3. Bulk Flow Distribution

Figure 3 shows contour maps of the vector representations of averaged stream-wise velocity, U, at ten cross-sections. It is obvious that the flow is symmetrical in the straight inflow reach (cross-sections −4.5 and −0.9), deflects to the inner bank in the bend (cross-section 10°) before transferring to the outer bank (cross-sections 60°, 90°, 120°, and 169°), and tends to reconstruct symmetry after exiting the bend in the outflow reach. Note that the flow resumes symmetry at cross-section 38°, a transition cross-section where the main flow achieves balance between the inner and outer banks.

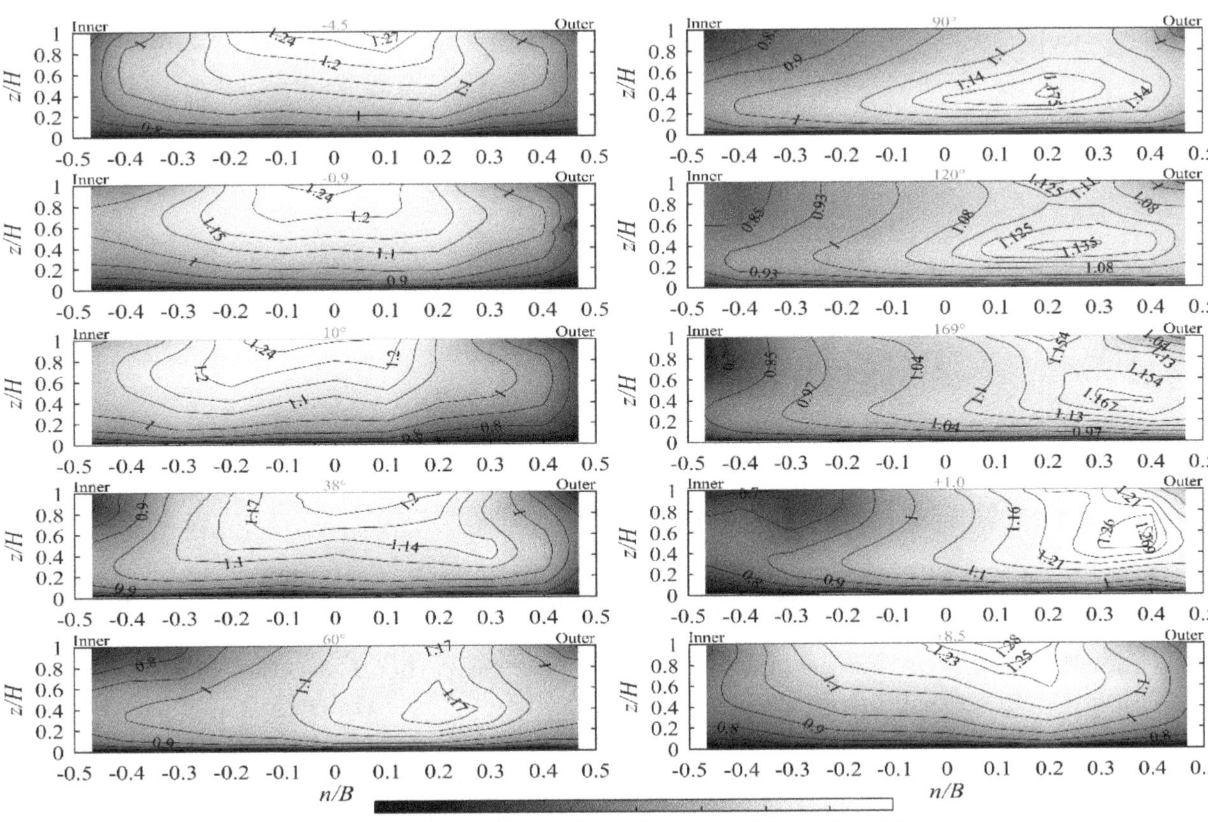

Figure 3. Stream-wise velocity distribution at ten cross-sections (−4.5, −0.9, 10°, 38°, 60°, 90°, 120°, 169°, +1.0, +8.5).

A detailed and quantitative description of the maximum stream-wise velocity, U_{max}, is given in Figure 4, in terms of both magnitude and position.

The magnitude of U_{max}, as can be seen in Figure 4a, shows a general decreasing trend when the flow enters the bend, reaches its minimum at cross-section 90°, and then turns to increase in the cross-sections well after the bend. At cross-section +1.0, however, U_{max} exhibits an unexpected rise over that in the straight inflow reach. This phenomenon may be due to the sudden change of the water level, which creates a positive stream-wise gradient near the outer bank and a negative one near the inner bank.

From the vertical position of U_{max} in Figure 4b, one can see that the maximum velocity occurs at the water surface in both the straight inflow and outflow reaches. In the channel bend, however, the flow sinks, resulting in a submersion of the maximum velocity well below the water surface. This is particularly noticeable in the central reach of the bend (60°–169°). After the bend, cross-section +1.0 witnesses a rise of water surface followed by a gradual decrease.

The change in the span-wise position of U_{max}, in Figure 4c, is consistent with established knowledge of curved flows. The flow starts to go toward the inner bank at cross-sections −0.9 and 10°, turns to the outer bank in the rest of the bend, and keeps left-deviated even in the straight outflow reach. It is interesting that the maximum deviation occurs at cross-section +1.0, i.e., 1.0 m away from the bend exit. This indicates that the flow needs a sufficiently long way to re-establish itself to uniform conditions.

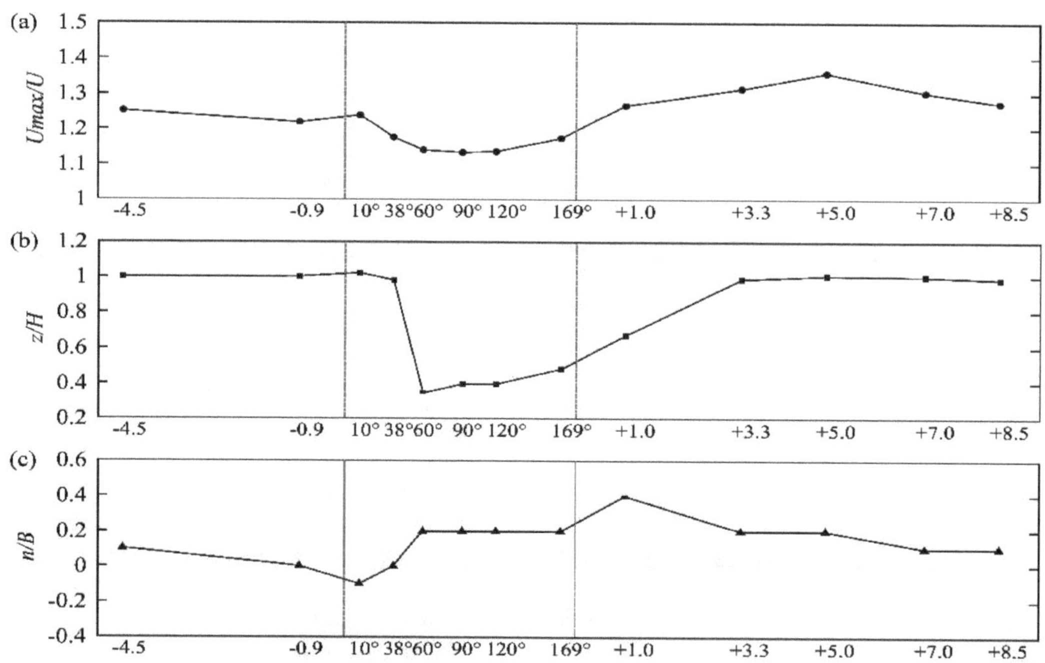

Figure 4. (**a**) Stream-wise evolution (U_{max}/U) of the magnitude of the stream-wise velocity; (**b**) Evolution of the vertical location of $U_{max}(z/H)$; (**c**) Evolution of the span-wise location (n/B) of U_{max}.

To further illustrate the asymmetry of the flow, we introduced another parameter, m, as the horizontal position where the flow discharge is split fifty−fifty. The result is shown in Figure 5. A comparison of Figures 4c and 5 indicates that the change of m in the stream-wise direction is similar to that of n/B, except in the reach from cross-section +3.3 to cross-section +5.0, where the increase of magnitude (Figure 4a) plays a more important role rather than the deflection of position (Figure 5).

Both Figures 4 and 5 reveal that the flow needs a long distance after exiting the bend to re-establish itself, and the curvature has a significant influence not only on the deflection, but also on the magnitude of U_{max}. For field engineering and numerical simulation of curved flows, such complexity needs in-depth consideration.

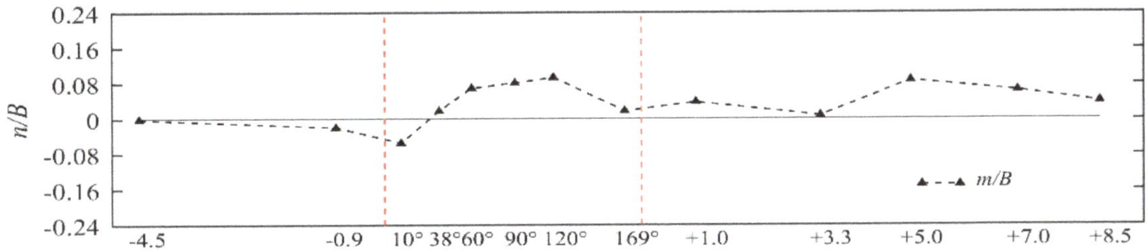

Figure 5. Stream-wise evolution of m along the flume (at 13 cross-sections).

4. Secondary Flow Characteristics

4.1. Surface Topography

Based on the measured water depth data from 25 cross-sections, the topography of the water surface in the entire channel was obtained, as shown in Figure 6. Note that the contour represents a subtraction of 6 cm from actual measurement data.

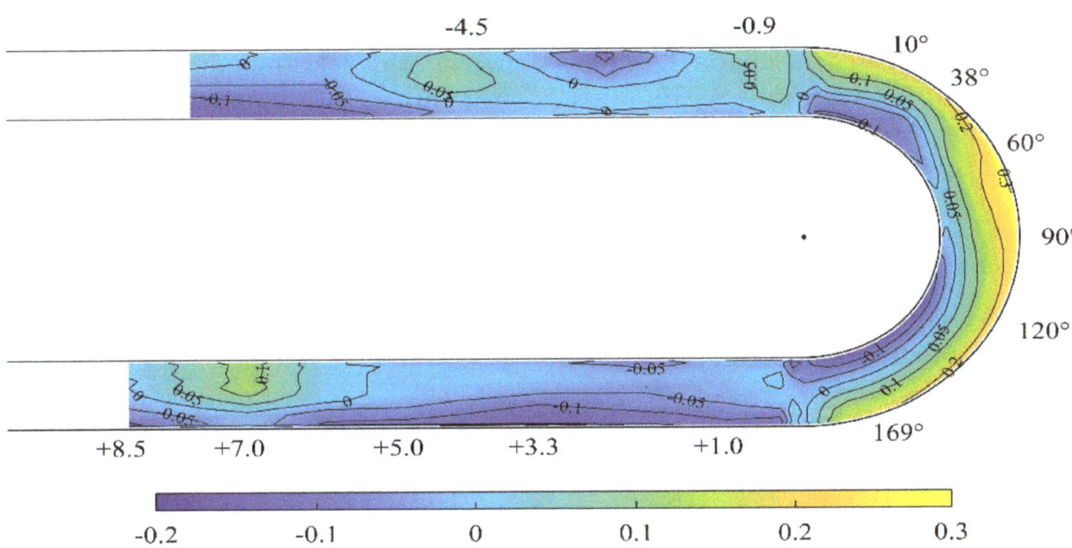

Figure 6. Topography of water surface in the channel derived from point gauge measurement (values are in centimeters and 6 cm defines '0' reference).

The presence of tilting water surface is a typical feature at channel bends. As expected, a transverse slope was found in the curved flow, i.e., it maintains a negligibly small presence in the straight inflow reach, becomes evident when the flow enters the bend, and reduces to nearly zero immediately after the bend. Within the bend, the outer banks witness a higher transverse slope than the inner bank. The difference in magnitude of transverse slope between the outer and inner banks diminishes in the last third of the bend as the water surface turns to be linear.

The transverse slope depends on a combination of factors, e.g., the incoming flow, the geometry of the bend, and the tail gate. When these factors vary, the transverse slope may display seemingly inconsistent characteristics. For instance, in their experimental and numerical study of a 90° bend with a short exit, Gholami et al. [21] reported a higher degree of surface tilting closer to the inner bank rather than the outer bank, a finding that is contradictory to the present result. In the straight reach of the curved open channel, Blanckaert [7] revealed a significantly larger transverse slope than the present study. Such differences indicate the dependence of flow characteristics on the bend geometry as well as the length of inflow/outflow reaches. If the inflow/outflow characteristics cannot be exactly specified, flow in the bend may exhibit significant variations which hinder verification and calibration of numerical schemes.

Figure 7 shows the change of the transverse slope along the channel. Note that for each cross-section the transverse slope (in degree) was calculated based on the water depth difference between the outer and inner banks.

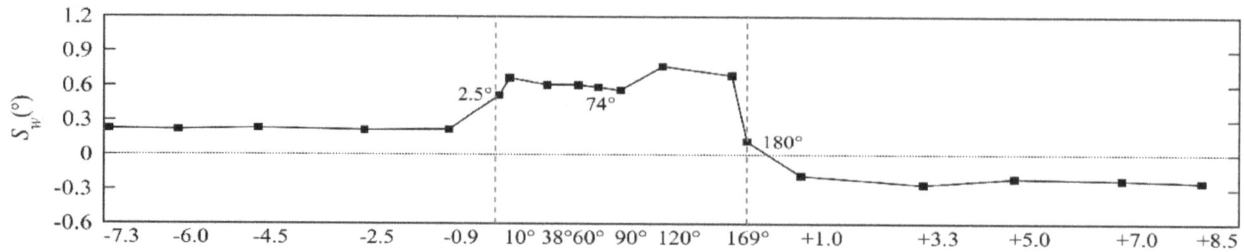

Figure 7. Evolution of the transverse slope along the flume.

In the bend, as we can see, the transverse slope increases quickly and remains at about 0.6 degrees until the end of the bend. At cross-section 120°, the transverse slope reaches its maximum, about 0.8 degrees.

In the straight reaches of the channel, as has been generally acknowledged, the transverse slopes are distinctly smaller. Interestingly, S_W remains positive in the inflow reach and becomes negative in the outflow reach. A positive S_W in the inflow reach is inconsistent with the water tilting in the bend. A negative S_W in the outflow reach is possibly due to the redistribution of flux and transfer of momentum.

4.2. *Secondary Flow and Vice Cells*

Vice cells of secondary circulation are visible in the experiments. This is critical in the investigation of the number and position of secondary flow.

The cross-section-averaged magnitude of vorticity along the flow is shown in Figure 8. Note that the vorticity is normalized by water depth H and cross-section-averaged stream-wise velocity U_f.

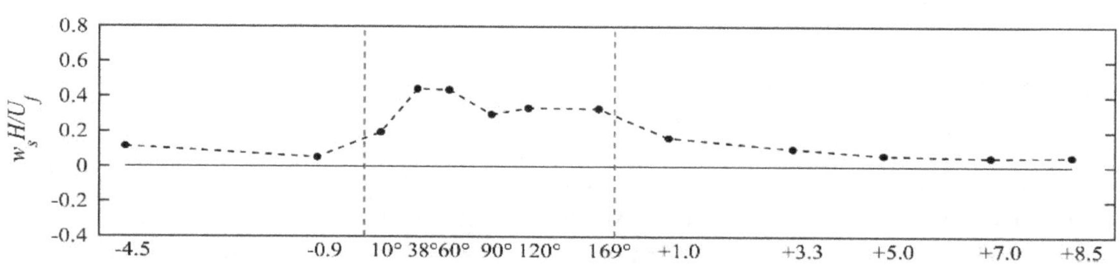

Figure 8. Evolution of dimensionless secondary flow strength.

In the bend, a gradual increase of transverse circulation is recognized, followed by a decrease. The maximum value occurs at cross-section 60°. The decrease is not distinct after cross-section +3.3. Similar findings have been reported by Rovovikii [42], Blanckaert [31], and Roca et al. [14].

It is interesting that the vorticity reaches its peak at cross-section 60° rather than at 90°, and it keeps roughly constant until 169°. Force analysis below helps to explain this phenomenon.

Within a curved channel, secondary circulation occurs as a result of local imbalance between the inward-directed pressure gradient force, $-1/(\rho \cdot \partial p/\partial n)$, and the outward-directed centrifugal force, u^2/r, as shown in Figure 9. The blue areas in Figure 9 represent the local force imbalance which corresponds to the last term in the simplified transverse momentum equation [12].

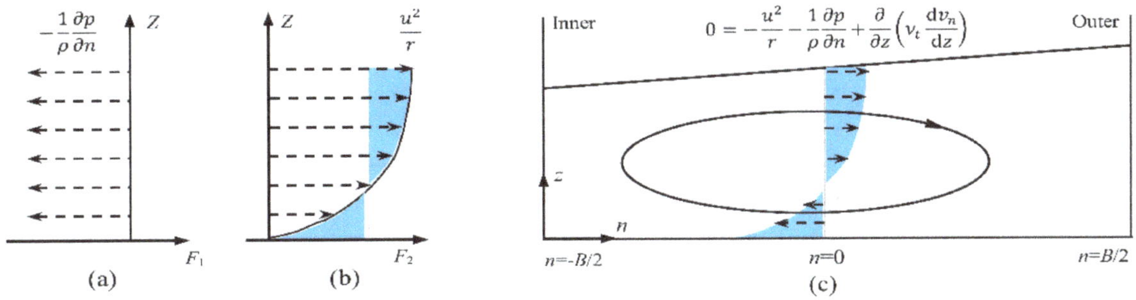

Figure 9. Schematic representation of curvature-induced secondary flow in a cross-section of a curved river reach; F1 represents the inward pressure difference; F2 represents the outward centrifugal force.

Obviously, the secondary flow is basically proportional to the local imbalance, which depends on the profile of the stream-wise velocity distribution along the vertical direction and the transverse pressure gradient [4]. The advective momentum transport by the cross-circulation cell brings down the stream-wise velocities in the upper part of the water column and enhances them in the lower part [4,16]. When the stream-wise velocity becomes more evenly distributed along the vertical, the outward centrifugal force will decrease in the upper part and increase in the lower side, as can be seen in Figure 9b. As a result, the intensity of circulation is reduced. There occurs, as it was, a "self-breaking" of circulation.

Figure 10 shows the pattern of the stream-wise vorticity and the cross-stream motions in ten cross-sections. The color map represents the vorticity in two senses of rotation, i.e., positive for anticlockwise and negative for clockwise. It is worth noting that the flow takes a relatively long distance for the circulation to decay after exiting the bend.

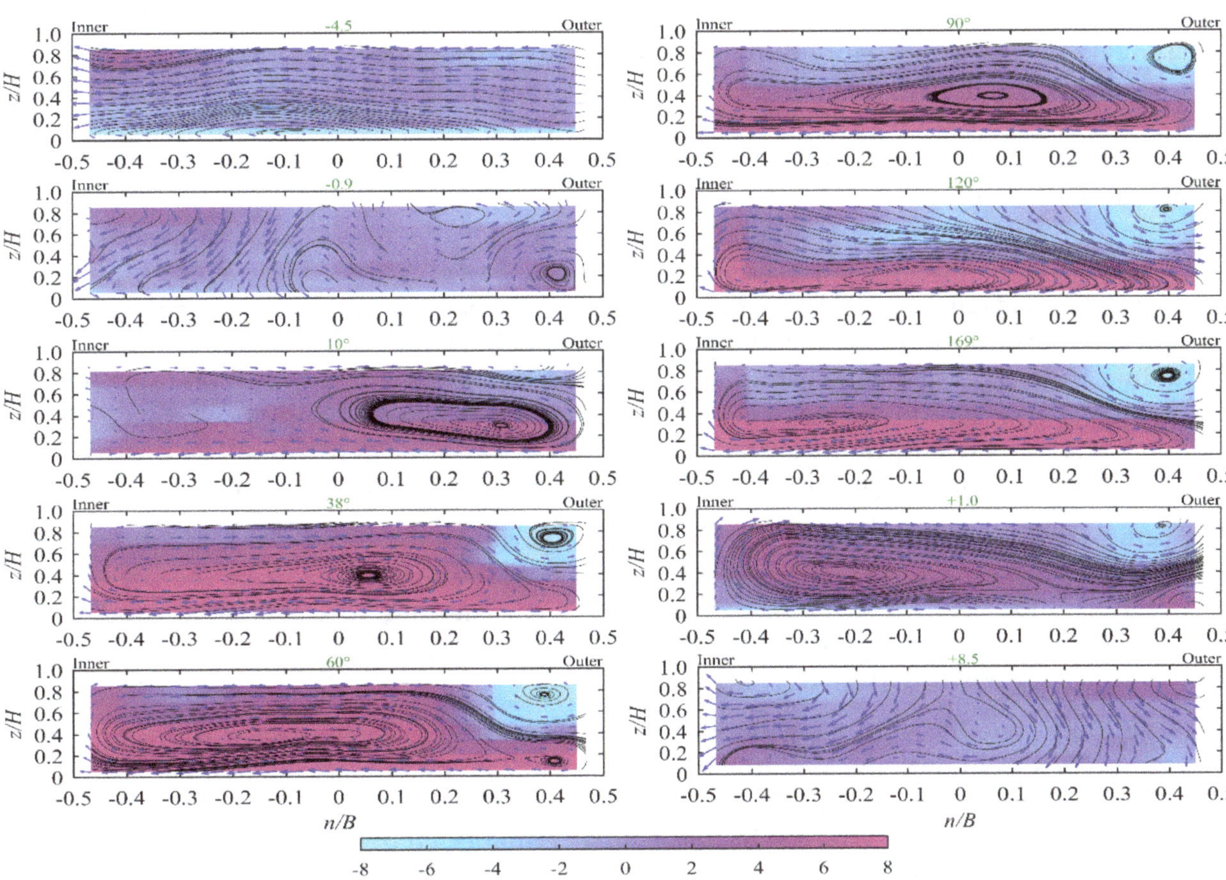

Figure 10. Secondary flow phenomena: velocity projection in the cross-sections (blue vectors); the 2D streamline patterns of the cross-flow (the black line); the dimensionless values of the vorticity (color).

In cross-section −4.5, one can see that the radial velocity components of the transverse circulation are all oriented to the inner bank. Similar results were also obtained by Gholami et al. [20] and Rozovskii [42]. This one-way radial flow leads to asymmetric velocity distributions, which are consistent with the inner-bank deflection of the stream-wise velocity (Figure 4). Accordingly, the high-velocity zone deviates toward the inner bank at the beginning of the bend.

The changing patterns of the secondary flow are illustrated by the streamlines in Figure 10. As can be seen, circulation appears before the flow enters the bend. The commencement of circulation can be recognized at cross-section −0.9 where the velocity vectors are generally oriented to the inner bank, but a visible circulation cell occurs in the outer-bank region near the bed. This finding is contradictory to the conclusion made by Gholami et al. [20], that no secondary flow exists at the entry of channel bends.

At cross-sections 38°, 60°, 90°, 120°, 169°, and +1.0, a large secondary cell presents across each cross-section and a counter-rotating weaker cell is also identified. These results are consistent with previous findings in both laboratory and field experiments, e.g., Einstein and Harder [43], Rozovskii [42], de Vriend [16], Bridge [44], and Blanckaert et al. [45]. When it rotates, the large secondary cell will transport sediment from the outer bank (where erosion occurs) to the inner bank (where deposition develops) of the bend. The presence of the small counter-rotating cell, in contrast, will attenuate such transportation process and help to protect the bank.

A comparison of cross-sections 10° and 38° reveals a dramatic change in the pattern of the rotating cell. This indicates a fast development of secondary flow in the entry of the bend due to the centrifugal acceleration. At cross-section 10°, the circulation is unique, with a large outer bank cell in the middle of the water depth. This is closely related to the fact that cross-section 10° is the only cross-section whose U_{max} is close to the inner bank and the core area of the contour (U) deflects to the inner bank (see Figure 3).

A closer look at cross-section 60° shows that near the outer bank, two small circulations occur with different rotation senses, one below the water surface (denoted the third cell) and the other close to the channel bed. This is probably due to the strong dip and deflection of the core of stream-wise velocity (Figures 3 and 4). The mechanism underlying the circulation is complicated by the effect of turbulence anisotropy near the outer bank. Actually, the U_{max} at cross-section 60° is located near the bed at $n/B = 0.3$ and $z/H = 0.4$ (Figure 3), where the big central region separates from the two small outer bank circulation cells.

The third cell occurring at cross-section 60° seems to decay in the following cross-sections and totally disappears at about 90°. This anisotropy generates a stream-wise vorticity of the additional secondary circulation moment, and thus enhances the tendency of the secondary circulation to split up and recur at cross-section +1.0. As the identification of such a small cell requires measurement with high spatial resolution, few previous studies have succeeded in providing detailed information.

The presence of an additional small cell at cross-sections 60° and +1.0 is closely related to the role of the kinetic energy fluxes from turbulence to the mean flow on the generation of the outer bank cell. The results in this paper may help to gain more insight into the kinetic energy transportation.

The outer-bank cell provides a buffer layer that protects the outer bank from any influence of the center-region cell and keeps the core of U_{max} a certain distance from the bank (Figure 3). At cross-section +1.0, the outer bank cell is very weak, failing to provide a good protection. This phenomenon agrees well with the stream-wise velocity distribution, in that the bulk flow deflects to the outer bank significantly (Figure 4c).

As general knowledge, the outer bank of a bend is more susceptible to erosion. However, the present results, consistent with those of Rovovikii [42], indicate that the most erosion-prone area is the concave bank near the exit from the bend at cross-section +1.0, where a sudden increase in U_{max} and an attenuation of the counter-rotating cell significantly enhances the erosional force of the flow.

It is surprising that at cross-section +8.5, some weak circulations still exist. This, again, indicates the importance of the length of the straight outflow reach for the flow to re-establish itself.

5. Conclusions

The study reported velocity measurements for the experiments at 25 cross-sections in a U-shaped open channel, based on PIV technique. The PIV system was designed as one which is able to obtain quasi-three-dimensional velocities. The observations focused not only on the bend, but also on the straight inflow/outflow reaches which were designed to be sufficiently long. The velocity distribution, the maximum velocity path, the flow surface topography, and the secondary flow have been evaluated and analyzed. The following conclusions may be drawn:

(1) Streamlines in cross-sections show secondary flows of different patterns along the entire channel. Both the curvature-induced secondary flow and weaker secondary flows, including vice cells, were recognized in the straight inflow/outflow reaches.

(2) The maximum secondary flow occurs at the first half of the bend (cross-section 60°), and a "self-breaking" process was observed, which is consistent with previous studies.

(3) The flow near the entry and exit of the bend is complex due to the sudden change of water surface. Flow redistribution also occurs in the straight inflow and outflow reaches, except the channel bend, indicating that the interaction between the main flow and secondary flow exists along the entire channel. If the inflow/outflow cannot be exactly specified, the flow in the bend may exhibit significant variations.

Author Contributions: R.B. and D.L. conceived and designed the experiments; R.B. collected the data; R.B. and H.C. designed the framework and analyzed the data of this study; D.L. and D.Z provided significant suggestions on the methodology and structure of the manuscript; R.B wrote the paper with the contribution of all co-authors.

Acknowledgments: The authors are grateful to Wang Xingkui for his help in design of the curved channel.

Abbreviations

ADV	Acoustic Doppler Velocity
ADVP	Acoustic Doppler Velocity Profiler
B	channel width (cm)
R	hydraulic radius (m)
R_c	radius of curvature of channel centerline (m)
S	stream-wise channel slope (−)
Q	water discharge (m^3/s)
H	averaged water flow depth along the flume (cm)
U_f	cross-averaged stream-wise velocity in the flume (m/s)
u*	shear velocity (cm/s)
Re	UR/ν is the Reynolds number
Re*	u*R/ν is the shear Reynolds number
Fr	the Froude number
g	gravitational acceleration (ms^{-2})
s	stream-wise reference coordinate (−)
n	span-wise reference coordinate; −15 cm represents the inner bank and 15 cm represents the outer one (−)
N	sample size (−)
z	vertical reference coordinate (−)
u	instantaneous stream-wise velocity (m/s)
v	instantaneous span-wise velocity (m/s)
w	instantaneous vertical velocity (m/s)

u'	instantaneous fluctuating stream-wise velocity (m/s)
v'	instantaneous fluctuating span-wise velocity (m/s)
w'	instantaneous fluctuating vertical velocity (m/s)
U	averaged stream-wise velocity (m/s)
V	averaged span-wise velocity (m/s)
W	averaged vertical velocity (m/s)
v_n	instantaneous depth-averaged span-wise velocity (m/s)
V_n	time-averaged value of v_n (m/s)
v^*	span-wise component of the secondary circulation, $v^* = v - V$ (m/s)
V^*	time averaged value of v^* (m/s)
Inner	Inner bank (−)
Outer	Outer bank (−)
θ	cross-section angle (°)
w_s	stream-wise vorticity, $w_s = \partial w/\partial n - \partial v/\partial z$ (s^{-1})
ρ	density of water (kg/m^3)
h	local water flow depth (cm)
U_{max}	the highest value of U in a cross-section (m/s)
m	the deviation of equipartition position (m)
N	sample size (−)
P	pressure (Pa)
S_w	transverse water surface slope (°)

References

1. Shiono, K.; Muto, Y.; Knight, D.W.; Hyde, A.F.L. Energy losses due to secondary flow and turbulence in meandering channels with overbank flows. *J. Hydraul. Res.* **1999**, *37*, 641–664. [CrossRef]
2. Edwards, B.F.; Smith, D.H. Critical wavelength for river meandering. *Phys. Rev. E* **2001**, *63*, 45304. [CrossRef] [PubMed]
3. Lancaster, S.T.; Bras, R.L. A simple model of river meandering and its comparison to natural channels. *Hydrol. Process.* **2002**, *16*, 1–26. [CrossRef]
4. Blanckaert, K.; Graf, W.H. Momentum Transport in Sharp Open-Channel Bends. *J. Hydraul. Eng.* **2004**, *130*, 186–198. [CrossRef]
5. Sin, K. *Methodology for Calculating Shear Stress in a Meandering Channel*; Colorado State University: Fort Collins, CO, USA, 2010.
6. Roca, M.; Martín-Vide, J.P.; Blanckaert, K. Reduction of Bend Scour by an Outer Bank Footing: Footing Design and Bed Topography. *J. Hydraul. Eng.* **2007**, *133*, 139–147. [CrossRef]
7. Blanckaert, K. Flow separation at convex banks in open channels. *J. Fluid Mech.* **2015**, *779*, 432–467. [CrossRef]
8. Kashyap, S.; Constantinescu, G.; Rennie, C.D.; Post, G.; Townsend, R. Influence of Channel Aspect Ratio and Curvature on Flow, Secondary Circulation, and Bed Shear Stress in a Rectangular Channel Bend. *J. Hydraul. Eng.* **2012**, *138*, 1045–1059. [CrossRef]
9. Silva, A.M.F.D.; Yalin, M.S. *Fluvial Processes*, 2nd ed.; CRC Press/Balkema: London, UK, 2001.
10. Roca, M.; Martín-Vide, J.P.; Blanckaert, K. Bend scour reduction and flow pattern modification by an outer bank footing. In Proceedings of the International Conference on Fluvial Hydraulic, Lisbon, Portugal, 6–8 September 2012; pp. 1793–1799.
11. Bai, Y.; Song, X.; Gao, S. Efficient investigation on fully developed flow in a mildly curved 180° open-channel. *J. Hydroinform.* **2014**, *16*, 1250–1264. [CrossRef]
12. Wei, M.; Blanckaert, K.; Heyman, J.; Li, D.; Schleiss, A.J. A parametrical study on secondary flow in sharp open-channel bends: Experiments and theoretical modelling. *J. Hydro-Environ. Res.* **2016**, *13*, 1–13. [CrossRef]
13. Vaghefi, M.; Akbari, M.; Fiouz, A.R. An experimental study of mean and turbulent flow in a 180 degree sharp open channel bend: Secondary flow and bed shear stress. *KSCE J. Civ. Eng.* **2015**, *20*, 1582–1593. [CrossRef]
14. Roca, M.; Blanckaert, K.; Martín-Vide, J.P. Reduction of Bend Scour by an Outer Bank Footing: Flow Field and Turbulence. *J. Hydraul. Eng.* **2009**, *135*, 361–368. [CrossRef]
15. Ramamurthy, A.S.; Han, S.S.; Biron, P.M. Characteristics of Flow around Open Channel 90° Bends with Vanes. *J. Irrig. Drain. Eng.* **2011**, *137*, 668–676.

16. De Vriend, H.J. Velocity redistribution in curved rectangular channels. *J. Fluid Mech.* **1981**, *107*, 423. [CrossRef]
17. Tang, X.; Knight, D.W. The lateral distribution of depth-averaged velocity in a channel flow bend. *J. Hydro-Environ. Res.* **2015**, *9*, 532–541. [CrossRef]
18. Abhari, M.N.; Ghodsian, M.; Vaghefi, M.; Panahpur, N. Experimental and numerical simulation of flow in a 90° bend. *Flow Meas. Instrum.* **2010**, *21*, 292–298. [CrossRef]
19. Termini, D.; Piraino, M. Experimental analysis of cross-sectional flow motion in a large amplitude meandering bend. *Earth Surf. Process. Landf.* **2011**, *36*, 244–256. [CrossRef]
20. Gholami, A.; Akhtari, A.A.; Minatour, Y.; Bonakdari, H.; Javadi, A.A. Experimental and numerical study on velocity fields and water surface profile in a strongly-curved 90° open channel bend. *Eng. Appl. Comput. Fluid Mech.* **2014**, *8*, 447–461. [CrossRef]
21. Farhadi, A.; Sindelar, C.; Tritthart, M.; Glas, M.; Blanckaert, K.; Habersack, H. An investigation on the outer bank cell of secondary flow in channel bends. *J. Hydro-Environ. Res.* **2018**, *18*, 1–11. [CrossRef]
22. Van der Kindere, J.W.; Laskari, A.; Ganapathisubramani, B.; de Kat, R. Pressure from 2D snapshot PIV. *Exp. Fluids* **2019**, *60*, 32. [CrossRef]
23. Scharnowski, S.; Bross, M.; Kähler, C.J. Accurate turbulence level estimations using PIV/PTV. *Exp. Fluids* **2019**, *60*, 1. [CrossRef]
24. Kalpakli Vester, A.; Sattarzadeh, S.S.; Örlü, R. Combined hot-wire and PIV measurements of a swirling turbulent flow at the exit of a 90° pipe bend. *J. Vis.* **2016**, *19*, 261–273. [CrossRef]
25. Nishi, Y.; Sato, G.; Shiohara, D.; Inagaki, T.; Kikuchi, N. A study of the flow field of an axial flow hydraulic turbine with a collection device in an open channel. *Renew. Energy* **2019**, *130*, 1036–1048. [CrossRef]
26. Lindken, R.; Merzkirch, W. A novel PIV technique for measurements in multiphase flows and its application to two-phase bubbly flows. *Exp. Fluids* **2002**, *33*, 814–825. [CrossRef]
27. Termini, D.; Di Leonardo, A. Efficiency of a Digital Particle Image Velocity (DPIV) Method for Monitoring the Surface Velocity of Hyper-Concentrated Flows. *Geosciences* **2018**, *8*, 383. [CrossRef]
28. Sarno, L.; Carleo, L.; Papa, M.N.; Villani, P. Experimental Investigation on the Effects of the Fixed Boundaries in Channelized Dry Granular Flows. *Rock Mech. Rock Eng.* **2018**, *51*, 203–225. [CrossRef]
29. Sarno, L.; Carravetta, A.; Tai, Y.C.; Martino, R.; Papa, M.N.; Kuo, C.Y. Measuring the velocity fields of granular flows—Employment of a multi-pass two-dimensional partical image velocimetry (2D-PIV) approach. *Adv. Power Technol.* **2018**, *29*, 3107–3123. [CrossRef]
30. Booij, R. Measurements and large eddy simulations of the flows in some curved flumes. *J. Turbul.* **2003**, *4*, 1–17. [CrossRef]
31. Blanckaert, K. *Flow and Turbulence in Sharp Open-Channel Bends*; EPFL: Lausanne, Switzerland, 2002.
32. Zeng, J.; Constantinescu, G.; Blanckaert, K.; Weber, L. Flow and bathymetry in sharp open-channel bends: Experiments and predictions. *Water Resour. Res.* **2008**, *44*. [CrossRef]
33. Blanckaert, K.; Graf, W.H. Mean flow and turbulence in open-channel bend. *J. Hydraul. Eng.* **2001**, *127*, 835–847. [CrossRef]
34. Odgaard, A.J.; Bergs, M.A. Flow processes in a curved alluvial channel. *Water Resour. Res.* **1988**, *24*, 45–56. [CrossRef]
35. Yen, B.C. On establishing uniform channel flow with tail gate. *Proc. Inst. Civ. Eng.-Water Marit. Eng.* **2003**, *156*, 281–283. [CrossRef]
36. Chen, H.; Zhong, Q.; Wang, X.; Li, D. Reynolds number dependence of flow past a shallow open cavity. *Sci. China Technol. Sci.* **2014**, *57*, 2161–2171. [CrossRef]
37. Zhong, Q.; Li, D.; Chen, Q.; Wang, X. Coherent structures and their interactions in smooth open channel flows. *Environ. Fluid Mech.* **2015**, *15*, 653–672. [CrossRef]
38. Moser, R.D.A.P. *Direct Numerical Simulation of Curved Turbulent Channel Flow*; National Aeronautics and Space Administration: Moffett Field, CA, USA, 1984.
39. Nezu, I.; Onitsuka, K. Turbulent structures in partly vegetated open-channel flows with LDA and PI V measurements. *J. Hydraul. Res.* **2001**, *39*, 629–642. [CrossRef]
40. Bradshaw, P. Turbulence secondary flows. *Fluid Mech.* **1987**, *19*, 53–74. [CrossRef]
41. Blanckaert, K.; De Vriend, H.J. Secondary flow in sharp open-channel bends. *J. Fluid Mech.* **1999**, *498*, 353–380. [CrossRef]
42. Rozovskii, I.L. *Flow of Water in Bends of Open Channels*, 2nd ed.; Academy of Sciences of the Ukrainian SSR: Jerusalem, Israel, 1957.

43. Einstein, H.A.; Harder, J.A. Velocity distribution and the boundary layer at channel bends. *Eos Trans. Am. Geophys. Union* **1954**, *35*, 114–120. [CrossRef]
44. Bridge, J.S. *Rivers and Floodplains*; Blackwell Publishing: Hoboken, NJ, USA, 2009.
45. Blanckaert, K.; Duarte, A.; Chen, Q.; Schleiss, A.J. Flow processes near smooth and rough (concave) outer banks in curved open channels. *J. Geophys. Res. Earth Surf.* **2012**, *117*. [CrossRef]

9

River Bathymetry Model based on Floodplain Topography

Ludek Bures [1,*], Petra Sychova [1], Petr Maca [1], Radek Roub [1] and Stepan Marval [2]

1 Department of Water Resources and Environmental Modeling, Faculty of Environmental Sciences, Czech University of Life Sciences Prague, Kamycka 1176, 165 21 Prague 6, Suchdol, Czech Republic; sychova@fzp.czu.cz (P.S.); maca@fzp.czu.cz (P.M.); roub@fzp.czu.cz (R.R.)

2 Research Institute for Soil and water Conservation, Zabovreska 250, 156 27 Prague 5, Czech Republic; marval.stepan@vumop.cz

* Correspondence: Buresl@fzp.czu.cz.

Abstract: An appropriate digital elevation model (DEM) is required for purposes of hydrodynamic modelling of floods. Such a DEM describes a river's bathymetry (bed topography) as well as its surrounding area. Extensive measurements for creating accurate bathymetry are time-consuming and expensive. Mathematical modelling can provide an alternative way for representing river bathymetry. This study explores new possibilities in mathematical depiction of river bathymetry. A new bathymetric model (Bathy-supp) is proposed, and the model's ability to represent actual bathymetry is assessed. Three statistical methods for the determination of model parameters were evaluated. The best results were achieved by the random forest (RF) method. A two-dimensional (2D) hydrodynamic model was used to evaluate the influence of the Bathy-supp model on the hydrodynamic modelling results. Also presented is a comparison of the proposed model with another state-of-the-art bathymetric model. The study was carried out on a reach of the Otava River in the Czech Republic. The results show that the proposed model's ability to represent river bathymetry exceeds that of his current competitor. Use of the bathymetric model may have a significant impact on improving the hydrodynamic model results.

Keywords: DEM; hydrodynamic modelling; river bathymetry; floods; Bathy-supp

---

1. Introduction

Knowledge of terrain morphology is crucial for the hydrodynamic modelling of floods. The accuracy and applicability of hydrodynamic models is driven by the nature, availability, and accuracy of source topographic data [1–3].

The digital elevation models (DEMs) are required as a main input for hydrodynamic modelling. Topographic mapping is conventionally conducted by ground surveying. The main advantage of such a method is its high accuracy. Among the major limitations of measured data acquisition are its high cost and time-consuming data collection. Therefore, ground mapping is increasingly being replaced by remote-sensing methods. Radar and laser altimetry are among the most commonly used remote-sensing techniques [4–6].

A description of the river channel and its surrounding area is necessary to create a DEM for purposes of hydrodynamic modelling. The DEM must have precise vertical accuracy and spatial resolution [1,7]. DEMs obtained from satellites are commonly used on a global scale, but the spatial resolution of these models often does not meet the requirements of precise hydrodynamic modelling. The most commonly cited DEMs used for hydrodynamic modelling are the Advanced Spaceborne Thermal Emission and Reflection Radiometer (ASTER), and the Shuttle Radar Topography Mission (SRTM) [8–11].

Aerial laser scanning (ALS) can be another source of input data for a DEM intended for hydrodynamic modelling. This data source can produce DEMs with high spatial resolution [4]. The method is based on Light Detection and Ranging (LiDAR) technology. ALS methods usually use an infrared laser beam, which is unable to penetrate the water surface or scan the river bed [12]. This problem can be solved by using dual LiDAR (DiAL) technology, which uses a combination of two laser beams [13–15]. However, in some cases, this technology fails because the green laser beam (used for scanning water areas) is not able to penetrate an aquatic environment characterized by high turbidity [16–18].

The most commonly used method for creating a DEM with correct bathymetry is to merge topographic (e.g., ALS) data with another source of bathymetric data. Often used for this purpose are acoustic Doppler current profilers (ADCP), Sound Navigation and Ranging (SONAR) techniques, theodolites, and Global Positioning System (GPS) stations [3,19,20]. The main advantage of these methods is their high accuracy of the acquired data. The main disadvantage lies in the cost of the data's acquisition.

Interpolation approaches have been developed in an attempt to reduce the volume of measured data needed to represent river bathymetry [21,22]. Some interpolation techniques for the creation of bathymetry use cross-sections with various spacings as input data [21,23,24]. Nevertheless, measured data are still required.

Another option for representing river bathymetry can be to use mathematical modelling methods. In works of Dutch and Wang [25] and James [26], the cross-section shape of fluvial sediment deposits was estimated using analytical curves. Moramarco et al. [27] have used entropy-based methods for estimation of the cross-section. Roub et al. [28] introduced a bathymetric model, which estimates the river channel on the basis of hydraulic parameters. The river channel is thereby schematized into a trapezoidal shape. The flow area of this channel is determined using Chezy's equation.

This paper explores new possibilities in mathematical representation of river bathymetry. A new model of river bathymetry based on topographic data describing the area surrounding the river channel is proposed. In this work, the model's ability to represent actual bathymetry is evaluated. A two-dimensional (2D) hydrodynamic model is used to evaluate the influence of the bathymetric model on the results of hydrodynamic modelling. Also presented is a comparison of the proposed model with another state-of-the-art model, the Cross-section Solver ToolBox (CroSolver) [28].

2. Materials and Methods

The proposed bathymetry model is based on analytical curves. The curves are bent into the shapes of the cross-sections. The most precise values of the model parameters are needed in order to obtain the best description of the river bathymetry.

2.1. *Input Data for the Bathy-Supp Model*

Three types of input data are needed for a river bathymetric model. The first is a DEM, describing the floodplain topography. Also extracted from the DEM are the terrain characteristics, the altitudes for height adjustment of the new cross-sections, and a definition for the aquifer area of the river channel. The second type of input data is the design flow rate. The design flow rate is the flow rate at the time of elevation data acquisition. The last, but not least, type of input data is Manning's roughness coefficient for the river channel.

2.2. *Construction of the Bathy-Supp Model*

The proposed bathymetric model is constructed in four main steps:

1. The user defines the number and location of the new cross-sections (location of cross-section endpoints) from which a new bathymetry model will be composed.

2. Computation of the spatial terrain characteristics (predictor variables) derived from the floodplain DEM, and estimation of the model parameters m_1 and m_2.
3. Cross-section construction and transformation.
4. River bed reconstruction.

2.2.1. Location of the New Cross-Sections

After displaying the input DEM, it is possible to identify the position of the river channel itself as a no-data region, which thereby divides the model into two (or more) parts. In this no-data region, the user defines the number of new cross-sections and their locations. The distance between the first and last cross-section defines that part of the river for which new bathymetry will be estimated. Each cross-section is defined by two endpoints. The distance between these endpoints defines the width of the channel. The endpoints have coordinates X and Y. A DEM point with the lowest altitude is searched in the circular space around each endpoint. The altitude of this lowest point is used as the Z coordinate of the endpoint. The radius of the circular space is called the lowest point search radius. An example of the cross-section location can be seen in Figure 1.

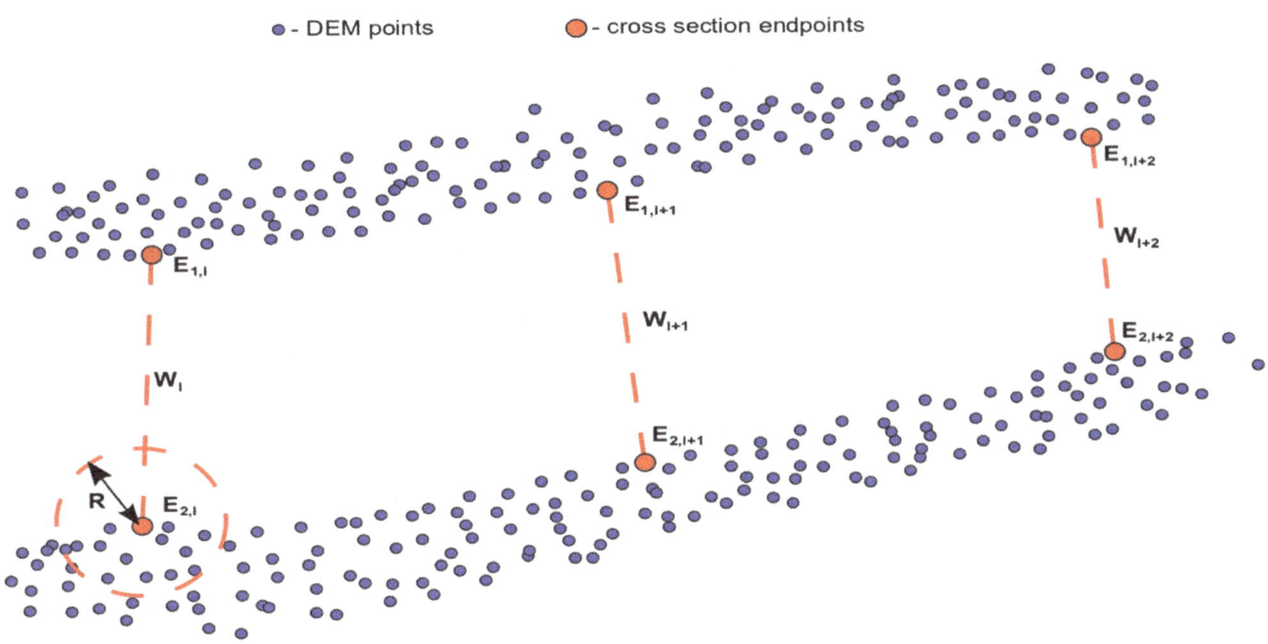

Figure 1. Defining the position of the new cross-sections. (width (W), endpoint (E), lowest point search radius (R)).

2.2.2. Explaining Model Parameters

Usually, the best-fit model parameters (m_1, m_2) are unknown. To establish them (without using an inverse problem-solving method), it is necessary to find the relationship between the search parameters and other explanatory variables. Spatial terrain characteristics can be used as possible explanatory variables [21,26]. The overall curvature, planar curvature, profile curvature, overall slope, slope in the x-direction, and slope in the y-direction were, in the present case, selected for predicting individual model parameters. The characteristics were determined as described by Zevenbergen and Thorne [29]. The source of terrain characteristics was the DEM input. Terrain characteristics were calculated for the all raster cells located behind the bank lines around the river. The terrain characteristics of the nearest raster cell were used to estimate the parameters of a given endpoint. As explanatory variables, the left bank terrain characteristics for model parameter m_1 and the right bank terrain characteristics for parameter m_2 were used.

Statistical learning methods were used for finding the dependence between model variables and terrain characteristics. Those statistical learning methods studied were multiple linear regression, extended linear regression, and random forest (RF).

A simple linear model (LM) is used for finding a linear relationship between a response and its predictor, and, in the case of multiple linear regressions, this relationship is based upon more than a single predictor. The LM is described in Equation (1):

$$y = \beta_0 + \sum \beta_i x_i + \varepsilon_i,\ \varepsilon_i \sim N\left(0, \sigma^2\right),\ i = 1, \ldots, M, \tag{1}$$

where βx are the linear parameters, x_i are predictor variables, ε_i is the error term, and M is the number of predictor variables. A mixed selection procedure, as described by Gareth et al. [30], was adopted for choosing the optimal number of variables.

An extended linear model with no random effects (GLS) was also used. This method extends linear regression with an ability to fit models with heteroscedastic and correlated within-group errors, but with no random effects [31]. The extended formula of the LM is described in Equation (2):

$$y = \beta_0 + \sum \beta_i x_i + \varepsilon_i,\ \varepsilon_i \sim N\left(0, \sigma^2 A^i\right),\ i = 1, \ldots, M, \tag{2}$$

where A_i are positive-definite matrices composed using variance and covariance matrices, β_x are the linear parameters, x_i are predictor variables, ε_i is the error term, and M is the number of predictor variables. Again, the mixed selection procedure, as described by Gareth et al. [30], was adopted for choosing the optimal number of variables.

Random forest (RF) is a combined machine learning method for classification and regression. This method is based on an ensemble of a regression tree (RT) algorithm. RT deals with tree structure by dividing the dataset into homogenous groups. That division is driven by some classification criterion, such as minimizing the variance of a given set of variables. In the case of RF, a dataset is divided into multiple sub-datasets by a bootstrap aggregating algorithm. For each sub-dataset, an RF of its own is constructed. This creates a group of random trees, termed RF. For each predictor variable, a measure of variable importance can be determined. Based on variable importance values, it is possible to decide which variables have significant impact for the response and which can be omitted [32].

All statistical analyses in this work were performed using statistical software R. The extended linear model with no random effects was applied using the package nlme [33], and the package randomForest [34] was used for random forests.

The coefficient of determination (R^2) was used to evaluate the reliability of model parameters. As a second quality assessment, vertical differences between models based on the best model parameters and a model based on estimated parameters were calculated. For this comparison, a similar approach is used in Section 2.1.

2.2.3. Cross-Section Construction and Transformation

Once the model parameters are estimated, new cross-sections can be constructed. A tested theoretical cross-section model Equation (3) is able to estimate the natural river cross-section on the basis of estimated parameters [35]. The studied theoretical model of the river cross-section is explained as:

$$z(d) = m_1 m_2 d^{m_1 - 1}\left(1 - d^{m_1}\right)^{m_2 - 1}, \tag{3}$$

where *z(d)* is the depth of water at a distance d from the left endpoint of the cross-section, while m_1 and m_2 are theoretical model parameters that are unique for each river cross-section. An example of the estimated shapes of the cross-section is shown in Figure 2.

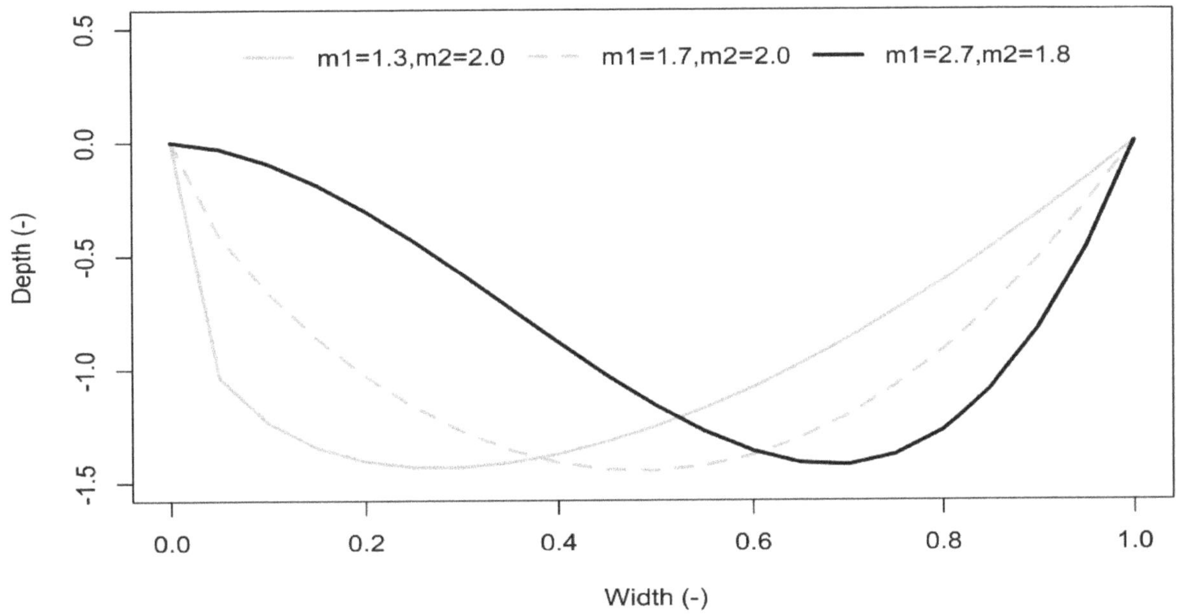

Figure 2. Example showing ability of the proposed bathymetric model to schematize a cross-section.

Due to the mathematical nature of the proposed model, a new cross-section width of 0–1 [35] must be selected in the first step. Note that the value 0 represents the left side of the cross-section. Additional points are inserted between the points 0 and 1. The user decides upon the number of points to insert. Once new stationing is defined, the depth value for each cross-section point is computed by applying Equation (3). New cross-sections produced by the proposed bathymetric model are in normative state (width and flow area are equal to (1)). Width transformation is simple. Stationing of the new cross-section is multiplied by the distance (W) between its endpoints (Figure 1). For the flow area transformation, an adequate flow area must be identified. This adequate flow area defines the flow area of the river channel required for the transfer of the design flow. The adequate flow area is determined using Chezy's equation on the basis of Manning's roughness coefficient, water surface slope, and design flow rate. This adequate flow area is compared with the area of the new cross-section. If the adequate flow area is smaller than the cross-sectional area, then the depths of the cross-section points are multiplied by the area multiplication parameter. This step is repeated until the adequate flow area is equal to or less than the area of the cross-section. Manning's coefficient, design flow rate, and area multiplication parameter are the input parameters. The water longitudinal surface slope and water surface elevation of each cross-section are extracted from the DEM. A similar approach had been used in the work of Roub et al. [28].

The final transformation step is to add the new cross-section into the coordinate system used. The XY coordinates of the first and last point of the cross-section correspond to the coordinates of the endpoints E_1 and E_2, for which the cross-section has been created. All internal points are placed on the line connecting the endpoints. Therefore, the coordinates of the internal points can be calculated based on the coordinates of the endpoints and its station value. The method of calculating the coordinates of internal points may vary depending on the coordinate system used. The lower of the altitudes of both endpoints is determined as the water level of this cross-section. The altitude (Z value) of each point is obtained by subtracting the water level and its depth. All cross-section points have coordinates X, Y, and Z and stationing after transformation.

2.2.4. River Bed Reconstruction

To create a three dimensional (3D) bathymetric model composed of isolated linear structures, such as cross-sections, spatial interpolation between these cross-sections must be used. Many different bed reconstruction algorithms are available in the literature [23,24,36,37]. In this contribution, we adopted the approach of Caviedes–Voullième [23]. This bed reconstruction algorithm is based upon cubic Hermite splines (CHS). Spline trajectory connects two depth points in two consecutive cross-sections, and it is driven by their normal vectors. The spatial interpolation (X, Y) of new bathymetric points is performed using CHS, and linear interpolation is performed for the height (Z). For a more detailed description, see the work of Caviedes–Voullième [23].

2.3. Model Suitability

Global optimization methods are used for estimating the best model parameters. The global optimization schemes are based on heuristics inspired by natural processes. Methods of differential evolution [38] are used for determining the solutions of inverse problems related to the parameters of a mathematical model of river bed surfaces. Differential evolution is a population-based stochastic optimization search algorithm that iteratively estimates a candidate solution with regard to a given measure of quality [38]. The analyzed objective function is a least squares method, determining the differences between the depths of the real cross-sections and the modelled cross-sections. We employed the *best1bin* algorithm in this work [38]. Analysis of the model's suitability was made in the C++ programming environment.

Model Suitability Evaluation

For evaluating the model's suitability, a vertical differences comparison between the model cross-sections (with the best model parameters) and the measured cross-sections (see Section 2.5.2) was made. The root mean square error (RMSE) and the mean absolute error (MAE) were calculated for this purpose. The equations for these evaluation criteria are as follows:

$$RMSE = \sqrt{\frac{1}{N}\sum_{i=1}^{N}(Elev_{MOD} - Elev_{REF})^2}, \quad (4)$$

$$MAE = \frac{1}{N}\sum_{i=1}^{N}|Elev_{MOD} - Elev_{REF}|, \quad (5)$$

where $Elev_{MOD}$ represents the elevation (m) obtained from each model cross-section and $Elev_{REF}$ represents the equivalent reference point obtained from the measured cross-section (see Section 2.5.2). N represents the number of the cross-section points.

2.4. Case Study Area

A part of the Otava River in the Czech Republic was chosen as an area of interest (Figure 3). The full length of the Otava River is 111.7 km with the basin area of 3840 km^2. The studied river reach is located into the lower part of the river (between 22.83 and 24.58 river km) and is 1.75 km long. The average depth of the river reach fluctuates around 1 m. The average bankfull width is between 22.8 and 52.7 m. The average annual flow rate is 23.4 m^3/s. The average water level is 354.84 m above sea level. The flow rates for N–year floods in the Otava River reach are shown in Table 1. $Q_{a.a.}$ denotes the average annual discharge.

Table 1. The flow rates for N-year floods in the Otava River reach [39].

N-year Floods	$Q_{a.a.}$	Q_1	Q_5	Q_{10}	Q_{50}	Q_{100}
Flow rate (m^3/s)	23.4	146	300	394	680	837

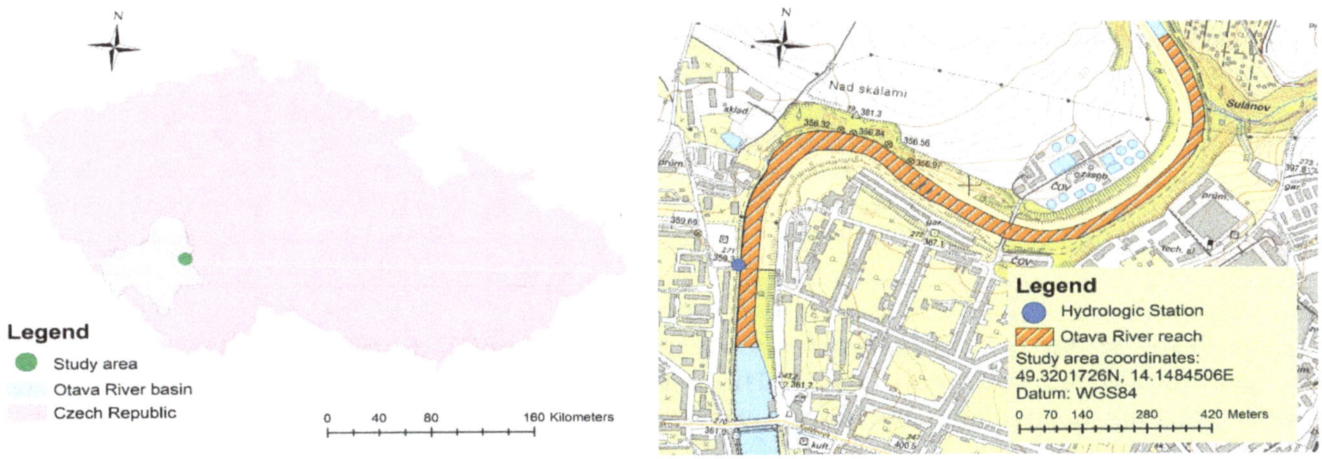

Figure 3. The study area: Otava River reach.

2.5. *Ground and Bathymetry Data*

This section of the paper describes the topographic source data and bathymetry data source used for identification of the model parameters and creating the DEMs.

2.5.1. Aerial Laser Scanning Data

The technology used for ALS data collection in this study was aerial laser scanning with one infrared laser beam. These ALS data were provided by the Czech Office for Surveying, Mapping, and Cadastre. Provided data were in the form of the list of elevation points. The point density per square meter varies depending on the slope of the terrain described. The point density starts with one point (flat area) and ends with dozens of points (steep area). Vertical data accuracy is +/−0.18 m in open areas and +/−0.3 m in forest areas. [40].

2.5.2. Acoustic Doppler Current Profiler Data

The data obtained by using an ADCP measurement device have the depth range 0.2–80 m and accuracy of +/−1% of the depth. The ADCP technology was successfully applied in recent works [18,41,42]. The data were sampled as a set of single cross-sections. The distance between these cross-sections was approximately 5 m. The distance between points within the cross-sections was less than 0.5 m. A total of 375 cross-sections were measured. These data were used for testing the suitability of the bathymetric model as well as for determining the best model parameters.

2.5.3. Compared DEMs

A total of four DEMs, based upon different sources of bathymetric data, were compared in this practical demonstration. The first DEM (ALS) was created only from the ALS data. It contains no bathymetric data and represents models of that sort produced by remote-sensing techniques.

The second DEM (CRO) was composed using ALS data, which describe the floodplain, and data from the Cross-section Solver ToolBox (CroSolver) bathymetric model. CroSolver is a software tool for improving an existing DEM in the riverbed area. The newly formed channel is schematized as trapezoidal or rectangular. The Chezy equation is used to determine the channel flow area for the design flow rate. The depth of the new channel is determined on the base of the size of the flow area, channel width, and selected channel schematization [28]. Table 2 provides the settings overview for the CroSolver model.

Table 2. The Cross-section Solver ToolBox (CroSolver) software specific settings.

Parameter	Value
Calculation method	Longitudinal gradient
Lowest point search radius (m)	5
Manning's roughness (s/m$^{1/3}$)	0.031
Bank slope 1:m	2
Flow rate (m^3/s)	36.8
Water surface slope calculation (m)	100

The third DEM (BAT) was composed using ALS data, which describes the floodplain and the Bathy-supp model proposed in this paper. The cross-section distance for the bathymetry creation was about 100 m. Table 3 provides the settings overview for the proposed model.

Table 3. Bathy-supp model specific settings.

Parameter	Value
Lowest point search radius (m)	5
Manning's roughness (s/m$^{1/3}$)	0.031
Flow rate (m^3/s)	36.8

The last DEM (ADP) is a model composed using ALS data and ADCP data. This model is used as a reference model because it is based on measured data (i.e., on data with assured accuracy). Table 4 provides background information on the DEMs being compared.

Table 4. Overview of compared digital elevation models (DEMs).

Digital Elevation Model	Floodplain Data	Bathymetry Data	Resolution (m)
ALS	LiDAR	–	0.5
CRO	LiDAR	CroSolver	0.5
BAT	LiDAR	Bathy-supp	0.5
ADP	LiDAR	ADCP	0.5

2.5.4. DEM evaluation

For evaluation of the DEM quality, the vertical differences between compared DEMs (ALS, CRO, BAT), and the reference DEM (ADP) were assessed. This comparison was made for all corresponding cells in the river channel area. Comparison of the models in the floodplain is not relevant. Identical data were used for its description. RMSE Equation (4) and MAE Equation (5) values were employed for evaluation.

2.6. *Hydrodynamic Modelling*

All hydrodynamic simulations were performed using a 2D HEC-RAS model [43]. HEC-RAS 2D allows computations for steady and unsteady flow conditions. The hydrodynamic results are determined on the base of solving of 2D Saint–Venant equations. Many authors successfully employed HEC-RAS 2D for hydrodynamic simulations [44,45].

In this study, hydrodynamic simulations for four different DEMs were made (see Section 2.5.3). For each DEM, four simulations were made with chosen N-year flow rates. Manning's roughness coefficients were set separately for each land cover category. Values in the range 0.035–0.10 s/m$^{1/3}$ were determined for inundations and the value for the main channel was 0.031 s/m$^{1/3}$. Selection of the Manning's values was verified by a calibration-verification process. The validation of model parameters was based on a flood event from December 2012, where, for discharges of 143 m^3/s, the recorded water level (hydrological station located in the model river reach) was equal to 356.43 m. The normal depth was used as the downstream boundary condition. The results of the basic hydrodynamic model (topography source was ADP DEM) for the given discharge provided a difference of 2 cm in the water level. That verified the accuracy of the model setup. All simulations began from a minimal start discharge, which then grew until eventually becoming steady [18,46,47]. Table 1 shows the selected N-year floods that were used as boundary conditions.

Water surface elevation (WSE) and Inundation areas (IAs) were evaluated to determine the influence of channel bathymetry representation. WSE was evaluated by comparing the differences between vertical measurements for the raster-based WSE (ALS, CRO, BAT) and those of the reference WSE (ADP). The RMSE (Equation (4) and MAE (Equation (5) values were used in the evaluation.

For IAs, the following compliance criterion was used:

$$IA_{dif} = \frac{|IA_{DEM} - IA_{REF}|}{IA_{REF}} \times 100 \tag{6}$$

where IA_{dif} represents the difference in inundation areas (%), IA_{DEM} represents the inundation area (km^2) of compared models (ALS, CRO, BAT), and IA_{REF} represents the inundation area of the ADP model (km^2).

3. Results and Discussion

3.1. Bathymetric Model Suitability

The ability of the proposed bathymetric model to schematize the measured cross-section was evaluated. The best model parameters were used for this purpose. The parameter range m_1 ranged from 1.0202 to 1.8219, and for the m_2 parameter from 1.0929 to 2.0376. Overall, 375 measured cross-sections were evaluated. The mean RMSE and MAE values were 0.16 m and 0.11 m, respectively. The variability of the RMSE and MAE values is shown in Figure 4.

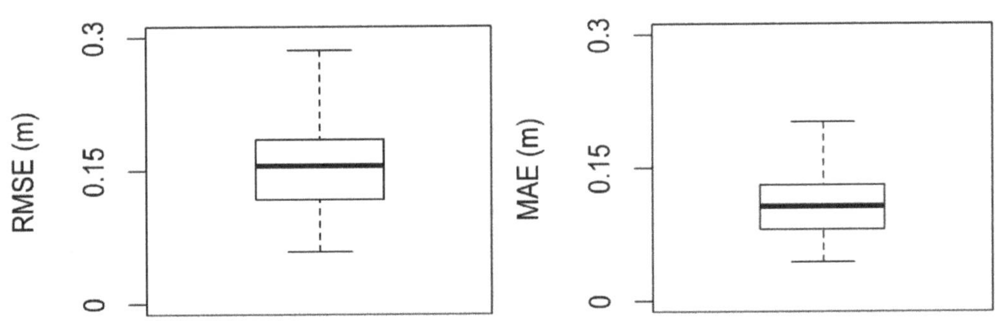

Figure 4. Comparing ability of the bathymetric model (with ideal parameters) to represent a measured cross-section. Shown are variances of root mean square error (RMSE) and mean absolute error (MAE) values.

3.2. Explaining Bathymetric Model Parameters

High-quality coefficient estimation is a prerequisite for correct bathymetric model creation. Therefore, various methods for its estimation were evaluated. For LM and GLS methods, overall curvature, planar curvature, overall slope, slope in the x direction, and slope in the y direction were considered as predictor variables. For the RF method, overall curvature, planar curvature, profile curvature, overall slope, slope in the x direction, and slope in the y direction were considered. LM and GLS methods provided similarly poor results. In contrast, the RF method produced a good result. Table 5 provides an overview of determination coefficients for the compared techniques. In view of these findings, the RT method was adopted for creating the bathymetric model.

Table 5. Coefficients of determination for the best-fit model parameters in comparing the estimation techniques linear model (LM), the extended linear model with no random effects (GLS), and the random forest (RF).

	LM	GLS	RF
R^2 (m_1)	0.145	0.161	0.918
R^2 (m_2)	0.085	0.118	0.914

Differences in parameter estimation quality may be due to nonlinear relationships in the data structure. More detailed analysis of the suitable regression structures and evaluation of its results are planned in follow-up research.

3.3. DEM Comparison

This comparison was made for cross-sections extracted from compared DEMs. The smallest divergence from the reference model (ADP) was achieved by the BAT model. The CRO model is unable to closely simulate the depth variability across the cross-section. This is due to the trapezoidal schematization of the channel. [28]. The ALS model almost completely neglects the riverbed area, which is specific for the sampling method used [4,12,40,48]. A graphical comparison is shown in Figure 5.

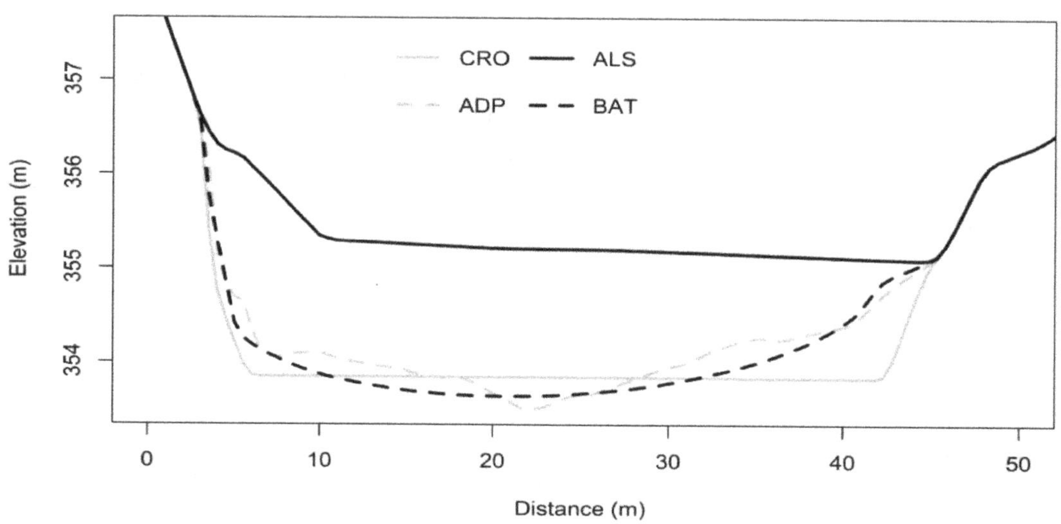

Figure 5. Visual comparison: Random cross-section derived from evaluated digital elevation models.

Table 6 describes the RMSE and MAE values achieved for the river channel (raster comparison). Both evaluated errors for the ALS model are >1. Note that the mean water depth in the channel at the time of acquiring ALS data was around 1 m. This suggests that the ALS model neglected almost the entire flow area of the river channel [16,20,28,42]. The RMSE and MAE values for the CRO model were 0.46 and 0.36 m, respectively. The smallest error values, RMSE 0.30 and MAE 0.23 m, were achieved by the BAT model. The raster cell difference variability is shown in Figure 6. Greater difference variability was achieved by the ALS model, and the smallest variability was achieved by the BAT model.

Table 6. RMSE and MAE errors for the compared DEMs (channel comparison).

	ALS	CRO	BAT
RMSE (m)	1.19	0.46	0.30
MAE (m)	1.06	0.36	0.23

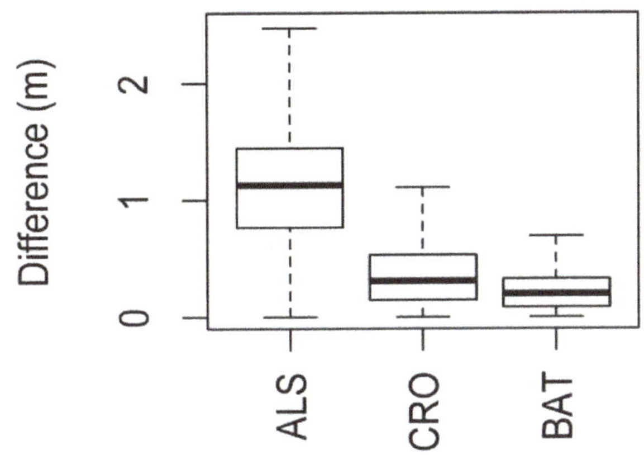

Figure 6. Variance of raster cell differences among compared digital elevation models.

Channel difference maps are shown in Figure 7. The biggest differences can be seen for the ALS model. The largest difference values are located along the thalweg (or stream centerline) location. Indeed, description of the river bottom was unrealistic. The error map presented for the CRO model identifies some areas with high cell difference values. These areas are located near embankment areas. The differences would probably be less significant if the cross-section were actually to have a trapezoidal shape (reflecting technical modification of the channel). For a natural river, the differences may be more significant. The BAT difference map provides the best results. Even here, however, it is possible to identify areas with a poor match. In this case, there was an excessive recess of one cross-section in the BAT model. The water surface slope derived from inundation data was slightly underestimated. This may be due to the accuracy of the data used to describe the floodplain [40].

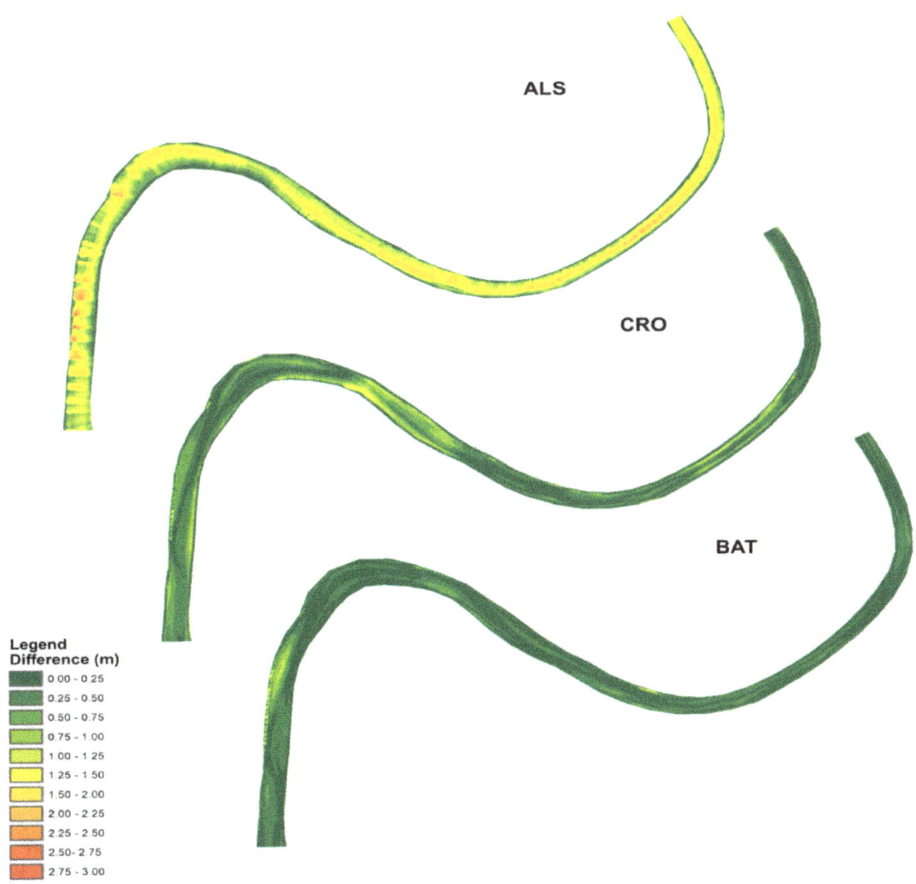

Figure 7. Error maps comparing evaluated digital elevation models (ALS, CRO, BAT) and the reference model (ADP).

3.4. Thalweg Comparison

The ALS model provided a poor match with the ADP model. Thalweg derived from the ALS model was shifted toward the water surface, where it fluctuated. Thus, it is evident that almost the entire flow area was neglected [16,20]. The CRO model had higher deviations relative to the BAT model. These deviations may occur in places where the channel narrows or expands locally. This is due to the fact that the local changes in flow velocity are not reflected in the CroSolver software [28]. The graphical comparison is shown in Figure 8, and the results of the numerical comparison of thalwegs are shown in Table 7.

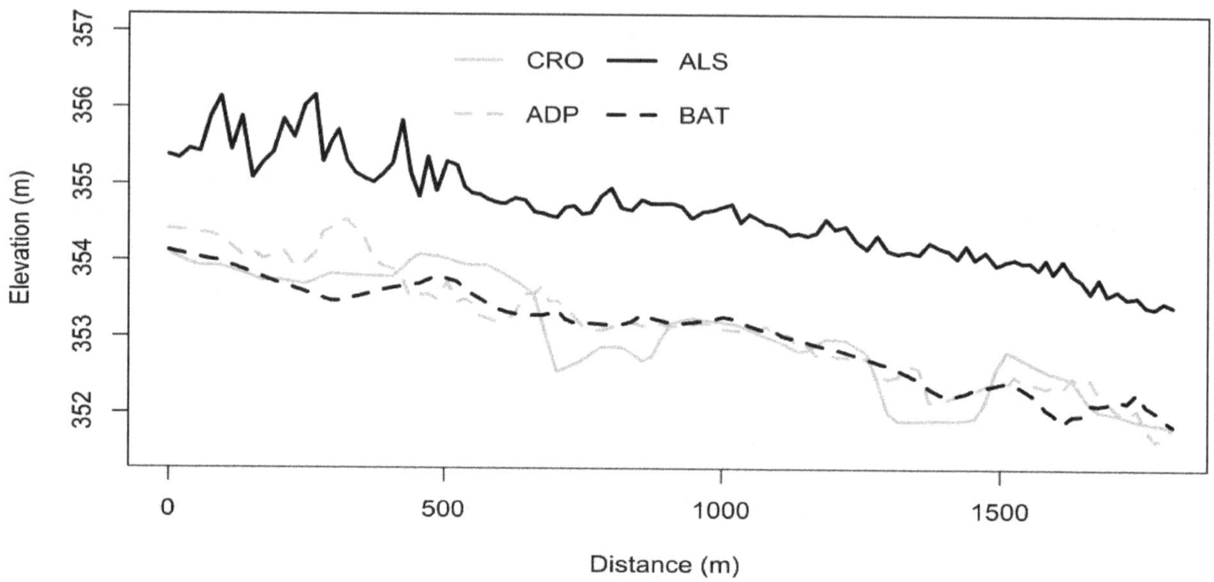

Figure 8. Visual comparison of thalwegs.

Table 7. RMSE and MAE values for the compared DEMs (thalweg comparison).

	ALS	CRO	BAT
RMSE (m)	1.52	0.37	0.30
MAE (m)	1.49	0.31	0.21

3.5. Water Surface Elevations Comparison

Figure 9 presents the variability in WSE errors between the compared DEMs (ALS, CRO, BAT) and the ADP DEM. The greatest deviations in WSE were seen in the ALS model, which manifested significant overestimation of WSE at all modelled flow rates. At flow $Q_{a.a.}$, the RMSE was more than 1.2 m, although the WSE variability was comparable to that for other models. For models CRO and BAT, the medians of the differences, as well as their variability, decreased with the increasing flow rate. The BAT model provided the smallest median values and the smallest variance for all rated flows. Table 8 presents RMSE and MAE values for WSE comparison.

Table 8. RMSE and MAE values for WSE comparison.

	Model	$Q_{a.a.}$	Q_1	Q_{10}	Q_{100}
RMSE (m)	ALS	1.24	0.87	0.80	0.85
	CRO	0.15	0.14	0.10	0.07
	BAT	0.13	0.06	0.04	0.03
MAE (m)	ALS	1.24	0.87	0.79	0.84
	CRO	0.13	0.12	0.09	0.05
	BAT	0.10	0.04	0.03	0.02

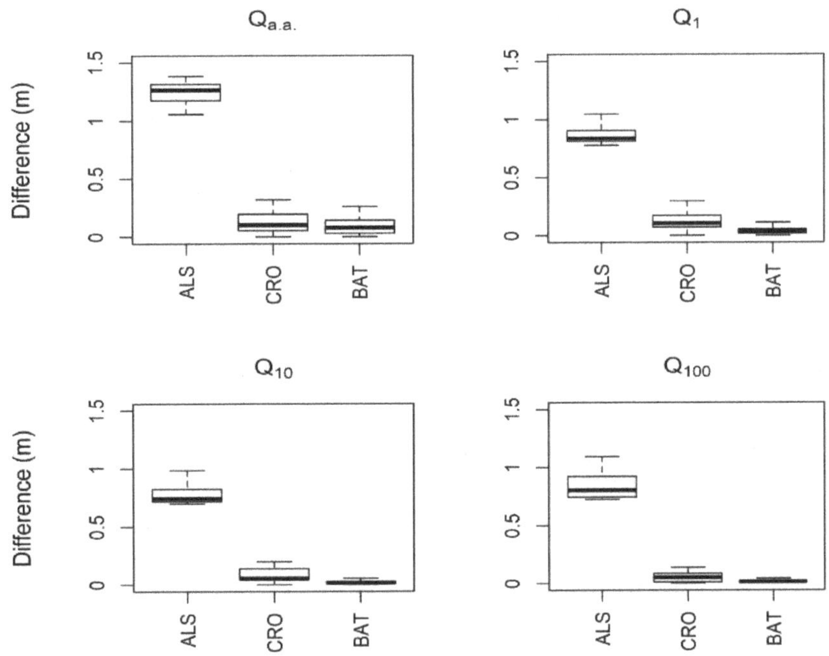

Figure 9. Variance of raster cell differences in water surface elevation between the compared digital elevation models (ALS, CRO, BAT) and the reference model (ADP).

3.6. *Inundation Areas Comparison*

Table 9 presents a comparison of inundation areas. The ALS model in all cases overestimated inundation vis-à-vis the reference ADP model. This overestimation was >50% in the case of $Q_{a.a.}$ These differences were diminishing with increasing flow, but nevertheless were >20% for the Q_{100} flow rate. Similar results had been presented in the works of Bures et al. [18] and Roub et al. [42]. In both those cases, software-modified DEMs (CRO, BAT) provided better results than the ALS model. The CRO model underestimated the ADP model values by as much as 4.5%. The BAT model underestimated them by as much as 3.8%. The fact that CRO and BAT consistently underestimate the results may reflect their similar model settings. The flow area for model cross-section transformation was set the same for the two models.

Table 9. Inundation area (IA) differences for compared DEMs.

	Model	$Q_{a.a.}$	Q_1	Q_{10}	Q_{100}
Inundation Area (km^2)	ADP	0.0540	0.0911	0.1288	0.1608
	ALS	0.0830	0.1182	0.1415	0.1953
	CRO	0.0536	0.0870	0.1277	0.1587
	BAT	0.0519	0.0898	0.1283	0.1598
Difference in Area (%)	ADP	-	-	-	-
	ALS	53.83	29.73	9.84	21.49
	CRO	0.6	4.5	0.87	1.27
	BAT	3.76	1.44	0.35	0.62

Overall, the ALS model showed the poorest results. That is because the technology used is unable to capture terrain under water. It was a LiDAR technology that, in our case, used an infrared laser beam [40]. Similar poor results could be expected from other remote-sensing methods [8–11]. The error produced by these models depends on the size of the neglected flow area of the river channel [18,42]. Although differential absorption LiDAR (DiAL) technologies may constitute exceptions to this general case [14,15], even using DiAL technology does not ensure any increase in accuracy. That is particularly the case when a high-turbidity river channel is measured [16–18,20].

The ALS model used in this study is a basic DEM, representative of the group of other remote-sensing models.

In the BAT model, the longitudinal water surface slope and water surface elevations were determined for all constructed cross-sections. Both of these parameters are derived from DEM inputs. The uncertainty in determining these parameters depends upon the accuracy of the data from which they are derived. It is possible, however, to assume that the development of remote-sensing techniques will reduce the uncertainty in determining water surface slope and elevations. Uncertainty in determining roughness parameters is the same for both models (CRO, BAT) and depends upon user experience. Further research in the field of model parameter estimation may bring improvement in this area. More appropriate regression relationships can be found, as can other explanatory variables. These variables might, for example, be the curvature of the flow centerline [21,38], mutual tilt of consecutive cross-sections [23], or ratio of the bank line distances.

In comparing the quality of the DEMs, we can see that the BAT model produced better results than the CRO model. The extent of the differences between the CRO and BAT models will probably depend upon the nature of the channel described. With respect to its adaptability, it can be assumed that the BAT model will produce better results in the case of natural channels. If the river channel has been technically adjusted (into trapezoidal shape), it can be assumed that the BAT model will produce similar results to a CRO model.

In the present bathymetric model, the interpolation mechanism introduced by Caviedes–Voullième [23] was used. Several interpolation methods can alternatively be used for this step [24,37,38]. Which of these methods will yield more reliable results remains an open question.

In the presented study, the importance of river bathymetry was evaluated mainly from the perspective of flood modeling. However, the proposed bathymetric model can find its place in other scientific disciplines, such as understanding the morphological changes of the river segment or understanding the hydrodynamic behavior of the river.

4. Conclusions

In this study, the new theoretical bathymetric model Bathy-supp was presented. The model is based on the analytical curves, which schematizes the river cross-sections. The shape of the analytical curves is driven by the floodplain topography. In the case study, the practical usability of the model was evaluated.

The vertical differences in the model's ability to represent the cross-sections were 0.16 m for RMSE and 0.11 m for MAE. For estimation of the model parameters, the three regression methods were used. The best parameter estimates were provided by the RF method. However, the regression relationships between model parameters and topographical characteristics need to be further investigated.

The study's results also show that, in the DEM quality assessment, the BAT model provides the best results in the river channel and thalweg representation. When assessing the WSE and IA, the BAT model provided better results than the CRO model, especially for high design flows rates.

Both DEMs created by merging ALS data with mathematical bathymetric models (CRO and BAT) provided significantly better results. The extent of this improvement was directly proportional to the size of the flow area neglected by using the ALS data itself.

The newly proposed Bathy-supp model was also compared with the state-of-the-art CroSolver model. The results showed that the Bathy-supp model describes the river bathymetry more accurately than the CroSolver model. This is because the analytical curves used in the Bathy-supp tool are more able to simulate the individual cross-sections of the channel than the trapezoidal shapes used by the CroSolver tool. Another advantage of the Bathy-supp model is in determining the flow area for each single cross-section from which the model is composed. This allows the model to take into account local narrows or expansions of the river channel.

The merging of bathymetric models with other types of remote-sensing data can significantly improve the DEMs, thereby enhancing their practical usefulness. Hence, the methods for mathematical schematizing of riverbeds should be further studied and improved.

The Bathy-supp model as an appropriate mechanism to improve DEM quality when other bathymetry measurements that are not available can be used. However, it will never be able to fully replace large bathymetric measurements.

Author Contributions: Conceptualization, L.B. and P.M.; methodology, L.B., P.M., and R.R.; validation, L.B., P.S., and R.R.; investigation, P.S.; data curation, L.B.; writing—original draft preparation, L.B.; writing—review and editing, P.S., S.M.; visualization, L.B. and S.M.; supervision, P.M. and R.R.; funding acquisition, R.R.

References

1. Horritt, M.S.; Bates, P.D. Effects of spatial resolution on a raster based model of flood flow. *J. Hydrol.* **2001**, *253*, 239–249. [CrossRef]
2. Horritt, M.S.; Bates, P.D.; Mattinson, M.J. Effects of mesh resolution and topographic representation in 2D finite volume models of shallow water fluvial flow. *J. Hydrol.* **2006**, *329*, 306–314. [CrossRef]
3. Merwade, V.; Cook, A.; Coonrod, J. GIS techniques for creating river terrain models for hydrodynamic modeling and flood inundation mapping. *Environ. Model. Softw.* **2008**, *23*, 1300–1311. [CrossRef]
4. Baltsavias, E.P. Airborne laser scanning: Basic relations and formulas. *ISPRS J. Photogramm. Remote Sens.* **1999**, *54*, 199–214. [CrossRef]
5. Lyzenga, D.R.; Malinas, N.R.; Tanis, F.J. Multispectral bathymetry using a simple physically based algorithm. *IEEE Trans. Geosci. Remote Sens.* **2006**, *44*, 2251–2259. [CrossRef]
6. Gao, J. Bathymetric mapping by means of remote sensing: Methods, accuracy and limitations. *Prog. Phys. Geogr.* **2009**, *33*, 103–116. [CrossRef]
7. Bates, P.D.; De Roo, A.P.J. A simple raster-based model for flood inundation simulation. *J. Hydrol.* **2000**, *236*, 54–77. [CrossRef]
8. Schumann, G.; Matgen, P.; Cutler, M.E.J.; Black, A.; Hoffmann, L.; Pfister, L. Comparison of remotely sensed water stages from LiDAR, topographic contours and SRTM. *ISPRS J. Photogramm. Remote Sens.* **2008**, *63*, 283–296. [CrossRef]
9. Ali, A.M.; Solomatine, D.P.; Di Baldassarre, G. Assessing the impact of different sources of topographic data on 1-D hydraulic modelling of floods. *Hydrol. Earth Syst. Sci.* **2015**, *19*, 631–643. [CrossRef]
10. Domeneghetti, A. On the use of SRTM and altimetry data for flood modeling in data-sparse regions. *Water Resour. Res.* **2016**, *52*, 2901–2918. [CrossRef]
11. Yan, K.; Di Baldassarre, G.; Solomatine, D.P.; Schumann, G.J.-P. A review of low-cost space-borne data for flood modelling: Topography, flood extent and water level. *Hydrol. Process.* **2015**, *29*, 3368–3387. [CrossRef]
12. Casas, A.; Benito, G.; Thorndycraft, V.R.; Rico, M. The topographic data source of digital terrain models as a key element in the accuracy of hydraulic flood modelling. *Earth. Surf. Proc. Landf.* **2006**, *31*, 444–456. [CrossRef]
13. Irish, J.L.; Lillycrop, W.J. Scanning laser mapping of the coastal zone: The SHOALS system. *ISPRS J. Photogramm. Remote Sens.* **1999**, *54*, 123–129. [CrossRef]
14. Hilldale, R.C.; Raff, D. Assessing the ability of airborne LiDAR to map river bathymetry. *Earth. Surf. Proc. Landf.* **2008**, *33*, 773–783. [CrossRef]
15. Mandlburger, G.; Hauer, C.; Wieser, M.; Pfeifer, N. Topo-Bathymetric LiDAR for Monitoring River Morphodynamics and Instream Habitats-A Case Study at the Pielach River. *Remote Sens.* **2015**, *7*, 6160–6195. [CrossRef]
16. Skinner, K.D. *Evaluation of Lidar-Acquired Bathymetric and Topographic Data Accuracy in Various Hydrogeomorphic Settings in the Lower Boise River, Southwestern Idaho*; US Geological Survey: Reston, VA, USA, 2007.
17. Bailly, J.S.; Le Coarer, Y.; Languille, P.; Stigermark, C.J.; Allouis, T. Geostatistical estimations of bathymetric LiDAR errors on rivers. *Earth. Surf. Proc. Landf.* **2010**, *35*, 1199–1210. [CrossRef]
18. Bures, L.; Roub, R.; Sychova, P.; Gdulova, K.; Doubalova, J. Comparison of bathymetric data sources used in hydraulic modelling of floods. *J. Flood Risk Manag.* **2018**, e12495. [CrossRef]
19. Allouis, T.; Bailly, J.S.; Pastol, Y.; Le Roux, C. Comparison of LiDAR waveform processing methods for very shallow water bathymetry using Ra-man, near-infrared and green signals. *Earth. Surf. Proc. Landf.* **2010**, *35*,

640–650. [CrossRef]

20. Laks, I.; Sojka, M.; Walczak, Z.; Wrozynski, R. Possibilities of Using Low Quality Digital Elevation Models of Floodplains in Hydraulic Numerical Models. *Water* **2017**, *9*, 283. [CrossRef]
21. Merwade, V.M.; Maidment, D.R.; Goff, J.A. Anisotropic considerations while interpolating river channel bathymetry. *J. Hydrol.* **2006**, *331*, 731–741. [CrossRef]
22. Jha, S.K.; Mariethoz, G.; Kelly, B.F. Bathymetry fusion using multiple-point geostatistics: Novelty and challenges in representing non-stationary bedforms. *Environ. Model. Softw.* **2013**, *50*, 66–76. [CrossRef]
23. Caviedes-Voullième, D.; Morales-Hernández, M.; López-Marijuan, I.; García-Navarro, P. Reconstruction of 2D river beds by appropriate interpolation of 1D cross-sectional information for flood simulation. *Environ. Model. Softw.* **2014**, *61*, 206–228. [CrossRef]
24. Schäppi, B.; Perona, P.; Schneider, P.; Burlando, P. Integrating river cross-section measurements with digital terrain models for improved flow modelling applications. *Comput. Geosci.* **2010**, *36*, 707–716. [CrossRef]
25. Deutsch, C.V.; Wang, L. Hierarchical object-based stochastic modeling of fluvial reservoirs. *Math. Geol.* **1996**, *28*, 857–880. [CrossRef]
26. James, L.A. Polynomial and power functions for glacial valley cross-section morphology. *Earth. Surf. Proc. Landf.* **1996**, *21*, 413–432. [CrossRef]
27. Moramarco, T.; Corato, G.; Melone, F.; Singh, V.P. An entropy-based method for determining the flow depth distribution in natural channels. *J. Hydrol.* **2013**, *497*, 176–188. [CrossRef]
28. Roub, R.; Hejduk, T.; Novák, P. Automating the creation of channel cross-section data from aerial laser scanning and hydrological surveying for modeling flood events. *J. Hydrol. Hydromech.* **2012**, *60*, 227–241. [CrossRef]
29. Zevenbergen, L.W.; Thorne, C.R. Quantitative analysis of land surface topography. *Earth. Surf. Proc. Landf.* **1987**, *12*, 47–56. [CrossRef]
30. Gareth, J. *An Introduction to Statistical Learning: With Applications in R*; Springer: New York, NY, USA, 2010.
31. Pinheiro, J.; Bates, D. *Mixed-Effects Models in S and S-PLUS*; Springer: New York, NY, USA, 2006.
32. Breiman, L. Random forests. *Mach. Learn.* **2010**, *45*, 5–32. [CrossRef]
33. Pinheiro, J.; Bates, D.; DebRoy, S.; Sarkar, D.; Package'nlme'. Linear and Nonlinear Mixed Effects Models. Available online: http://www.cran.r-project.org/web/packages/nlme/nlme.pdf (accessed on 18 June 2018).
34. Liaw, A.; Wiener, M.; Package'randomForest'. Breiman and Cutlers Random Forests for Classification and Regression; Version 4.6–12. Available online: http://www.cran.r-project.org/web/packages/randomForest/randomForest.pdf (accessed on 18 June 2018).
35. Kumaraswamy, P. A generalized probability density function for double-bounded random processes. *J. Hydrol.* **1980**, *46*, 79–88. [CrossRef]
36. Merwade, V. *Creating River Bathymetry Mesh from Cross-Sections*; School of Civil Engineering, Purdue University: West Lafayette, Indiana, 2017. Available online: https://web.ics.purdue.edu/~vmerwade/research.html#river (accessed on 10 February 2019).
37. Dysarz, T. Development of RiverBox—An ArcGIS Toolbox for River Bathymetry Reconstruction. *Water* **2018**, *10*, 1266. [CrossRef]
38. Storn, R.; Price, K. Differential evolution—A simple and eficient heuristic for global optimization over continuous spaces. *J. Global. Optim.* **1997**, *11*, 341–359. [CrossRef]
39. CHMI. Evidence Card for Profile No. 127. Available online: http://hydro.chmi.cz/hpps/popup_hpps_prfdyn.php?seq=307049 (accessed on 8 February 2019).
40. Brazdil, K.; Belka, L.; Dusanek, P.; Fiala, R.; Gamrat, J.; Kafka, O. *The Technical Report to the Digital Elevation Model 5th Generation DMR 5G*; VGHMÚř Dobruška: ZÚ Pardubice, Czech Repubilc, 2012.
41. Nihei, Y.; Kimizu, A. A new monitoring system for river discharge with horizontal acoustic Doppler current profiler measurements and river flow simulation. *Water Resour. Res.* **2008**, *44*, 1–15. [CrossRef]
42. Roub, R.; Kurkova, M.; Hejduk, T.; Novak, P.; Bures, L. Comparing a hydrodynamic model from fifth generation dtm data and a model from data modified by means of crosolver tool. *AUC Geogr.* **2016**, *51*, 29–39. [CrossRef]
43. Brunner, G.W. *HEC-RAS River Analysis System 2D Modeling User's Manual*; US Army Corps of Engineers—Hydrologic Engineering Center: Davis, CA, USA, 2016. Available online: https://www.hec.usace.army.mil/software/hec-ras/downloads.aspx (accessed on 10 February 2019).
44. Quirogaa, V.M.; Kurea, S.; Udoa, K.; Manoa, A. Application of 2D numerical simulation for the analysis of

the February 2014 Bolivian Amazonia flood: Application of the new HEC-RAS version 5. *Ribagua* **2016**, *3*, 25–33. [CrossRef]

45. Maskong, H.; Jothityangkoon, C.; Hirunteeyakul, C. Flood hazard mapping using on-site surveyed flood map, HEC-RAS V. 5 and GIS tool: A case study of Nakhon Ratchasima municipality, Thailand. *Int. J. Geomate* **2019**, *16*, 1–8. [CrossRef]
46. Cook, A.; Merwade, V. Effect of topographic data, geometric configuration and modeling approach on flood inundation mapping. *J. Hydrol.* **2009**, *377*, 131–142. [CrossRef]
47. Di Baldassarre, G.; Schumann, G.; Bates, P.D.; Freer, J.E.; Beven, K.J. Flood-plain mapping: A critical discussion of deterministic and probabilistic approaches. *Hydrol. Sci.* **2010**, *55*, 364–376. [CrossRef]
48. Cavalli, M.; Tarolli, P. Application of LiDAR technology for rivers analysis. *Ital. J. Eng. Geol. Environ.* **2011**, *11*, 33–44. [CrossRef]

10

Capturing the Motion of the Free Surface of a Fluid Stored within a Floating Structure

Roman Gabl [1,2,*], Jeffrey Steynor [1], David I. M. Forehand [1], Thomas Davey [1], Tom Bruce [1] and David M. Ingram [1]

1 School of Engineering, Institute for Energy Systems, FloWave Ocean Energy Research Facility, The University of Edinburgh, Max Born Crescent, Edinburgh EH9 3BF, UK; Jeff.Steynor@ed.ac.uk (J.S.); D.Forehand@ed.ac.uk (D.I.M.F.); Tom.Davey@flowave.ed.ac.uk (T.D.); Tom.Bruce@ed.ac.uk (T.B.); David.Ingram@ed.ac.uk (D.M.I.)

2 Unit of Hydraulic Engineering, University of Innsbruck, Technikerstraße 13, 6020 Innsbruck, Austria

* Correspondence: Roman.Gabl@ed.ac.uk

Abstract: Large floating structures, such as liquefied natural gas (LNG) ships, are subject to both internal and external fluid forces. The internal fluid forces may also be detrimental to a vessel's stability and cause excessive loading regimes when sloshing occurs. Whilst it is relatively easy to measure the motion of external free surface with conventional measurement techniques, the sloshing of the internal free surface is more difficult to capture. The location of the internal free surface is normally extrapolated from measuring the pressure acting on the internal walls of the vessel. In order to understand better the loading mechanisms of sloshing internal fluids, a method of capturing the transient inner free surface motion with negligible affect on the response of the fluid or structure is required. In this paper two methods will be demonstrated for this purpose. The first approach uses resistive wave gauges made of copper tape to quantify the water run-up height on the walls of the structure. The second approach extends the conventional use of optical motion tracking to report the position of randomly distributed free floating markers on the internal water surface. The methods simultaneously report the position of the internal free surface with good agreement under static conditions, with absolute variation in the measured water level of around 4 mm. This new combined approach provides a map of the free surface elevation under transient conditions. The experimental error is shown to be acceptable (low mm-range), proving that these experimental techniques are robust free surface tracking methods in a range of situations.

Keywords: free surface measurement; optical motion capturing; tank test; wave gauge; sloshing

---

1. Introduction

Designs for floating structures that hold internal fluids with a free surface must be capable of withstanding both the internal and external fluid pressures. The movement and stability of the structure is also affected by both fluid bodies. To investigate these effects, researchers must quantify sloshing within a free floating structure, which requires new methods of free surface tracking that are robust and repeatable. It is important that these measurement techniques are benchmarked and quantitatively understood if employed to quantify the free surface in experimental investigations that will be used to validate numerical methods for predicting such interactions, especially with larger motions.

Comparable research questions on the interaction between two separated water bodies can be found in oscillating water column (OWC) wave energy converters [1,2] as well as in ship design [3–5], especially in connection with the transport of liquefied natural gas (LNG) [6]. Sloshing experiments are an essential part of the design process, a problem typically explored using closed tanks mounted on motion platforms [7–9]. In most cases, such investigations focus on the impact forces exerted on the walls with respect to different filling levels. Hence, pressure transducers are the key measurement instrument [10–12].

For very large floating structures (VLFS), experimental data and numerical solutions are available, which investigate the deformation of VLFS depending on different mooring types [13–15]. Lee et al. [16] included multiple inner tanks filled with water in a VLFS and presented a numerical approach based on finite element method. This approach exhibits good agreement with experimental data but is limited to small motions. An example application of the full-scale structure would be as buoyant energy storage [17,18]. The size of such a device is comparable to ships and large platforms.

The motion of the free surface is a key indicator used in the comparison of experimental results and numerical simulations. The precision and accuracy of the free surface measurement is important in the validation of the numerical results. In general, a high spatial density of measurement points from two different and ideally independent systems is preferable to provide redundancy and reduce uncertainty [19]. It is also important that any additional mass on the model be minimised so that mass balance is not affected by the instruments. Finally, the movement of the floating tank should not be impeded by the experimental and measurement systems.

Classical approaches to detect and follow free surfaces are based on surface piercing wave gauges [20–25] or ultrasonic/acoustic probes, which measure the delay of the ultrasonic signal echo caused by the density change at the free surface [26,27]. Submerged acoustic instruments can measure the fluid velocity beneath the free surface at laboratory scale [28,29] as well as for field applications [30]. Such point measurement devices are very robust and precise. The observation of visible markers is an option for laboratory experiments [31,32] and can also be applied for field measurements [33].

Gomit et al. [34] present a method to detect the free surface depending on the 2D velocity field in a towing tank. In that case, a horizontal laser sheet is observed by up to three cameras and this requires optical access from the bottom of the tank. Similarly, vertical laser sheets used for particle image velocimetry (PIV) can be applied to measure the velocities as well as to detect free surfaces [35–40]. Belden and Techet [41] combined this method for air and water to detect the surface. These mentioned approaches are limited to a two-dimensional section, which results in a single slice through the surface and delivers results similar to a Light Detection And Ranging (LIDAR) measuring system [42,43]. Weitberecht et al. [44] evaluate the accuracy of surface PIV, which use the free water surface as measurement plane without a laser. In this case the water depth is comparably shallow (approximately 4.6 cm) and the measurement of velocity vector field base on particles seeded continuously onto the water surface. Comparable big motion of the free surface in sloshing tanks leads to a high rate of mixing, which makes it harder to maintain the particle distribution. Akutina et al. [45] capture the movement of neutrally buoyant particles and address the problem of the measurement validation as well as reduce the influence of optical distortion.

Evers and Hager [46,47] present an approach to capture the curvature of the water surface caused by an impulse water wave in a tank. In their method, a light grid is projected on the free surface and the deformation is captured with cameras. The 2D images of the grid deformation are then transformed into a calculated 3D surface. This approach requires the addition of pigment to the water [48,49], which is unlikely to be a viable option when extending the approach to a large wave basin. Additionally, the floating structure would shadow the free surface and hence measurement near the structure walls would not be possible.

This paper presents a new approach to capture the free surface motion inside a floating tank with bespoke floating markers using a 3D motion capturing system. Comparable wave gauges have been previously used in wave run-up experiments on a slope [38,50,51] or on fixed cylinders [52] but in these cases the wave gauges were always stationary. Wave gauges made of copper tape are also investigated to determine if the water run-up within or around a floating body can be captured accurately even if the body orientation changes in relation to its initial configuration. Experiments with an open-top clear acrylic cylinder filled with water (Figure 1) under static and transient conditions have been conducted. The aim of this work is to evaluate these two approaches for a use in combination with a floating structure.

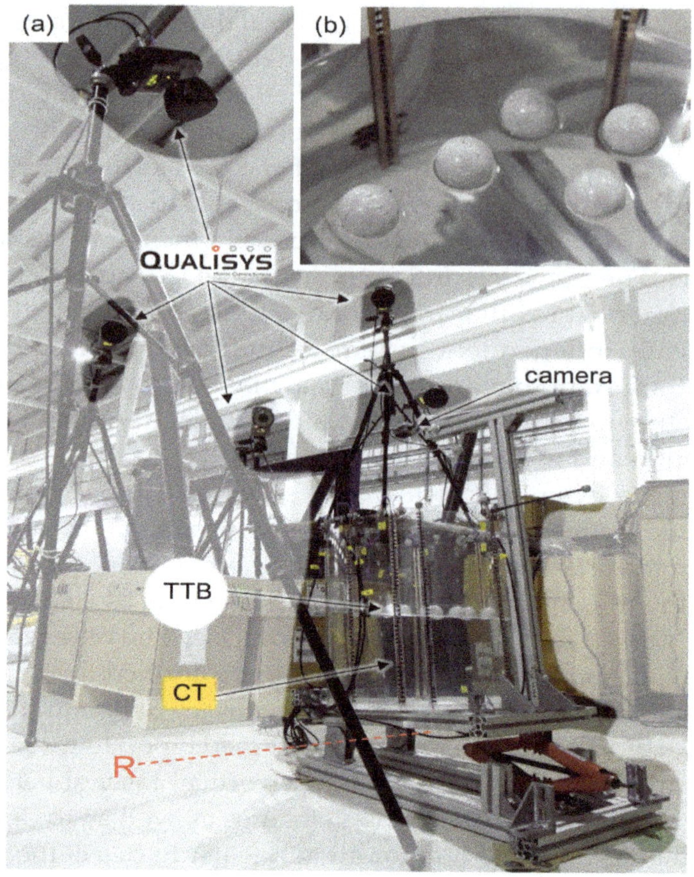

Figure 1. Experimental set-up—acrylic cylindrical tank mounted on an inclinable framework including table tennis balls (TTB) and wave gauges made of copper tape (CT)—rotation axis *R*—(**a**) overview of the inclined structure and (**b**) detail of the inner water surface.

2. Experimental Method

2.1. Overview

A cylindrical tank with an inner diameter D of 0.49 m is investigated (Figure 1). The transparent acrylic structure has a material thickness of 5 mm for the cylinder walls and 7 mm for the floor plate. If the cylinder is filled to its maximum height H of 0.5 m, it contains 94.28 L of water, ignoring any errors in circularity. For the presented experiments, the tank is fixed on a framework (rather than free floating), which allows an inclination around one fixed rotation axis R (Figures 1 and 2). The position of the structure is determined by a four camera Qualisys motion capturing system.

The water level in the tank is given by a measurement scale on the cylinder wall and captured by a camera, which is mounted at a constant position relative to the opening of the cylinder (Figure 1). This is used as a reference for the two measurement methods under investigation for the free surface motion: (a) floating infra-red reflective markers (Section 2.2) and (b) wave gauges made of conductive copper tape (Section 2.3). The first approach is based on the motion capturing system and is a new implementation, which enables this system to also capture a changing water surface. The investigation also aims to demonstrate that wave gauges made of light weight copper tape can be bonded to a moving body and deliver reliable measurements under different inclinations without the need for a recalibration.

2.2. Floating Infra-Red Reflective Markers

The movement of the containing cylindrical tank is captured with a commercial optical motion capture system from Qualisys. All six degrees of freedom (DoF) are determined based on the relative position of the markers, which are mounted on the rigid acrylic structure [53–56]. The Qualisys system is also capable of simultaneously tracking multiple independently moving reflective markers in the three translational degrees of freedom. Hence, the possibility of capturing the free water surface with additional floating markers randomly located inside the structure can be explored.

All known commercially available passive spherical reflective markers are either not suitable for wet conditions (i.e., they lose their reflectivity) or are not buoyant. Consequently, a bespoke solution is chosen to make each floating reflective marker from a standard table tennis ball. The table tennis ball (mass of 2.7 g, diameter 40 mm) serves as the basic body, which has enough buoyancy and visibility for the application. Reflective tape is applied to provide the reflectivity to infra-red light. This set-up is comparable to the taped underwater markers offered by Qualisys, but much lighter, water resistant and buoyant. A preliminary feasibility study showed that such a marker ball can be tracked by the Qualisys system whilst afloat on the surface and subject to wave motions in the University of Edinburgh's Curved Wave Tank [57] and in the FloWave basin, which has a diameter of 25 m and is also located at the University of Edinburgh [58–63].

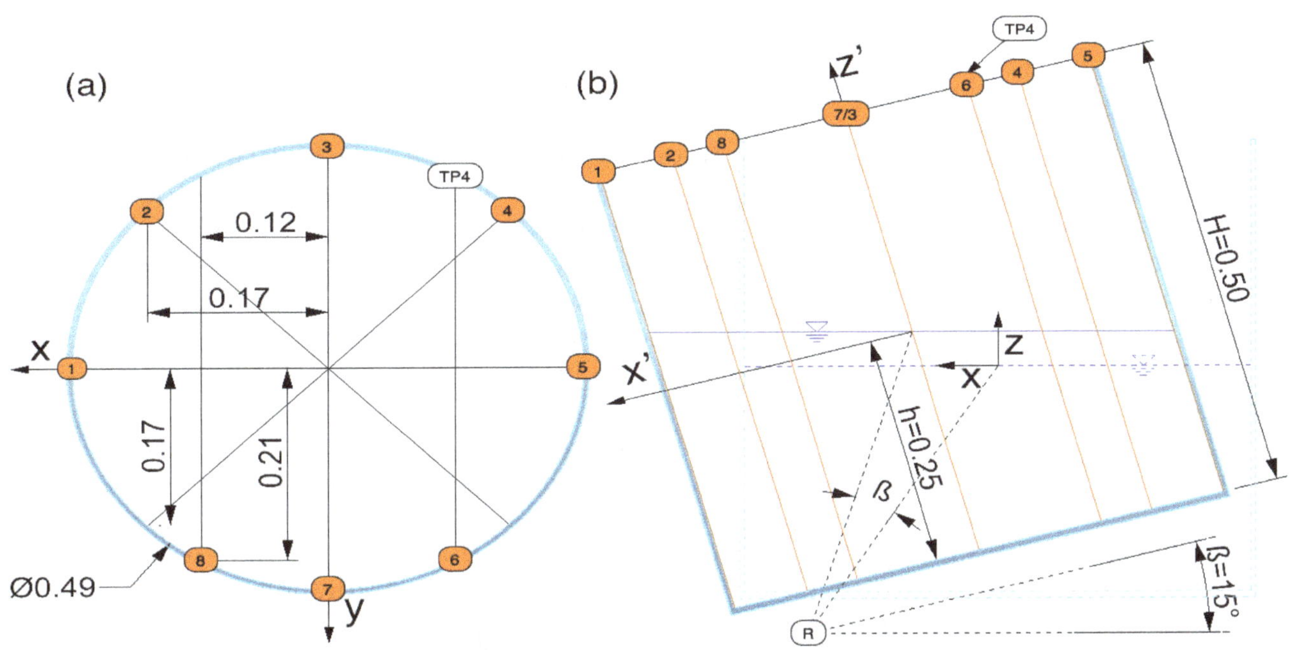

Figure 2. Basic geometry (**a**) plan view of the upright cylinder including the numbering of the wave gauges and (**b**) side view of the inclined geometry with an example pitch angle β—R is the rotation axis and $TP4$ the reference point introduced in Section 3.2—all dimensions in [m].

For the experiments, 12 marker balls are added to the water in the circular tank (Figure 1). They are left to freely float on the surface and are captured simultaneously with the rigid body markers. By distributing the balls well over the entire surface, nearly no clustering effects were observed. In the rare case that the balls do cluster, the system is capable of identifying correctly up to three markers, which may be in contact with each other. In post-processing, the markers have to be separately identified as the software cannot automatically recognise them. This is only possible for rigid bodies with a fixed distance between multiple markers. The identification of the marker balls has to be done only once for each measurement and further tracking is done automatically by the software. Alternatively, all unidentified trajectories can be exported and processed with additional software (for example Matlab). In this case, the initial fill level inside the cylinder can be chosen to autonomously identify the markers and separate from those, which are attached to the body.

As shown in Figure 1, the Qualisys system includes one Oqus 210c for video capturing and four Oqus 300+ cameras for marker tracking. The capture rate was set to 100 Hz for all tests.

Two sets of results from the motion capture system are presented here: (a) definition of the 6 DoF motion of the rigid body and (b) the coordinates of the free floating markers, which are recorded relative to the global coordinate system.

2.3. Copper Tape Resistive Wave Gauges

Wave gauges are a very reliable and accurate measurement system to detect and record the motion of water free surfaces. They are standard measurement instruments in tank testing [64] and can also be used under very complex conditions [38,50,59,60,65] as well as near submerged structures [21,22,66,67]. This measurement system correlates the resistance between two parallel rods with the submerged water depth. A direct electric current would lead to an anode/cathode arrangement leading to removal of ions in the water surrounding the gauges [20]. Hence the amplifier sends an AC signal with a typical excitation frequency of 10 kHz [68]. In general, the specification of such a wave probe depends on the application as well as the maximum expected wave amplitude. Typically heights vary between 0.3 to 1 m with a rod diameter of 3 to 4 mm [68], but thinner wire can be utilised if necessary. Andresen and Frigaard [69] applied wave gauges made of a wire with 1 mm near a cylinder under wave conditions and even thinner, namely 0.13 and 0.4 mm, are used by Liu et al. [70] in a wind-wave tank.

In the case of a tank test the typical accuracy of the conductive probes are in the range of ±0.25–0.5 mm respectively 0.1% [64,68]. Similar ranges of accuracy, namely ±0.3 mm, can be achieved by for ultrasonic sensors for experimental conditions as presented by Longo [28]. In the case of aerated and breaking wave this range has to be expanded. An accuracy of 1 mm is found by Gomit et al. [34] for the measurement of the crest with a resistive probe made by HR Wallingford. Petti and Longo [71] corrected the systematic error in the swash zone between the actual front on the beach in relation to the measurement by the run up wire with a separate video observation. A fixed offset of 5 mm was found as a good compromise to reduce the geometrical error without excessive signal error [71]. Whittaker et al. [51] quantify the accuracy for runup experiments with ±2 mm in the vertical direction based on effects of surface tension and calibration.

At the outset, the possibility of using conductive paint was investigated as part of preliminary tests. Those proved that it is possible to fabricate wave gauges by drawing on a surface with conductive paint (MG Chemicals 842AR-P Conductive Pen Silver). The hand drawn wave gauge was connected to the standard Edinburgh Designs' wave gauge box. Despite that the application with this method is inaccurate, the technique is very flexible and can be used in the future for specific applications that limit the use of tapes and rods. When dry, the conductive paint is water-resistant but remains very delicate and subject to mechanical damage. The variable cross section of the conductive paint leads to a non linear variation in gauge resistance that is not easily quantified, and therefore is only recommended for discrete water level measurements.

Based on their robust and exact mounting in comparison to the conductive paint, wave gauges made of copper tape were chosen [51,52]. Previous applications of such a tape can found as a switch [72,73] to indicate discrete water levels. In another study, Heller and Spinneken [38] bonded stainless steel strips on the surface of the inclined walls. For the current project, the scale for the water level inside the tank was used as an alignment for the conductive tape on each side. Separations of 30 mm, 24 mm, 20 mm, 15 mm, 10 mm, 5 mm, 4 mm and 1 mm between the 6.25 mm wide copper tapes were investigated. Preliminary dip tests showed that very small spacings between the two tapes lead to a delayed response to water level changes due to residual water films reducing the gauge resistance. Greater separations reduced the potential for stray conductive paths resulting from the residual water films that had led to erroneous measurements. Smaller separations provided a better signal to noise ratio and a reduced error in detecting oblique water surfaces run up. The 5 mm separation was chosen as the distance between the two tapes, resulting in a centerline separation of $w = 11.35$ mm. Whittaker et al. [51] used a bigger spacing of 20 mm. As part of the current tests, the durability of wave gauges based on conductive tape could be shown for normal laboratory conditions.

One Edinburgh Designs WG8USB Wave Gauge Controller is used for the presented work. The sample rate is 128 Hz [68]. Figure 2 shows the location of each wave gauge. In total, all the available 8 gauges are used, which is the capacity of a single controller. The rotation axis R for the following presented experiments (Section 3.2) is parallel to the y-axis. The wave gauges are arranged in four pairs with equal distance to the y-axis. In addition to the quadrants, the two wave gauges $WG2$ and $WG4$ are located at octants and have the equal $|x|$ and y distance of $D/(2 \cdot \sqrt{2}) = 0.173$ m. With an $|x|$ coordinate of $D/4$ for the $WG6$ and $WG8$, the y-value is $D \cdot \sqrt{3}/4 = 0.212$ m. For the evaluation of the inclination tests an additional marker $TP4$ (tank point, Figure 2) is needed. The $[x, y, z]$-coordinates of this point are $[-D/4,\ D \cdot \sqrt{3}/4,\ H - h + dh]$ with an offset dh of approximately 10 mm and height of the cylinder H of 0.5 m. The origin of the local coordinate system is located in the centre of the tank and on the initial water level h of 0.25 m. The measurements of the Qualisys system are based on a different global coordinate system, which is defined as part of the calibration. The results are transformed into a local coordinate system, which has its origin at the tank point $TP4$ (Figure 2).

3. Results

3.1. *Influence of Variable Inner Water Depth h*

In the first phase of the investigation, different water levels are measured with both methods to quantify the differences between the two approaches. For this set of tests, the location of the cylinder is fixed in the stationary position and the volume of fluid inside the tank is changed. This investigation is similar to the required calibration of the WGs and to the application of wave run-up experiments [38,51,52]. Consequently, the measurements at the WG can be used to verify the new approach based on the floating markers.

A total of 50 different steady states are investigated. The observed value is the change in water depth Δh inside the tank. The zero/reference value is set to the initial water depth $h = 0.25$ m. Both investigated measurement systems are calibrated independently and all depths are verified with a scale inside the tank captured by a camera.

All measurements for this variation of the water level, as well as the inclination tests in Section 3.2, are repeated three times and include 20 s recording time for each. One data point in Figure 3 represents the mean value over the entire test duration for all of the repetitions, for all eight wave gauges. The same averaging techniques are applied to the z-coordinate results of the Qualisys system for all 12 markers.

In Figure 3a the measurements are directly compared (in mm). Ideally, both measurement systems would report the same results. The absolute differences in Figure 3b indicates that the measurement based on the balls has a tendency to underestimate the height. For most of the cases, the differences are in the range of 4 mm and only a smaller water depth results in an increase up to around 7 mm. It is assumed that the influence of potentially biased measurements through the transparent cylinder is increased with a lower water level. The repetition of the Δh equal 0 mm allows checking for such a potential time-dependent effect. Theses repetitions were conducted spread over the whole experimentation by starting with this water level, which is first increase up to maximum, than reduce to the minimum level and back to the starting value of Δh equal 0 mm. A detailed analysis of those runs indicates two very good agreements near to the calibration process and the following repetitions have an increased difference. A change of the air temperature over the experimental time may cause an additional uncertainty in the water temperature of the comparably small volume in the cylinder and the used set-up (Figure 1).

In the case of the wave gauges made of conductive tape, the minimum standard deviation (of all eight wave gauges and three repetitions each 20 s long) can be found around the zero value. For both increasing and decreasing water levels, the standard deviation value increases up to an approximate value of ±3 mm (Figure 3c). For the standard deviation of the measurements based on the Qualisys system (Figure 3d), such a clear trend cannot be found in relation to Δh_Q-values, hence there is no connection between the system calibration and its origin. The reduction of the standard deviation for the highest water depths in the tank can be explained by improved marker visibility. In general, the standard deviation is in the same range as the observed absolute differences between the two measurement systems.

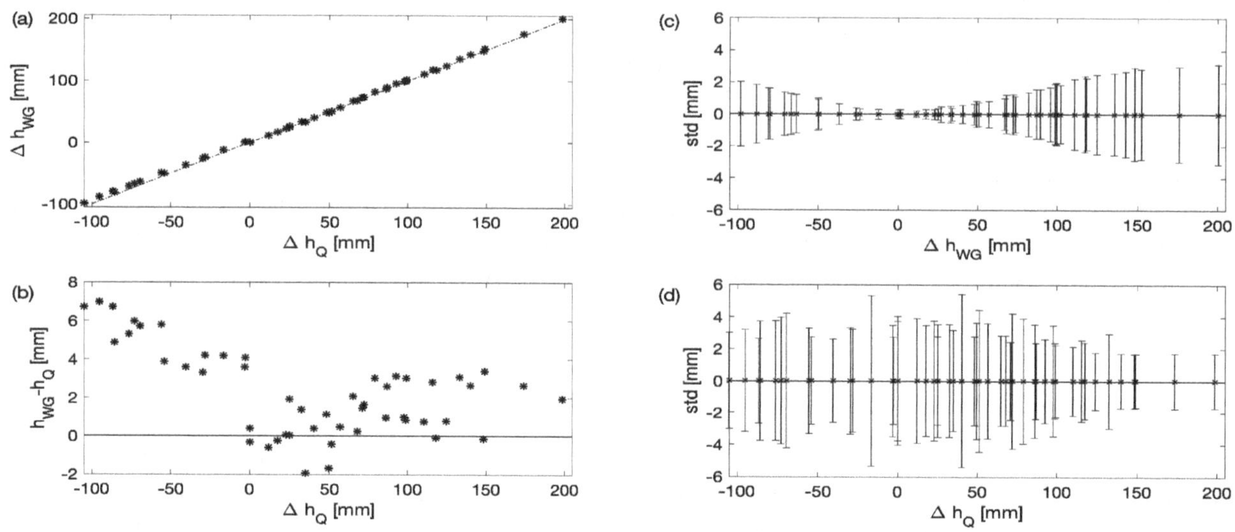

Figure 3. Comparison of the methods with a variable water depth h inside the cylinder—(**a**) comparison between water depth captured with the marker balls via Qualisys (Q) and wave gauges (WG); (**b**) absolute differences between the methods; (**c**,**d**) standard deviation (std) for each method.

Whittaker et al. [51] quantified the accuracy of this method with approximately ±2 mm in the vertical direction for a run-up experiment, which seem to be also applicable for the wave gauges made of copper tape. It seems to be reasonable to double this value for the floating balls based on the bigger standard deviation as well as the absolute differences between the two methods. On this basis, the next step is to investigate an inclined body in Section 3.2.

3.2. Influence of Variable Inclination Angle of the Tank

A fundamental difference between the two measurement systems is their point of reference. The locations of the infra-red markers are captured and reported based on a global reference frame which is independent of the position of the cylinder. The wave gauges are directly connected to the cylinder and thus measure the free surface in the local frame of reference of the cylinder. In contrast to the previous applications of onshore wave run-up probes [38,51,52], the orientation of the WGs changes according to the movement of the body on which they are bonded. Those tests allow to explore whether this effect on the measurement is significant or not.

It is necessary to transform the position of the floating markers to the local frame of reference of the cylinder or transform the wave gauge results into the global coordinate system. The required rotation matrix and the position vector is provided by the motion capturing system.

Figure 4a shows the mean value of the measurements of the two wave gauges laying on the y-axis ($WG3$ and $WG7$) vary with the rigid body's pitch angle as captured by Qualisys (β_Q). In the case of a rotation around the R-axis, which is parallel to the rotation axis y, those values should remain at 0 mm as well as the global mean value over all eight wave gauges. The other pairs of wave gauges (namely $WG1$ & $WG5$, $WG2$ & $WG4$, $WG6$ & $WG8$) are analysed in Figure 4b. The absolute distances in the x-direction for each pair is equal. Consequently, the mean value of each pair should also be

0 mm. Both analyses indicate that with an increasing β_Q-angle, a bigger difference to the theoretical position also occurs. This is possibly due to deformation of the structure resulting in a decrease in cross-sectional area, in turn increasing the height of the fluid in the structure. This will not be an issue for structures tested in the wave tank, as the surrounding water pressure will counteract the inner pressure.

The results of the three pairs of wave gauges with an x offset are used to back calculate the inclination of the structure β_{WG}, which should be identical to the pitch angle β_Q measured by the Qualisys system. Figure 4c presents the direct comparison of those values and Figure 4d the differences. A calculation of the inclination based on the wave gauge data would result in an under prediction of approximately 1°.

For the analysis of the measurements based on the marker balls, the vertical transition caused by the rotation has to be included. Basic geometric dependences are shown in Figure 2 for the centre point on the water surface. A theoretical height increase $z_{Q,t}$ is calculated to compare the observation z_Q of the markers. The value $z_{Q,t}$ depends on the measured pitch angle β_Q of the cylinder and the exact position of the rotation axis R. It is assumed that the rotation axis R is parallel to the y-axis, and to specify the location of the axis R, a circle is fitted through the measured x- and z-coordinates of the tank point $TP4$.

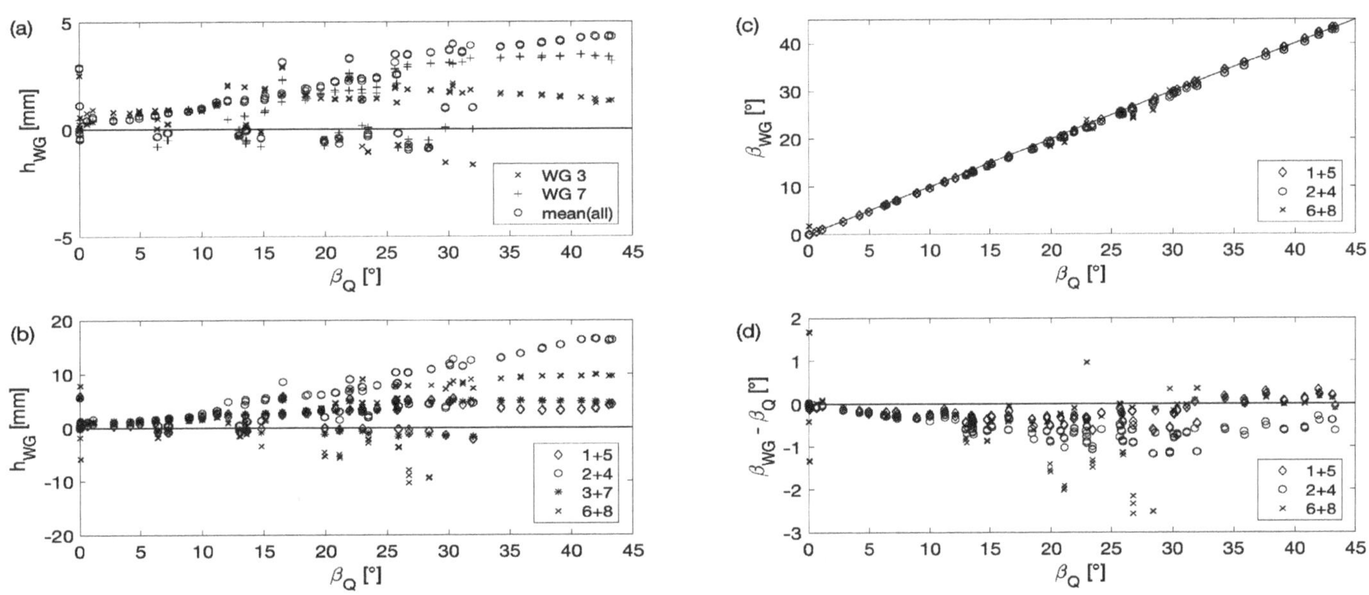

Figure 4. Influence of variable pitch angle β on the wave gauges (WG)—(**a**,**b**) analysis of the measured changes in the free surface; (**c**,**d**) comparison between the measured angle β_Q and the calculated β_{WG} based on the wave gauges pairs.

Figure 5a allows a comparison of the theoretical and measured values for the water surface and tank point $TP4$ in relation to the angle β_Q of the body. All values are transformed so that 0 mm indicates the initial water surface. The absolute differences between calculated and measured values are shown in Figure 5d. Both graphs indicate that, in a range of β_Q up to 30°, the theory over-predicts the measurement and beyond that point it under-predicts.

For a further analysis, the same data-set is plotted in relation to the measured roll angle α_Q in Figure 5c and the yaw angle γ_Q Figure 5e. A comparable behaviour to the angle β_Q could be found for α_Q around 1°, but not for γ_Q. This indicates that the complete system is not completely rigid and includes a small inclination around the x-axis and a part of the differences can be explained as a result. The standard deviation of the measurements of the marker balls (Figure 5b) show the same range as found by changing the water level h inside the tank (Figure 3d; Section 3.1).

The results of both stationary cases, namely the set of measurements with changing inner water level and variable inclination of the tank, show a reasonably small error of a few millimetres for each measurement system. The possible over-prediction of the water level in the case of inclined wave gauge in relation to the water surface should be considered for further comparisons with numerical simulations. Circularity measurements may verify if the increase in water level is due to deformation of the structure or a wave gauge error.

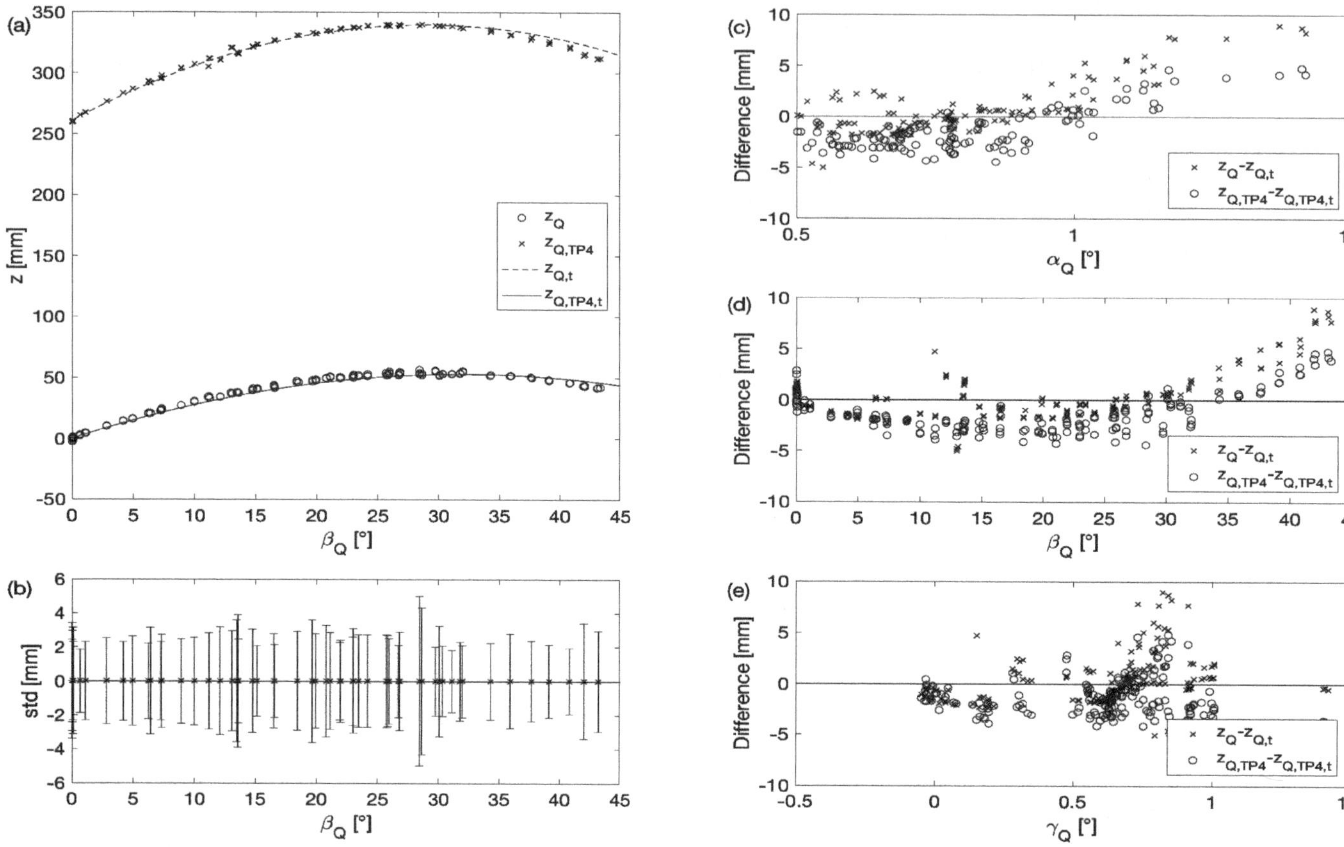

Figure 5. Influence of variable pitch angle β on the floating markers—(**a**) comparison of the theoretical with the measured z-coordinate of the reference point $TP4$ and the free surface; (**b**) standard deviation of the measured balls; (**c**–**e**) absolute difference of the z-coordinate in relation to all three rotation axes.

3.3. Transient Behaviour

The previous investigations assume steady-state conditions and focused on establishing a proof of concept for the adapted measurement system. However, the goal is to investigate the free floating structure motions and inner free water surface when influenced by periodic wave conditions in a wave tank. In this case the free water surface inside the cylinder will be non stationary. Different drop tests are conducted to show that both measurement concepts are capable of capturing transient water surfaces inside of the tank. To achieve this, the cylinder is fixed in an inclined position and the supporting mechanism is suddenly removed. Hence, the tank falls onto an end stop and the oscillation of the free surface inside the tank is stimulated.

The presented experiment is chosen because a small overtopping volume is provoked near the wave gauge $WG1$. Figure 6 shows the rotation from an initial pitch angle β_Q of 21.1° to 6.4°. For the other two rotations, the observed differences $\Delta\alpha$ and $\Delta\gamma$ to the starting condition (inclined cylinder) are very small. All systems are synchronised so that the approximated start of the movement is at $t = 0.25$ s. The end position of the structure has an inclination in pitch caused by the end stop, which is added to the framework. The first impact compresses this end stop and a nearly stable situation of the cylinder is reached after about one second. A small periodic movement of the structure is also observed, which is caused by the moving water body in the tank combined with the soft support of the end stop material.

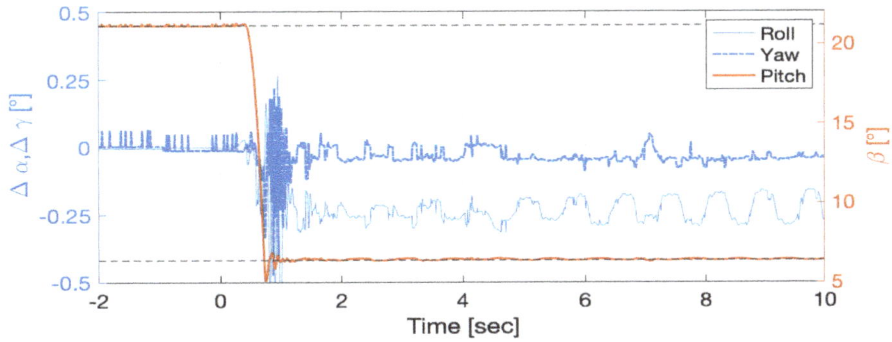

Figure 6. Measured angles of the rigid body caused by the sudden removal of the support structure.

Simultaneously two videos are captured, which give a very good overview of the behaviour of the water stored in the structure. The measured position of the markers can be directly connected with the video from the Oqus 210c. In contrast to this fixed position of observation, the second camera on the framework moves with the structure. This allows comparison of the visible scale inside the cylinder with the captured free surface changes [7,31]. Different random samples are analysed as part of the plausibility check and demonstrate a very good agreement (typically in the low range of a few millimetres) between both experimental methods. Therefore, the global coordinates of the marker balls had to be transformed into the local coordinate system of the cylinder.

Figure 7 presents the results of the wave gauges for the drop test. In each graphic, each pair of wave gauges with the same x offset are grouped, and the normalised water surface displacement η/h is plotted as the y-value. This non-dimensional elevation is also used for the second and third row in Figure 8 (two different side views) for the following analysis of the floating markers. As mentioned, the initial water height $h = 0.25$ m is equal to half of the total height H of the cylinder. The η/h value equal to 1 [-] in the first big peak of the wave gauge $WG1$ indicates the overtopping at this part of the cylinder. A wave period of approximately 0.7 s is observed for the introduced oscillation. The final conditions also include an inclination due the end stop material, which is added to the support structure, which alters the resting position of the structure, affecting the final value of all wave gauges except $WG3$ and $WG7$ because they lie on the y-axis.

The results show that the main oscillation direction is at first aligned with the movement of the cylinder in the x-z-plane and changes after a few seconds to the orthogonal direction in the y-z-plane. The $WG1$ and $WG5$ show a damped oscillation with a nearly constant decrease of the amplitude. The peak values for η/h at the $WG3$ and $WG7$ increase over the same time frame (Figure 7).

In addition to the wave gauge results, the elevation of each of the 12 free surface marker balls can be plotted. Figure 8 uses a different approach to show the three-dimensional motion of the markers. Four different time periods were chosen. For each period of record, the measured coordinates of all the 12 points were plotted in one graph. The top view (upper row) shows that the balls have the tendency to group together. Nevertheless, a good coverage is maintained. The side view of the first 5 s time period also captures the overflow with a η/h-value near to 1 [-], which fortunately did not result in the loss of a marker. The change of the main oscillation direction can also be observed in the measured data of the marker balls (going from 3rd to 4th column in Figure 8).

This simplified transient experiment showed the capability of both methods to capture the free surface inside the cylinder. The advantages and limitations of each approach are discussed in the following section.

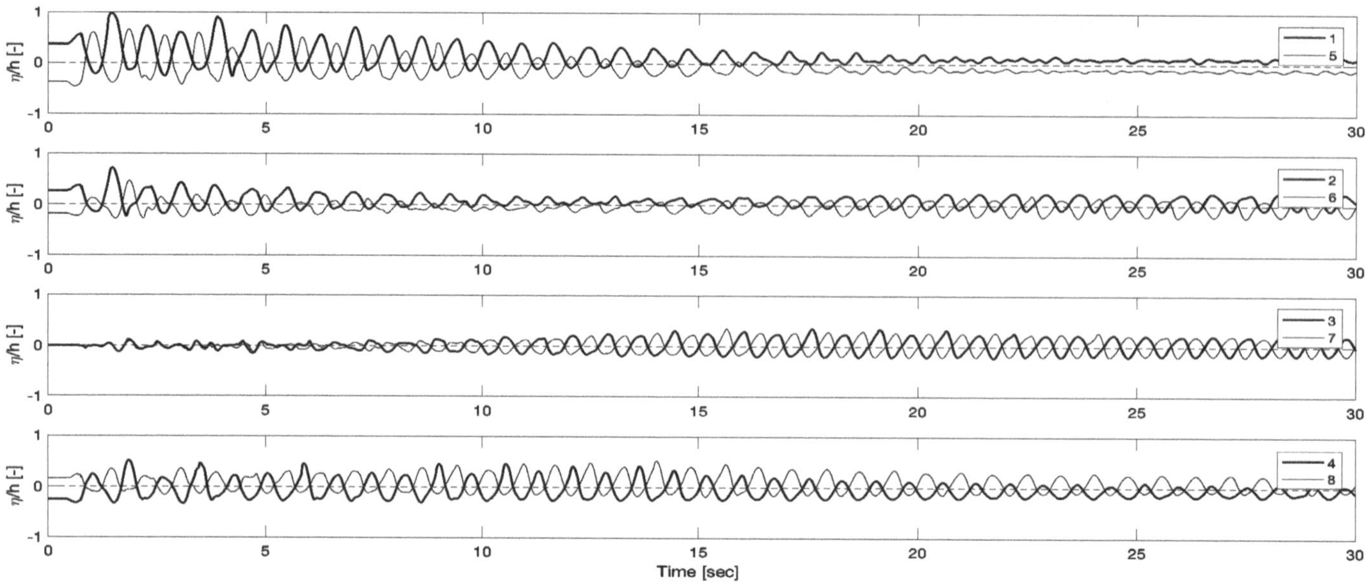

Figure 7. Results of the transient experiment—wave gauges (WG) in pairs as defined in Figure 2—normalised surface elevation η / h.

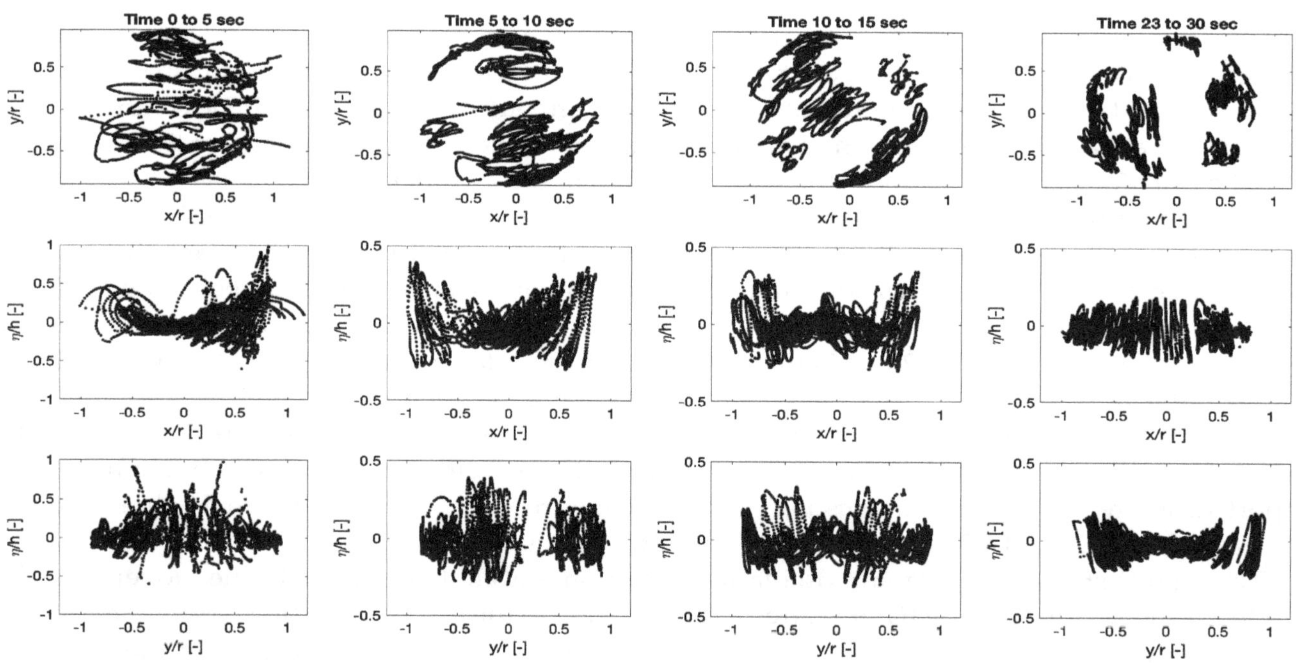

Figure 8. Results of the transient experiment—3D location of the measured marker balls; four different time periods (columns) and three different view directions (rows).

4. Discussion

Both measurement systems investigated here fulfill the basic requirements mentioned in Section 1 for the future use of capturing the motion of the free water surface inside a floating structure in a wave tank.

The addition of the bespoke marker balls into the tank is simple to implement, and makes use of the motion capture systems which are typical equipment in a laboratory with wave basins. Camera positions may have to be adapted and more carefully chosen, so that the inside of the floating cylinder is easily captured. A motion capture camera view from the top would be very advantageous for 3D capture of the markers, but this may not always be practical to realise.

The floating markers are capable of surviving very rough conditions in the cylinder and are very light. It is thus assumed that the influence of adding floating markers to the free surface is very small, but it cannot be excluded completely. Specific experiments with and without the markers should be compared for the individual applications in the future. For the current project, 12 markers were utilized, which covered approximately 8% of the available surface inside the cylinder. More floating markers could be applied, but as a result the interaction between the markers would also increase. The ideal number and size of markers should be optimised for each project.

Under certain conditions the capturing of one individual marker's position is interrupted because fewer than two cameras detect it. This can be caused by obscuring components (such as the mounted camera at the top) or extreme sloshing, including breaking waves. There is also the possibility that one marker is washed out of the structure. This is an abort criterion because the water volume inside the tank is reduced in such a case. Overtopping of the structure cannot be captured with this measurement system.

The novel characteristic of this system is that the markers cover the free surface inside the cylinder (Figure 8), but the exact location cannot be predicted beforehand and is not repeatable. This situation can be improved with a splitting of the available volume inside the tank, for example with a vertical mesh. This can limit the area which each marker can cover but also has a further effect on the fluid dynamics. Conventional wave gauges could be a better option instead of such highly restricted markers.

The wave gauges made of the copper tape are fixed to the cylinder structure and no additional support structure is needed. This prevents unnecessary mass additions and subsequent changes in the centre of mass. An electrical connection is still required, but the cable can be added to the mooring lines required to stabilise the floating body. This solution requires an insulating and robust surface to apply the copper tape. A minimum distance between two pairs has to be maintained so that no interference occurs (similar to normal wave gauges). The capillary effect between the two tapes on the wall may lead to erroneous measurement. Based on a preliminary investigation, a distance of 5 mm was chosen for this investigation (Section 2.3). Such a suitable choice of the distance between the two electrodes helps to minimise this effect. Furthermore, maximum local water heights in the floating structure are caused by run up on the walls which is quantifiable with this method. If data analysis of the wave gauges is carried out in the local coordinates of the cylinder, no additional post processing is needed to use the measured data. The major disadvantage of video recording the scale inside the tank is the post processing required to attain useful data.

It is well known that the measurement method of wave gauges is highly dependent on the electrical properties of the water. The water for the tests herein was taken from the 2.4 million litre basin at FloWave and not fresh from the main water supply, thus minimising any temperature adjustments or other changes. In comparison to the wave tank, the stored water volume inside the cylinder (Figure 1) is very small. The changes of the air temperature over a working day also influenced the temperature of the water in the tank. Consequently, some of the observed errors may be caused by this change in environmental conditions and the (water) temperature should be measured in comparable investigations in the future. The temperature changes of the water in the wave tank under normal conditions is nearly negligible and hence the conditions for a submerged structure are far better than in the present configuration of being supported on a framework.

5. Conclusions

This paper presents two different measurement methods for capturing the motion of the free water surface inside a floating structure by re-purpose existing systems: (a) floating markers, which are captured in addition to the movement of the rigid body and (b) wave gauges made of conductive copper tape fixed to the walls of the cylinder. The additional mass of both systems is very small compared to the structure. Both systems are tested under laboratory conditions and their durability is shown.

The conductive tape wave gauge proved very accurate and consistent in measuring the water height at the walls, where the maximum heights occur due to water run-up. It can be shown that the conductive tape wave gauges can also be applied to moving objects and that a change in the inclination in comparison to the calibration has a small effect on the accuracy. The standard deviation of the measurement system was found to be equal or smaller than 2 mm (variable water height, Section 3.1). Those measurements are directly connected to the movable structure, or local coordinate system. In contrast to this, the marker balls can freely float inside the water body and the measurement is in the global coordinate system. Hence the movement of the rigid body is captured in global coordinates, so that the free floating markers can be transformed into local coordinates for comparison with the copper tape wave gauge measurements. The standard deviation for this measurement system is in the range of 4 mm using markers with a diameter of 40 mm (Sections 2.2 and 3.1). A reduction of the marker size can result in a higher accuracy and should be considered in the future.

Different stationary and transient tests proved that both methods are capable of delivering measurements with a very high accuracy in low mm-range. The combination of both systems allows one to compare the two independent measurement methods and use the advantages of each approach. This has a significant benefit in providing useful data sets as a validation experiment for future numerical studies of free floating objects.

Author Contributions: R.G., J.S., D.I.M.F., T.D., T.B. and D.M.I. are responsible for the conceptualisation of the experimental investigation. R.G., J.S. and T.D. measured the data and analysed the data. R.G. and J.S. wrote the initial draft and D.I.M.F., T.D., T.B. and D.M.I. reviewed and edited the paper.

Acknowledgments: Further thanks go to the support team of Edinburgh Designs and Qualisys for their help.

Abbreviations

The following abbreviations are used in this manuscript:

LNG	liquefied natural gas
Q	Qualisys
VLFS	very large floating structures
WG	wave gauge

References

1. Chaplin, J.R.; Heller, V.; Farley, F.J.M.; Hearn, G.E.; Rainey, R.C.T. Laboratory testing the Anaconda. *Philos. Trans. R. Soc. A* **2012**, *370*, 403–424.
2. Farley, F.J.M.; Rainey, R.C.T.; Chaplin, J.R. Rubber tubes in the sea. *Philos. Trans. R. Soc. A* **2012**, *370*, 381–402. [CrossRef] [PubMed]
3. Hirdaris, S.E.; Bai, W.; Dessi, D.; Ergin, A.; Gu, X.; Hermundstad, O.A.; Huijsmans, R.; Iijima, K.; Nielsen, U.D.; Parunov, J.; et al. Loads for use in the design of ships and offshore structures. *Ocean Eng.* **2014**, *78*, 131–174. [CrossRef]
4. Jiao, J.; Ren, H.; Chen, C. Model testing for ship hydroelasticity: A review and future trends. *J. Shanghai Jiaotong Univ.* **2017**, *22*, 641–650. [CrossRef]
5. Sharma, R.; Kim, T.W.; Storch, R.L.; Hopman, H.; Erikstad, S.O. Challenges in computer applications for ship and floating structure design and analysis. *CAD Comput. Aided Des.* **2012**, *44*, 166–185. [CrossRef]
6. Bureau Veritas (BV). *Design Sloshing Loads for LNG Membrane Tanks—Guidance Note NI 554 DT R00 E*; Bureau Veritas (BV): Paris, France, 2010.
7. ITTC—Seakeeping Committee of the 28th ITTC. ITTC Quality System Manual Recommended Procedures and Guidelines—Procedure Sloshing Model Tests, Revision 00, Section: 7.5-02-07-02.7. Available online: https://ittc.info/media/7625/75-02-07-027.pdf (accessed on 27 December 2018).

8. Lyu, W.; Moctar, O.E.; Potthoff, R.; Neugebauer, J. Experimental and numerical investigation of sloshing using different free surface capturing methods. *Appl. Ocean Res.* **2017**, *68*, 307–324. [CrossRef]
9. Song, Y.K.; Chang, K.A.; Ryu, Y.; Kwon, S.H. Experimental study on flow kinematics and impact pressure in liquid sloshing. *Exp. Fluids* **2013**, *54*, 1592. [CrossRef]
10. Ariyarathne, K.; Chang, K.-A.; Mercier, R. Green water impact pressure on a three-dimensional model structure. *Exp. Fluids* **2012**, *53*, 1879–1894. [CrossRef]
11. Chuang, W.L.; Chang, K.A.; Mercier, R. Impact pressure and void fraction due to plunging breaking wave impact on a 2D TLP structure. *Exp. Fluids* **2017**, *58*, 68. [CrossRef]
12. Kim, S.Y.; Kim, K.H.; Kim, Y. Comparative study on pressure sensors for sloshing experiment. *Ocean Eng.* **2015**, *94*, 199–212. [CrossRef]
13. Kim, K.-T.; Lee, P.-S.; Park, K.C. A direct coupling method for 3D hydroelastic analysis of floating structures. *Int. J. Numer. Methods Eng.* **2013**, 96, 842–866. [CrossRef]
14. Wei, W.; Fu, S.; Moan, T.; Lu, Z.; Deng, S. A discrete-modules-based frequency domain hydroelasticity method for floating structures in inhomogeneous sea conditions. *J. Fluids Struct.* **2017**, *74*, 321–339. [CrossRef]
15. Yoon, J.S.; Cho, S.P.; Jiwinangun, R.G.; Lee, P.S. Hydroelastic analysis of floating plates with multiple hinge connections in regular waves. *Mar. Struct.* **2014**, *36*, 65–87. [CrossRef]
16. Lee, K.H.; Cho, S.; Kim, K.T.; Kim, J.G.; Lee, P.S. Hydroelastic analysis of floating structures with liquid tanks and comparison with experimental tests. *Appl. Ocean Res.* **2015**, *52*, 167–187. [CrossRef]
17. Klar, R.; Steidl, B.; Sant, T.; Aufleger, M.; Farrugia, R.N. Buoyant Energy—Balancing wind power and other renewables in Europe's oceans. *J. Energy Storage* **2017**, *14*, 246–255. [CrossRef]
18. Klar, R.; Steidl, B.; Aufleger, M. A floating energy storage system based on fabric. *Ocean Eng.* **2018**, *165*, 328–335. [CrossRef]
19. Gabl, R.; Achleitner, S.; Neuner, J.; Aufleger, M. Accuracy analysis of a physical scale model using the example of an asymmetric orifice. *Flow Meas. Instrum.* **2014**, *36*, 36–46. [CrossRef]
20. Hughes, S.A. *Physical Models and Laboratory Techniques in Coastal Engineering*; Advanced Series on Ocean Engineering; World Scientific: Singapore, 1993. [CrossRef]
21. Lykke Andersen, T.; Frigaard, P.; Damsgaard, M.L.; De Vos, L. Wave run-up on slender piles in design conditions - Model tests and design rules for offshore wind. *Coast. Eng.* **2011**, *58*, 281–289. [CrossRef]
22. Mavrakos, S.A.; Chatjigeorgiou, I.K.; Lentziou, D.M. Wave Run-Up and Second-Order Wave Forces on a Truncated Circular Cylinder Due to Monochromatic Waves. In Proceedings of the 24th International Conference on Offshore Mechanics and Arctic Engineering, Halkidiki, Greece, 12–17 June 2005; pp. 231–238.
23. Teh, H.M.; Venugopal, V.; Bruce, T. Hydrodynamic characteristics of a free-surface semicircular breakwater exposed to irregular waves. *J. Waterway Port Coast. Ocean Eng.* **2012**, *138*, 149–163. [CrossRef]
24. Tripepi, G.; Aristodemo, F.; Veltri, P. On-Bottom Stability Analysis of Cylinders under Tsunami-Like Solitary Waves. *Water* **2018**, *10*, 487. [CrossRef]
25. Peruzzo, P.; De Serio, F.; Defina, A.; Mossa, M. Wave Height Attenuation and Flow Resistance Due to Emergent or Near-Emergent Vegetation. *Water* **2018**, *10*, 402. [CrossRef]
26. Chiapponi, L.; Longo, S.; Tonelli, M. Experimental study on oscillating grid turbulence and free surface fluctuation. *Exp. Fluids* **2012**, *53*, 1515–1531. [CrossRef]
27. Caron, P.A.; Cruchaga, M.A.; Larreteguy, A.E. Study of 3D sloshing in a vertical cylindrical tank. *Phys. Fluid* **2018**, *30*, 082112.
28. Longo, S. Experiments on turbulence beneath a free surface in a stationary field generated by a Crump weir: Free-Surface characteristics and the relevant scales. *Exp. Fluids* **2010**, *49*, 1325–1338.
29. Longo, S.; Liang, D.; Chiapponi, L.; Jimenez, L.A. Turbulent flow structure in experimental laboratory wind-generated gravity waves. *Coast. Eng.* **2012**, *64*, 1–15.
30. Sellar, B.; Harding, S.; Richmond, M. High-resolution velocimetry in energetic tidal currents using a convergent-beam acoustic Doppler profiler. *Meas. Sci. Technol.* **2015**, *26*, 085801. [CrossRef]
31. Bonakdar, L.; Oumeraci, H.; Etemad-Shahidi, A. Run-up on vertical piles due to regular waves: Small-scale model tests and prediction formulae. *Coast. Eng.* **2016**, *118*, 1–11. [CrossRef]
32. Tosun, U.; Aghazadeh, R.; Sert, C.; Özer, M.B. Tracking free surface and estimating sloshing force using image processing. *Exp. Thermal Fluid Sci.* **2017**, *88*, 423–433. [CrossRef]
33. Bechle, A.J.; Wu, C.H. Virtual wave gauges based upon stereo imaging for measuring surface wave characteristics. *Coast. Eng.* **2011**, *58*, 305–316. [CrossRef]

34. Gomit, G.; Chatellier, L.; Calluaud, D.; David, L. Free surface measurement by stereo-refraction. *Exp. Fluids* **2013**, *54*, 1540. [CrossRef]
35. Bregoli, F.; Bateman, A.; Medina, V. Tsunamis generated by fast granular landslides: 3D experiments and empirical predictors. *J. Hydraul. Res.* **2017**, *55*, 743–758. [CrossRef]
36. Chang, K.-A.; Liu, P.L.-F. Pseudo turbulence in PIV breaking-wave measurements. *Exp. Fluids* **2000**, *29*, 331–338. [CrossRef]
37. Fritz, H.M.; Hager, W.H.; Minor, H.E. Landslide generated impulse waves. 1. Instantaneous flow fields. *Exp. Fluids* **2003**, *35*, 505–519. [CrossRef]
38. Heller, V.; Bruggemann, M.; Spinneken, J.; Rogers, B.D. Composite modelling of subaerial landslide-tsunamis in different water body geometries and novel insight into slide and wave kinematics. *Coast. Eng.* **2016**, *109*, 20–41. [CrossRef]
39. Weigand, A. Simultaneous mapping of the velocity and deformation field at a free surface. *Exp. Fluids* **1996**, *20*, 358–364.
40. Park, J.; Im, S.; Sung, H.J.; Park, J.S. PIV measurements of flow around an arbitrarily moving free surface. *Exp. Fluids* **2015**, *56*, 56. [CrossRef]
41. Belden, J.; Techet, A.H. Simultaneous quantitative flow measurement using PIV on both sides of the air-water interface for breaking waves. *Exp. Fluids* **2011**, *50*, 149–161. [CrossRef]
42. Blenkinsopp, C.E.; Turner, I.L.; Allis, M.J.; Peirson, W.L.; Garden, L.E. Application of LiDAR technology for measurement of time-varying free-surface profiles in a laboratory wave flume. *Coast. Eng.* **2012**, *68*, 1–5. [CrossRef]
43. Montano, L.; Li, R.; Felder, S. Continuous measurements of time-varying free-surface profiles in aerated hydraulic jumps with a LIDAR. *Exp. Thermal Fluid Sci.* **2018**, *93*, 379–397. [CrossRef]
44. Weitbrecht, V.; Kühn, G.; Jirka, G.H. Large Scale PIV Measurements at the Surface of Shallow Water Flows. *Flow Meas. Instrum.* **2002**, *13*, 237–245. [CrossRef]
45. Akutina, Y.; Mydlarski, L.; Gaskin, S.; Eiff, O. Error analysis of 3D-PTV through unsteady interfaces. *Exp. Fluids* **2018**, *59*, 53.
46. Evers, F.M.; Hager, W.H. Videometric Water Surface Tracking: Towards Investigating Spatial Impulse Waves. In Proceedings of the 36th IAHR World Congress, Delft, The Netherlands, 28 June–3 July 2015; pp. 6618–6623.
47. Evers, F.M.; Hager, W.H. Spatial impulse waves: Wave height decay experiments at laboratory scale. *Landslides* **2016**, *13*, 1395–1403. [CrossRef]
48. Mignot, E.; Moyne, T.; Doppler, D.; Riviere, N. Clear-water scouring process in a flow in supercritical regime. *J. Hydraul. Eng.* **2016**, *142*, 04015063-1. [CrossRef]
49. Przadka, A.; Cabane, B.; Pagneux, V.; Maurel, A.; Petitjeans, P. Fourier transform profilometry for water waves: How to achieve clean water attenuation with diffusive reflection at the water surface? *Exp. Fluids* **2012**, *52*, 519–527. [CrossRef]
50. Heller, V.; Spinneken, J. On the effect of the water body geometry on landslide-tsunamis: Physical insight from laboratory tests and 2D to 3D wave parameter transformation. *Coast. Eng.* **2015**, *104*, 113–134.
51. Whittaker, C.N.; Fitzgerald, C.J.; Raby, A.C.; Taylor, P.H.; Orszaghova, J.; Borthwick, A.G.L. Optimisation of focused wave group runup on a plane beach. *Coast. Eng.* **2017**, *121*, 44–55. [CrossRef]
52. Niedzwecki, J.M.; Duggal, A.S. Wave runup and forces on cylinders in regular and random waves. *J. Waterway Port Coast. Ocean Eng.* **1992**, *118*, 615–634. [CrossRef]
53. Kang, Z.; Ni, W.; Zhang, L.; Ma, G. An experimental study on vortex induced motion of a tethered cylinder in uniform flow. *Ocean Eng.* **2017**, *142*, 259–267. [CrossRef]
54. O'Connell, K.; Thiebaut, F.; Kelly, G.; Cashman, A. Development of a free heaving OWC model with non-linear PTO interaction. *Renew. Energy* **2018**, *117*, 108–115. [CrossRef]
55. Sjökvist, L.; Wu, J.; Ransley, E.; Engström, J.; Eriksson, M.; Göteman, M. Numerical models for the motion and forces of point-absorbing wave energy converters in extreme waves. *Ocean Eng.* **2017**, *145*, 1–14. [CrossRef]
56. Usherwood, J.R. The aerodynamic forces and pressure distribution of a revolving pigeon wing. *Exp. Fluids* **2009**, *46*, 991–1003. [CrossRef] [PubMed]
57. Gyongy, I.; Richon, J.-B.; Bruce, T.; Bryden, I. Validation of a hydrodynamic model for a curved, multi-paddle wave tank. *Appl. Ocean Res.* **2014**, *44*, 39–52. [CrossRef]

58. Draycott, S.; Noble, D.; Davey, T.; Bruce, T.; Ingram, D.; Johanning, L.; Smith, H.; Day, A.; Kaklis, P. Re-creation of site-specific multi-directional waves with non-collinear current. *Ocean Eng.* **2018**, *152*, 391–403. [CrossRef]
59. Draycott, S.; Sutherland, D.; Steynor, J.; Sellar, B.; Venugopal, V. Re-creating waves in large currents for tidal energy applications. *Energies* **2017**, *10*, 1838. [CrossRef]
60. Draycott, S.; Davey, T.; Ingram, D.M. Simulating Extreme Directional Wave Conditions. *Energies* **2017**, *10*, 1731.
61. Ingram, D.; Wallace, R.; Robinson, A.; Bryden, I. The design and commissioning of the first, circular, combined current and wave test basin. In Proceedings of the Oceans 2014 MTS/IEEE Taipei, Taiwan, 18–21 May 2015.
62. Sutherland, D.R.J.; Noble, D.R.; Steynor, J.; Davey, T.; Bruce, T. Characterisation of current and turbulence in the FloWave Ocean Energy Research Facility. *Ocean Eng.* **2017**, *139*, 103–115. [CrossRef]
63. Draycott, S.; Sellar, B.; Davey, T.; Noble, D.R.; Venugopal, V.; Ingram, D. Capture and Simulation of the Ocean Environment for Offshore Renewable Energy. *Renew. Sustain. Energy Rev.* **2019**. under review.
64. MARINET (2012) Work Package 2: Standards and Best Practice—D2.1 Wave Instrumentation Database. Revision: 05. Available online: http://www.marinet2.eu/wp-content/uploads/2017/04/D2.01-Wave-Instrumentation-Database.pdf (accessed on 23 November 2018).
65. Heller, V.; Chen, F.; Brühl, M.; Gabl, R.; Chen, X.; Wolters, G.; Fuchs, H. Large-scale experiments into the tsunamigenic potential of different iceberg calving mechanisms. *Sci. Rep.* **2018**. [CrossRef]
66. Mai, T.; Greaves, D.; Raby, A.; Taylor, P.H. Physical modelling of wave scattering around fixed FPSO-shaped bodies. *Appl. Ocean Res.* **2016**, *61*, 115–129. [CrossRef]
67. Wienke, J.; Oumeraci, H. Breaking wave impact force on a vertical and inclined slender pile—Theoretical and large-scale model investigations. *Coast. Eng.* **2005**, *52*, 435–462. [CrossRef]
68. EDL-Edinburgh Designs Ltd. Available online: http://www.edesign.co.uk/product/wavegauges/ (accessed on 23 November 2018).
69. Andersen, T.L.; Frigaard, P. Horns Rev II, 2D-Model Tests: Wave Run-Up on Pile. In *DCE Contract Reports, No. 3*; Department of Civil Engineering, Aalborg University: Aalborg, Denmark, 2006.
70. Liu, H.-T.; Katsaros, K.B.; Weissman, M.A. Dynamic response of thin-wire wave gauges. *J. Geophys. Res.* **1982**, *87*, 5686–5698. [CrossRef]
71. Petti, M.; Longo, S. Turbulence experiments in the swash zone. *Coast. Eng.* **2001**, *43*, 1–24.
72. Bruce, T.; van der Meer, J.W.; Franco, L.; Pearson, J.M. Overtopping performance of different armour units for rubble mound breakwaters. *Coast. Eng.* **2009**, *56*, 166–179. [CrossRef]
73. Pullen, T.; Allsop, W.; Bruce, T.; Pearson, J. Field and laboratory measurements of mean overtopping discharges and spatial distributions at vertical seawalls. *Coast. Eng.* **2009**, *56*, 121–140. [CrossRef]

11

Estimation of Precipitation Evolution from Desert to Oasis using Information Entropy Theory

Xiangyang Zhou [1,2], Zhipan Niu [2,3,*] and Wenjuan Lei [4]

1 College of Resource and Environmental Engineering, Guizhou University, Guiyang 550025, China; xyzhou6@gzu.edu.cn
2 State Key Laboratory of Hydraulics and Mountain River Engineering, Sichuan University, Chengdu 610065, China
3 Institute for Disaster Management and Reconstruction, Sichuan University, Chengdu 610065, China
4 College of Architecture & Environment, Sichuan University, Chengdu 610065, China; leixiaojuan333@126.com
* Correspondence: niuzhipan@sina.com.

Abstract: The cold-wet effect of oasis improves the extreme natural conditions of the desert areas significantly. However, the relationship between precipitation and the width of oasis is challenged by the shortage of observed data. In this study, the evolution of annual precipitation from desert to oasis was explored by the model establishment and simulation in Tarim Basin of northwestern China. The model was developed from the principle of maximum information entropy, and was calibrated by the China Meteorological Forcing Dataset with a high spatial resolution of 0.1° from 1990 to 2010. The model performs well in describing the evolution of annual precipitation from the desert to oasis when the oasis is wide enough, and the R^2 is generally more than 0.90 and can be up to 0.99. However, it fails to simulate the seasonal precipitation evolution because of the non-convergence solved by nonlinear fitting and the unfixed upper boundary condition solved by the least square method. Through the simulation with the parameters obtained from the nonlinear fitting, the basic patterns, four stages of precipitation evolution with the oasis width increasing, are revealed at annual scale, and the current stages of these oases are also uncovered. Therefore, the establishment of the model and the simulated results provide a deeper insight from the perspective of informatics to understand the regional precipitation evolution of the desert–oasis system. These results are not only helpful in desertification prevention, but also helpful in fusing multisource data, especially in extreme drought desert areas.

Keywords: evolution of precipitation; model simulation; information entropy theory; desert–oasis areas; Tarim Basin

1. Introduction

Oasis serves to improve the extreme natural conditions of the arid regions by affecting the regional hydrometeorological factors [1–7], and the oasis–desert interactions are important for the stable co-existence of oasis and desert ecosystems [8] and water resources management [9]. Hence, quantifying the spatial evolution of precipitation with the oasis width increase is crucial to the regional eco-environmental security in the arid areas.

In fact, the spatial evolutions of precipitation are affected by the cold-wet effect of oasis. The cold effect is mainly resulted from the greater absorption of latent heat through evapotranspiration [10,11] and the higher surface albedo of the vegetation than the desert surface [3]. The wet effect is because of more water vapor source from the evapotranspiration of oasis [3,4]. As a result, the local water vapor can account for up to 20% of the precipitation in arid and semiarid regions [12,13] and 20–50% in humid regions [14–17]. In fact, the extreme arid environment leads to the much larger difficulties of data observation. Many studies are based on the field observation at a low spatial resolution [1–6,12–17], although some multiscale datasets are available in specific areas such as Heihe Basin located in the northwestern China [18]. Thus, generally, exploring the relationship between the precipitation and oasis width quantitatively is still challenged by the resolution of data.

Merging multisource data is a useful tool to solve the problem of data shortage and low resolution, especially fusing the calibrated remote sensing data with the land surface observed data. After the early attempts in the 1980s [19,20], the resolution of the assimilation data has improved substantially. For example, as for the ocean system, the current resolution is about 1/4–1/6 degrees [21,22]. In China, the current resolution is 0.1 degrees spatially [23,24], and its good spatial continuity has been demonstrated by several studies. These studies have uncovered the spatial distribution of precipitation and evaporation in the Nam Co basin in Tibetan Plateau [25,26], and the spatial patterns of permafrost based on the climatic factors of this dataset [27]. The higher resolution of fusion dataset provides an alternative source to analyze the spatial evolution of precipitation.

Entropy, combining the micro and macro status, has been widely used to simulate the evolution of a system. Information entropy is a better measure of variability than the variance and coefficient of variation when the probability distribution is not symmetric [28,29], because entropy may be related to higher order moments of a distribution [30]. Hence, it has been employed to evaluate the complexity of typical chaos [31] and uncertainty of precipitation and the potential water resources ability at different scales [32–35]. Spatially, entropy has been widely employed to evaluate the spatial distribution of rainfall gauge net [36,37], quantify the soil water dynamic processes [38–40], explore the multifractal generation [41,42] and simulate the shallow water and solution transportation based on the molecule collision [43,44]. These applications above indicate that entropy is not only a good index to describe the uncertainty of a time series, but also a good method to describe the spatial continuity.

Therefore, the objective of this study was to quantitatively explore the evolution of precipitation from desert to inner oasis based on the Tarim Basin, northwestern China. A model describing precipitation evolution was developed from the information entropy theories, and then the parameters of model were fitted by the China Meteorological Forcing Dataset with the spatial resolution of 0.1 degrees. Through the simulation, the typical precipitation evolution patterns weare revealed, and the current stage of these oases were also identified. The establishment of the model and the simulated results provide a deeper insight from the perspective of informatics to understand the regional precipitation evolution of the desert–oasis system. These results are useful for the desertification prevention and multisource data fusion, especially in extreme drought desert areas.

2. Study Area, Data and Methodology

2.1. Study Area

The Tarim Basin, covering an area of approximately 560,000 km^2, is located in the south of Xinjiang Province in northwestern China [45,46], as shown in Figure 1. The Taklamakan Desert, which formed at least 5.3 Ma years ago, is in the center of the Tarim Basin [47]. The total area of the oasis is approximately 103,900 km^2 [11,48]. In the desert region, the annual precipitation is from less than 15 to 60 mm, increasing from the east to the west [49], and the increase rate is 10.15, 6.29, and 0.87 mm per decade in the mountain, oasis and desert areas, respectively, based on the observed data from 1960 to 2010 [50]. However, the annual potential evaporation is over 3200 mm based on the evaporation from water surface, and most of the basin is a generally unsuitable or extremely unsuitable area for human settlement [51]. Currently, the irrational reclamation of land and overuse of natural resources have led to the destruction of vegetation and shrinkage of the water area [52–54] and have threaten the security and sustainable development of oases [48].

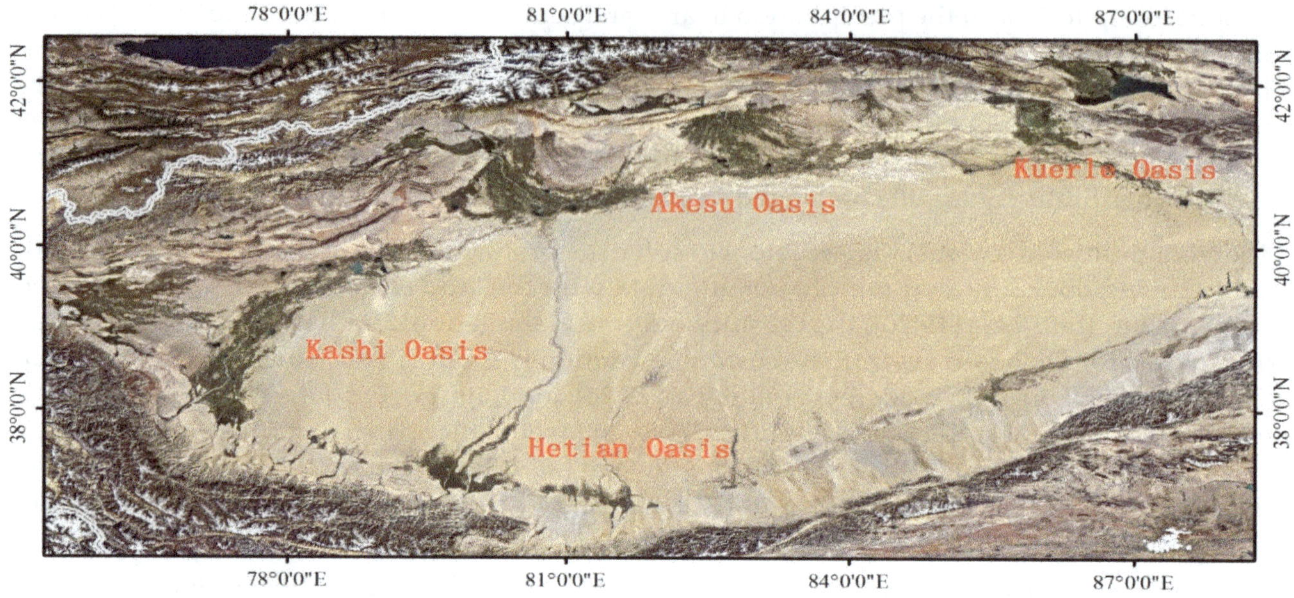

Figure 1. Spatial distribution of oasis in Tarim Basin, obtained from Google Earth. The linkage for Google's permissions is: https://www.google.ca/permissions/geoguidelines.

2.2. Data

In this study, the China Meteorological Forcing Dataset, developed by the Data Assimilation and Modeling Center for Tibetan Multispheres, Institute of Tibetan Plateau Research, Chinese Academy of Sciences [23,24], was used to test the performance of the model and to calibrate the parameters of the precipitation evolution form the desert toward oasis. The dataset was produced by merging a variety of data sources. More details on the dataset are given in the user's guide of the Dataset [55]. The spatial and temporal resolutions of the dataset are 0.1 degrees and 3 h, respectively, and the data length is 21 years (1990–2010). Based on the dataset, the spatial distributions of annual precipitation in Tarim basin are shown in Figure 2.

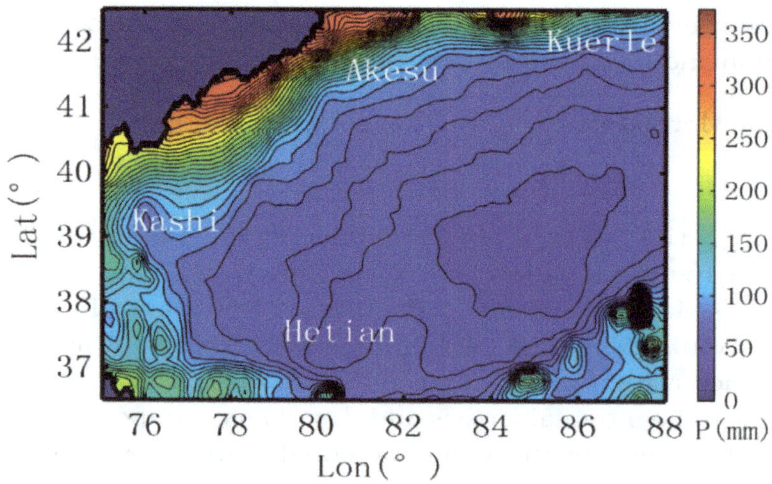

Figure 2. Spatial distribution of annual precipitation in Tarim Basin based on The China Meteorological Forcing Dataset.

2.3. Information Entropy and Principle of Maximum Entropy

Assuming the probability density function of a continuous variable x is expressed by $f(x)$, and the information entropy is calculated as [56,57]:

$$H(x) = -\int_a^b f(x)\ln(x)dx \tag{1}$$

$$\int_a^b f(x)dx = 1 \tag{2}$$

where a and b and the lower and upper boundaries of the variable x, respectively.

In addition, the known information of the variable represented by $g_i(x)$, such as mean and standard deviation, can be given by the following formula:

$$\int_a^b g_i(x)f(x) = N_i \tag{3}$$

Hence, the general solution of $f(x)$ can be obtained by solving the conditional extreme values using the method of Lagrange multipliers: the objective function is the maximum information entropy, and the constraint is the known information in Equation (3). More details about the mathematical derivation is presented in the literature [40,58]. The general form of $f(x)$ is given as:

$$f(x) = e^{\sum_{i=1}^{n} g_i(x)} \tag{4}$$

3. Establishment of the Model

Assuming the precipitation at the boundary area between oasis and desert is P_0, the precipitation evolution from desert toward the oasis is represented by the integration of a function $p(z)$, as given in Equation (5):

$$P(z) = P_0 + \int_0^Z p(z)dz \tag{5}$$

where z represents the distance from the desert–oasis boundary to a certain point of the oasis. $P(z)$ represents the precipitation at the location with distance z. Equation (5) can also be transformed into the following form:

$$P(z) - P_0 = \int_0^Z p(z)dz \tag{6}$$

Usually, $P(z)$ reaches a constant approximately when the oasis is wide enough. Here, Equation (6) can be written as:

$$\int_0^{+\infty} p(z)dz = C \tag{7}$$

where the constant C represents the maximum increment of precipitation with the increase of oasis width. Dividing both sides of the equation by the constant C, Equation (7) becomes:

$$\frac{1}{C}\int_0^{+\infty} p(z)dz = 1 \tag{8}$$

Hence, the item of $p(z)/C$ can be considered as a probability density function, and its information entropy is calculated by Equations (9) and (10):

$$H = -\int_0^Z \frac{1}{C}p(z)\ln\frac{p(z)}{C}dz \tag{9}$$

$$H = -\frac{1}{C}\int_0^Z p(z)\ln p(z)dz + \frac{1}{C}\ln C \quad (10)$$

As for a specific oasis, the spatial distribution can be considered as stable approximately in a mid-long term. Correspondingly, the arithmetic mean and geometric mean of precipitation evolution can be considered as constant, as given by Equations (11) and (12), respectively.

$$\int_0^\infty \frac{z}{C}p(z)dz = \mu_z \quad (11)$$

$$\int_0^\infty \frac{\ln(z)}{C}p(z)dz = \nu_z \quad (12)$$

where μ_z and ν_z represent the arithmetic mean and geometric mean. Here, the Lagrange function is obtained based on the information entropy of precipitation evolution with the constraints of Equations (8), (11) and (12), as given by Equation (13).

$$\begin{aligned} L = -\tfrac{1}{C}\textstyle\int_0^Z p(z)\ln p(z)dz + \tfrac{1}{C}\ln C + \lambda_0(\int_0^Z \tfrac{1}{C}p(z)dz - 1) \\ +\lambda_1(\textstyle\int_0^Z \tfrac{z}{C}p(z)dz - \mu_z) + \lambda_2(\int_0^Z \tfrac{\ln(z)}{C}p(z)dz - \nu_z) \end{aligned} \quad (13)$$

Let $\frac{\partial L}{\partial p(z)} = 0$, then the general solution of $p(z)/C$ is obtained by solving the conditional extreme values, as given in Equation (14):

$$p(z)/C = e^{\lambda_0 - 1 + \lambda_1 z + \lambda_2 z \ln z} \quad (14)$$

Equation (14) can also be written as:

$$p(z)/C = e^{\lambda_0 - 1}e^{\lambda_1 z}z^{\lambda_2} \quad (15)$$

Let $\lambda_2 = \alpha - 1$, $\lambda_1 = -1/\beta$. It is easy to obtain that $e^{\lambda}{}_0{}^{-1} = -{}^{\alpha}/\Gamma(\alpha)$. Thus, the probability density function $p(z)/C$ can also be expressed as:

$$p(z)/C = \frac{1}{\Gamma(\alpha)\beta^\alpha}z^{\alpha-1}e^{-z/\beta} \quad (16)$$

Substituting Equation (16) into Equation (5), the evolution of precipitation with the oasis width increase is:

$$P(z) = P_0 + C\int_0^z \frac{1}{\Gamma(\alpha)\beta^\alpha}z^{\alpha-1}e^{-\frac{z}{\beta}}dz \quad (17)$$

In fact, the right-hand term of Equation (16) is the probability density function of a gamma distribution. Therefore, Equation (17) shows that the precipitation evolution model is based on the scaled cumulative distribution function of a gamma distribution with an initial precipitation. In other words, the model is based on the linear transformation of the cumulative distribution function of a gamma distribution. The parameters of the model include the shape factor α, scale factor β, zooming constant C and initial precipitation P_0 at the boundary between desert and oasis.

Assuming the initial precipitation P_0 at the boundary between desert and oasis is 50 mm/year and the zooming constant C is 50, the illustration of the precipitation evolution are shown in Figure 3 with different shape and scale factors.

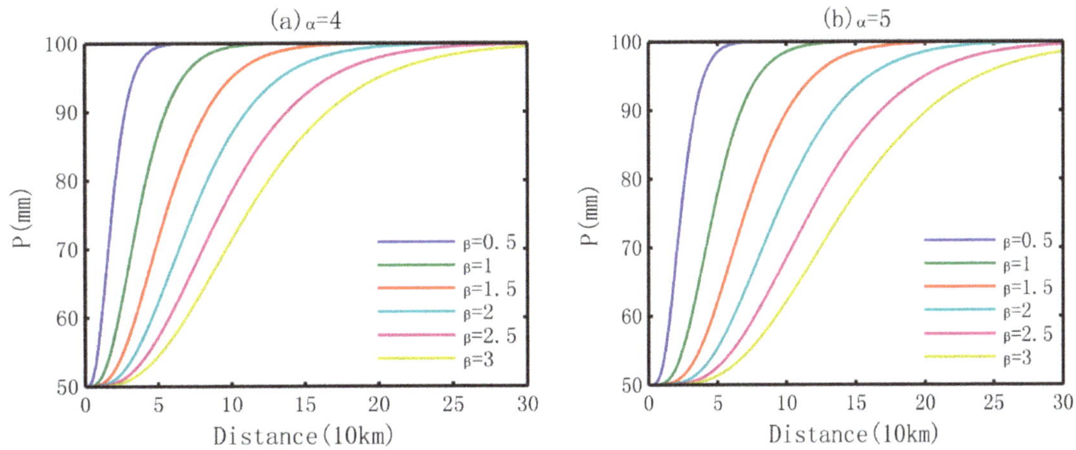

Figure 3. Illustration of precipitation evolution from desert to inner oasis. Assuming the initial precipitation P_0 at the boundary between desert and oasis is 50 mm/year and the zooming constant C is 50, the precipitation evolution is shown with different shape factor and different scale factor.

4. Calibration of the Model and the Simulated Results

4.1. Calibration of the Model

4.1.1. Performance of the Model

In this study, four main oases were used to test the performance of the model. These oases are Kashi Oasis located in the western basin, Akesu Oasis in the northern basin, Kuerle Oasis in the northeastern basin and Hetian Oasis in the southwestern basin. The evolution routes of precipitation from desert to oasis are shown in Figure 4. Generally, these routes are located in the middle of these oases.

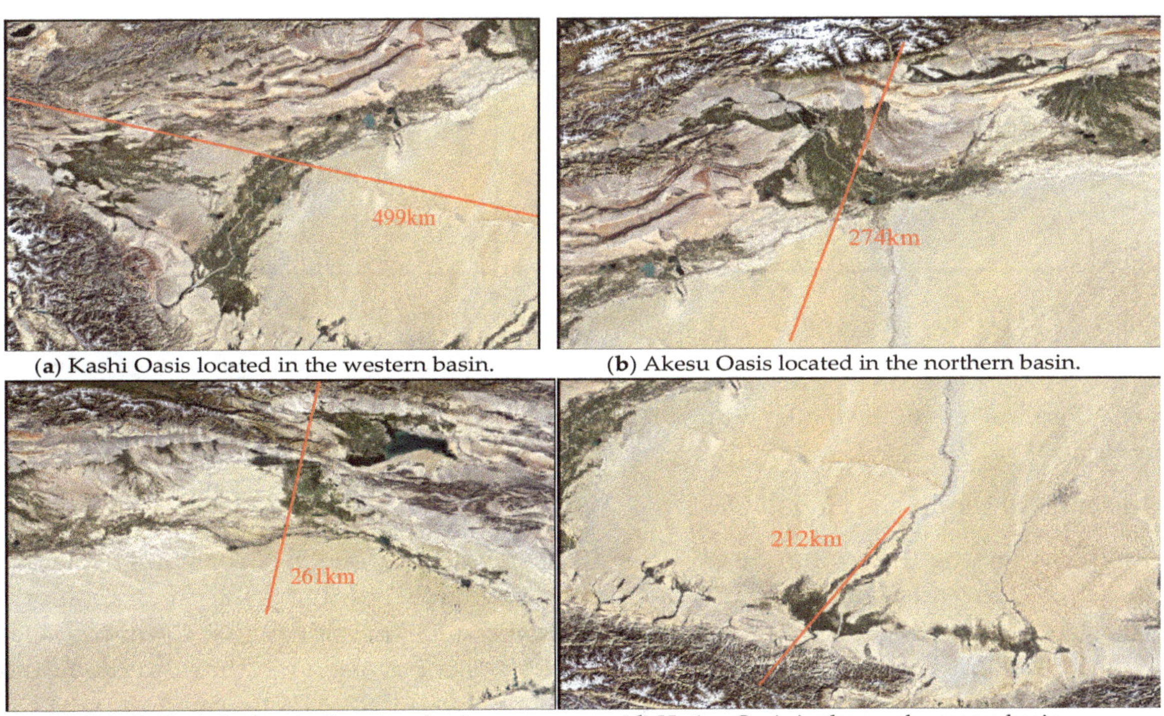

(**a**) Kashi Oasis located in the western basin. (**b**) Akesu Oasis located in the northern basin. (**c**) Kuerle Oasis in the northeastern basin (**d**) Hetian Oasis in the southwestern basin.

Figure 4. Selected routes of precipitation transition from the desert to oasis in different location of the Tarim Basin, obtained from Google Earth. The permission of Google to use Google Earth Periodicals is given at: https://www.google.ca/permissions/geoguidelines.

Following the routes shown in Figure 4, the evolution of precipitation from desert to oasis is obtained, and the results are scattered in Figure 5. Here, the parameters of the model are obtained by nonlinear fitting. The correlation coefficient is used to evaluate the performance of the model given by Equation (17). The fitted results and the square of correlation coefficient, R^2, of the four selected routes are also shown in Figure 5.

The model performs well in three oases: Kashi Oasis, Akesu Oasis and Kuerle Oasis, corresponding to R^2 of 0.99, 0.93 and 0.97, respectively. However, the model does not perform well in Hetian Oasis and the R^2 is 0.47. To improve the accuracy trend of precipitation evolution, the model should be calibrated further for the Hetian Oasis.

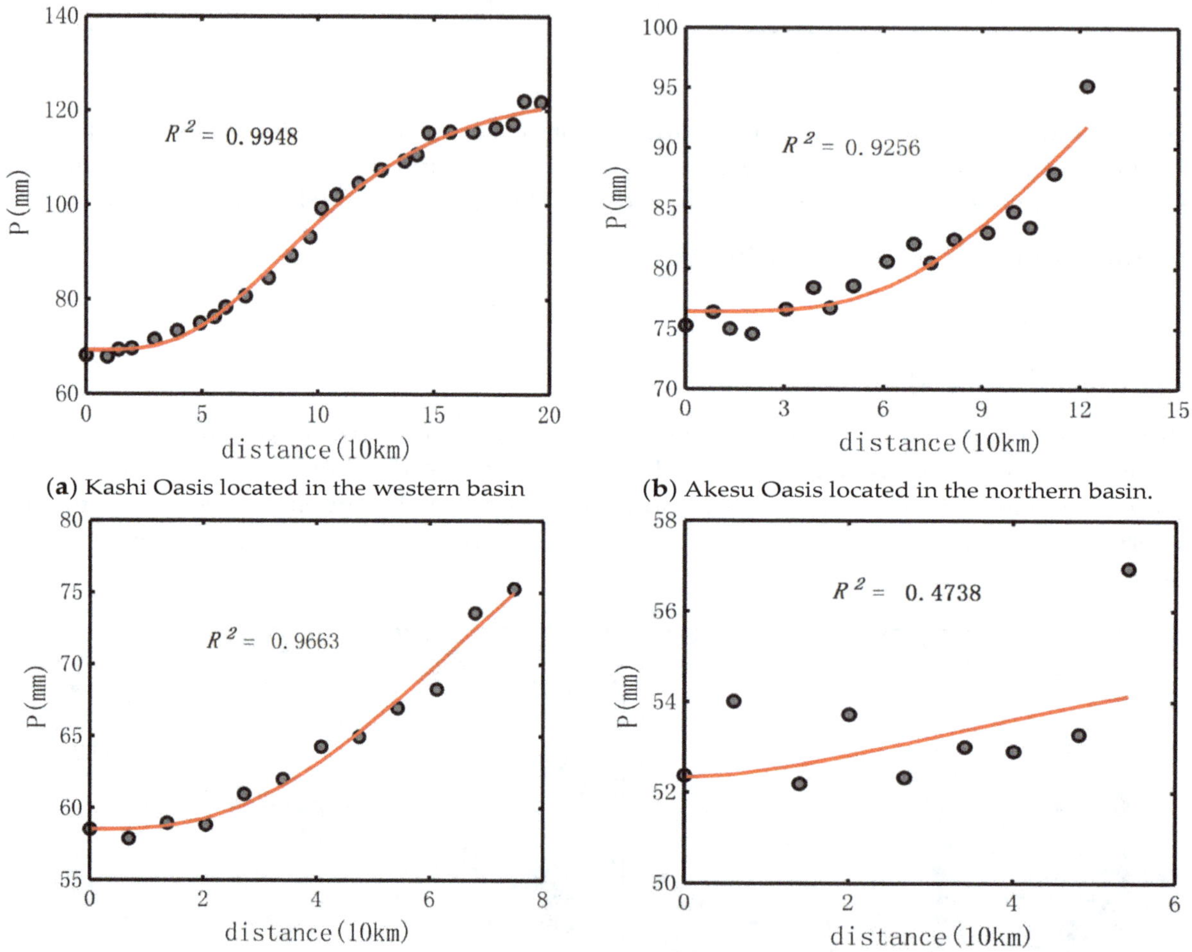

(**a**) Kashi Oasis located in the western basin

(**b**) Akesu Oasis located in the northern basin.

(**c**) Kuerle Oasis in the northeastern basin.

(**d**) Hetian Oasis in the southwestern basin.

Figure 5. Performance of the model simulating precipitation from desert to oasis in different areas of Tarim Basin.

4.1.2. Calibration of Model for Hetian Oasis

Through combining the spatial distribution of oasis and the local hydrological cycle, two main reasons were found to lead to the poor performance of the model. The first is the much smaller width of Hetian Oasis. As shown in Figure 5, the Hetian Oasis is about 55 km wide, while Kashi Oasis is 200 km, Akesu Oasis is 120 km and Kuerle Oasis is 80 km. The second reason is that the prevailing wind direction of the Hetian Oasis is opposite to the cold effect of oasis [59–61]. Our previous study revealed that the opposite direction leads to water vapor from the local evapotranspiration accumulating over the buffer zone located in the down wind direction of the oasis [49], as shown in Figure 6a. Hence, the evolution of precipitation in the Hetian Oasis should be divided into two phases: the buffer zone with abnormal high precipitation and the regular pattern of the precipitation evolution after the buffer zone, as shown in Figure 6b.

By overlapping the spatial distribution of water vapor content and oasis layout, it was found that the first four grids should be removed to calibrate the model. After their removal, the calibrated model performs much better, and the R^2 increased from 0.47 to 0.87, as shown in Figure 6b.

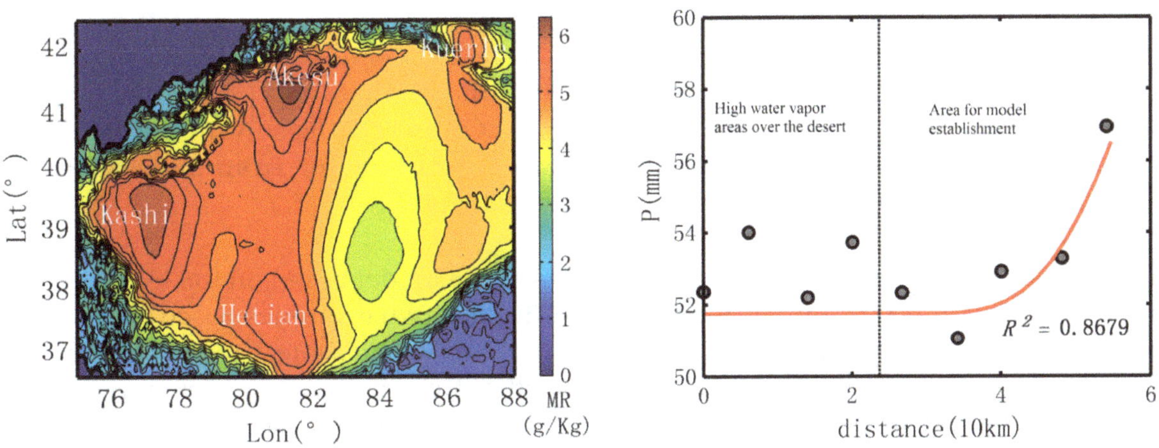

(**a**) Spatial distribution of water vapor. (**b**) Calibrated model in Hetian Oasis.

Figure 6. Spatial distributions of water vapor in Tarim Basin and the calibrated model in Hetian.

4.1.3. Parameter Inversion

Based on the results obtained by nonlinear fitting, the parameters of the model for all four oases are obtained, as shown in Table 1. According to Equation (17) and Figure 3, the meanings of these parameters are clear. P_0 is the initial value that represents the precipitation at the boundary area between oasis and desert. Here, AKesu Oasis exhibits largest value, followed by Kashi, Kuerle and Hetian respectively. The zooming constant, C, represents the maximum potential increment of precipitation with the oasis width increase. The results of zooming constant C suggest the maximum promotion on local precipitation in Kashi, middle level in AKesu and Kuerle, and the minimum promotion in Hetian Oasis. Shape and scale factors represent the increase rate of the precipitation. A larger value is usually accompanied with a smaller increase rate, and vice versa (Figure 3).

Table 1. Performance of the model and the parameters inversion for precipitation evolution from desert to oasis in different locations of the Tarim Basin.

Main Oases	Initial Precipitation P_0 (mm)	Zooming Constant C (-)	Shape Factor α (-)	Scale Factor β (-)	Correlation R^2
Kashi Oasis (Western Basin)	69.31	54.24	4.64	2.32	0.99
Akesu Oasis (Northern Basin)	76.46	43.04	5.16	3.00	0.93
Kuerle Oasis (Northeastern Basis)	58.54	44.60	3.36	3.00	0.97
Hetian Oasis without calibration (Southwestern Basin)	52.34	3.34	2.00	3.00	0.47
Hetian Oasis after calibration (Southwestern Basin)	51.76	32.57	1.02	5.41	0.87

4.1.4. Influence of the Local Terrain

Apart from the cold-wet effect of oasis, the significant uplift of the terrain would lead to a sharp decrease in temperature and result in a substantial increase in precipitation at the local scale. As given in our previous study [49], the increment of the elevation is about 80–100 m. Meanwhile, the widths of Kashi, Akesu, Kuerle and Hetian Oases are approximately 210, 120, 80 and 55 km, respectively, as shown in Figure 5. This indicates that the increase rate is ordered by: Hetian > Kuerle > Akesu > Kashi.

However, this trend is opposite to the increment and increase rate of precipitation from desert to oasis. Hence, the increase of precipitation from desert to oasis is not dominated by the local terrain.

4.1.5. Performance of the Model at Seasonal Scale

The performance of the model at seasonal scale was also analyzed. The first problem confronted is that the result does not converge when solved by nonlinear fitting using MATLAB (MathWorks, Natick, MA, USA). Here, the results are based on least square method and the results are shown by Figure 7. Generally, the model seems to perform well with several exceptions, including the winter of Kuerle and Hetian Oases, and the summer of Hetian Oasis. The reason can be attributed to the transportation of water vapor. As discussed above, the wind blowing from the oasis to the desert would decrease the cold-wet effect of the oasis substantially [49].

However, the second problem is that the solutions of the model based on the least square method are substantially affected by the upper boundary conditions. Two examples based on different upper boundary conditions are given to illustrate this problem. As shown in Table 2, the initial precipitation P_0 is not affected by the upper boundary conditions. The other three parameters, however, are influenced significantly. For example, the difference of the parameter C estimated for the summer of the Kuerle Oasis can be up to 140 mm, although the R^2 is 0.99 for both solutions. Here, the failure of the model may be resulted from the resolution of the data or the more complex water vapor transportation. Hence, further studies should be carried out to explore the simulation of the season precipitation evolution from desert to the oasis.

Table 2. Comparisons on the sensitivity of parameter with different upper boundary conditions (UBC) at seasonal scale. R^2 are indicated in italics and Bold.

Oasis	Season	UBC: $P_{0u} = 10$; $C_u = 60$, $\alpha_u = 8$, $\beta_u = 8$					UBC: $P_{0u} = 10$; $C_u = 200$, $\alpha_u = 20$, $\beta_u = 20$				
		P_0 (mm)	C (-)	α (-)	β (-)	R^2	P_0 (mm)	C (-)	α (-)	β (-)	R^2
(a) Kashi	Spring	16.87	30.80	2.47	8.00	*0.99*	16.53	57.55	1.78	20.00	*0.99*
	Summer	31.81	21.93	8.00	1.12	*0.98*	31.81	21.93	9.11	0.99	*0.98*
	Autumn	11.20	11.59	1.91	8.00	*0.99*	10.94	18.18	1.33	20.00	*0.99*
	Winter	8.44	7.39	5.49	2.55	*0.99*	8.44	7.39	5.49	2.55	*0.99*
(b) Akesu	Spring	15.14	60.00	8.00	2.37	*0.83*	15.24	32.22	20.00	0.65	*0.89*
	Summer	35.03	14.36	0.74	8.00	*0.97*	34.94	19.32	0.60	20.00	*0.98*
	Autumn	13.59	60.00	8.00	2.38	*0.87*	13.66	200.00	10.08	2.12	*0.88*
	Winter	9.19	2.00	1.42	2.80	*0.86*	9.19	2.00	1.42	2.80	*0.86*
(c) Kuerle	Spring	12.35	11.05	1.91	8.00	*0.98*	12.33	28.21	1.65	20.00	*0.98*
	Summer	27.10	60.00	3.32	4.11	*0.99*	27.03	199.94	2.71	9.94	*0.99*
	Autumn	11.65	4.41	1.47	8.00	*0.94*	11.64	5.87	1.36	11.83	*0.94*
	Winter	7.01	0.65	8.00	8.00	*0.02*	7.01	2.91	17.15	16.99	*0.01*
(d) Hetian	Spring	10.74	60.00	8.00	1.41	*0.96*	10.76	199.35	8.15	1.73	*0.96*
	Summer	23.76	60.00	8.00	1.92	*0.09*	23.76	200.00	20.00	0.54	*0.16*
	Autumn	8.11	60.00	6.02	3.03	*0.78*	8.11	200.00	5.60	4.52	*0.78*
	Winter	9.76	60.00	8.00	2.33	*0.01*	9.73	200.00	20.00	0.57	*0.05*

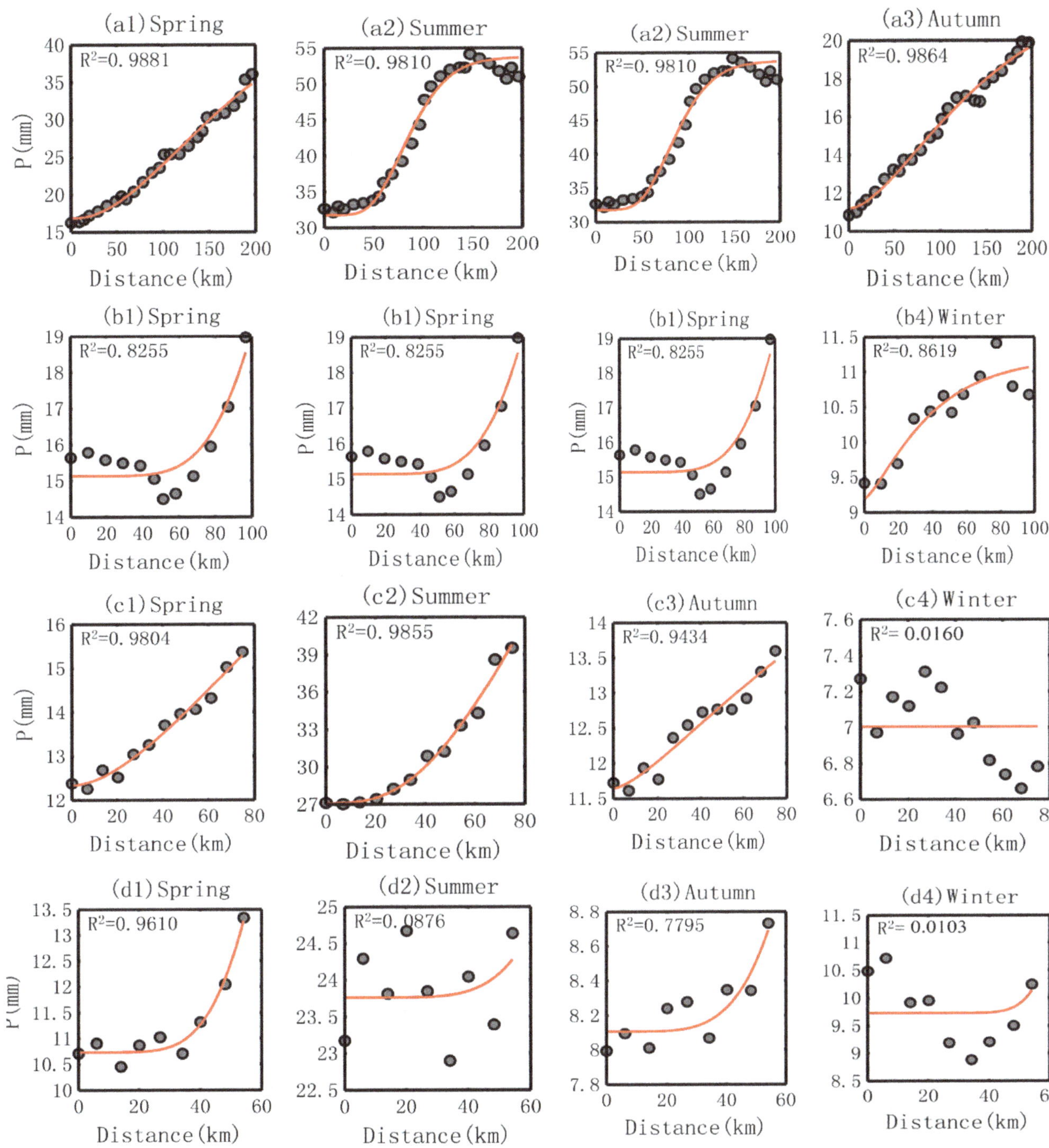

Figure 7. Performance of the information entropy based model at seasonal scale: (**a**) Kashi Oasis located in the western basin; (**b**) Akesu Oasis located in the northern basin; (**c**) Kuerle Oasis in the northeast basin; and (**d**) Hetian Oasis in the southwestern basin.

4.2. Simulated Results

Based on the performance of the model at different scales, the parameters are reasonable at annual scale. The evolution of precipitation with the increase of oasis width was calculated by Equation (17) using the parameters given in Table 1. The results are shown in Figure 8, from which the basic patterns of precipitation evolution are revealed and the current stage of these oases are also obtained. More details on the two aspects are as follows.

4.2.1. Basic Patterns of Precipitation Evolution

According to the simulated results (Figure 8), it could be found that the basic patterns of precipitation evolution exhibit four stages: a small increase at first, then a much faster increase after a threshold of oasis width, followed by a small increase rate again when the oasis width reaches a relatively large width, and finally the increase rate decreases to close to zero when the oasis is wide enough. The threshold values of the four stages are also obtained (Table 3).

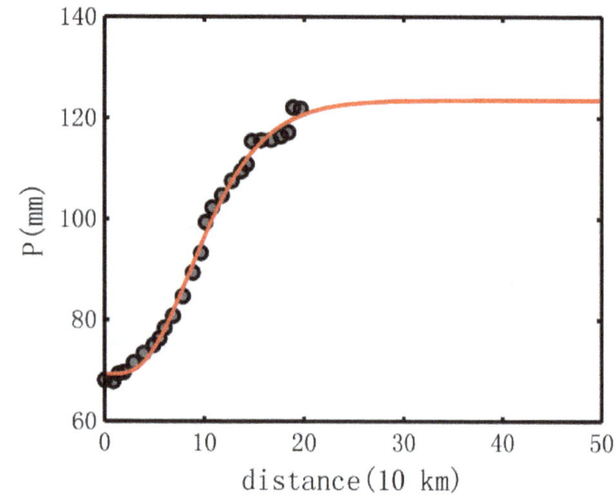

(**a**) Kashi Oasis located in the western basin.

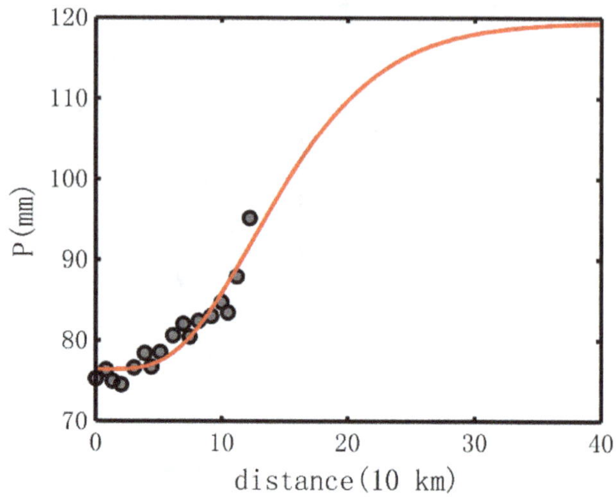

(**b**) Akesu Oasis located in the northern basin.

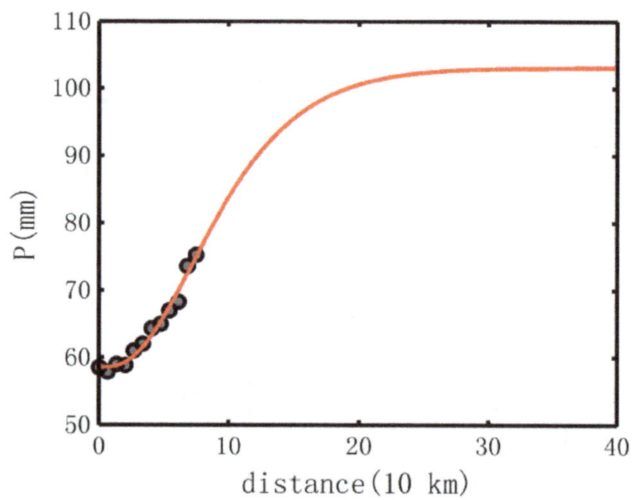

(**c**) Kuerle Oasis in the northeastern basin.

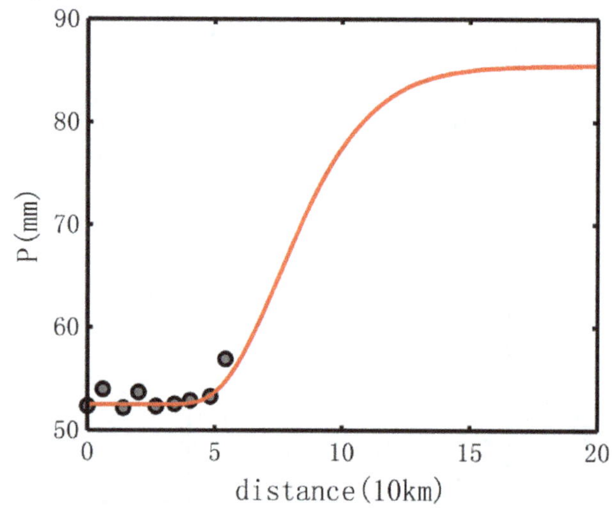

(**d**) Hetian Oasis in the southwestern basin.

Figure 8. The simulated precipitation evolution from desert to oasis with the oasis width increase.

1. The first stage: Small increase rate. The oasis width of the first stage (small increase rate) is about 30–50 km. This means that, if the oasis width is less than the threshold value, the promotion of oasis' cold-wet effect on local precipitation is not substantial. More specifically, this width is about 40, 50, 40 and 55 km for Kashi, Akesu, Kuerle and Hetian Oases, respectively.
2. The second stage: Large increase rate. The second threshold value is about 150–185 km. In this period, the increase rate of precipitation is much larger than before. Generally, precipitation exhibits a linear trend with the oasis width increase. Hence, the increase rate of precipitation is approximately constant. As shown in Table 3, the increase rate varies substantially in the four oases, with the value of 3.32, 2.86, 1.76 and 0.65 mm/10 km corresponding to Kashi Oasis, Akesu Oasis, Kuerle Oasis and Hetian Oasis, respectively.

3. The third stage: Small increase rate again. The threshold value for the third stage is about 200–300 km. Two oases, Kashi and Hetian, display smaller value of about 200 km. The other two oases, Akesu and Kuerle, display larger values of 265 and 300, respectively. This means that Kashi and Hetian Oases would reach the maximum promotion on local precipitation easier than the other oases because of the smaller threshold value. However, the opposite direction between the cold effect of oasis and water vapor transportation lead to Hetian Oasis experiencing a longer first stage than the other three oases, as well as smaller precipitation locally.
4. The fourth stage: Stable status. When the oasis width is larger than the threshold value of the third stage, the precipitation evolution reaches the fourth stage, which is almost constant.

Table 3. Typical patterns of precipitation evolution with the increase of oasis width in different oases of the Tarim Basin based on the simulated results.

Stages	Threshold Value and Average Slope	Kashi Oasis	Akesu Oasis	Kuerle Oasis	Hetian Oasis
(1) Small Increase Rate	Threshold value	0–40 km	0–50 km	0–40 km	0–55
	Average slope	Not substantial	Not substantial	Not substantial	Not substantial
(2) Large Increase Rate	Threshold value	40–185 km	50–185 km	40–180 km	55–155 km
	Average slope	3.32 mm/10 km	1.76 mm/10 km	2.86 mm/10 km	0.65 mm/10 km
(3) Small Increase Rate Again	Threshold value	185–220 km	185–300 km	180–265 km	155–200 km
	Average slope	Not substantial	Not substantial	Not substantial	Not substantial
(4) Stable Status	Threshold value	>220 km	>300 km	>265 km	>200 km
	Average slope	Close to 0	Close to 0	Close to 0	Close to 0

4.2.2. Current Stages of Different Oases

The widths of Kashi, Akesu, Kuerle and HetianOases are 200 km, 135 km, 80 km and 60 km, respectively, as shown in Figure 5. By contrast with Figure 8, it can be obtained that the four oases belong to different stages that are manifested by the following aspects.

1. Kashi Oasis. Kashi Oasis is at the beginning of the fourth stage, in which the precipitation is stable with the oasis width increase. This implies that the annual precipitation will be the same as the maximum value (about 120 mm/year) if the oasis expands towards the desert.
2. Akesu Oasis. The width of the Akesu Oasis is about 135 km. Hence, the evolution of precipitation in Akesu Oasis is at the first half of the second stage currently. The second stage means largest increase rate of precipitation with oasis width increase but the middle level precipitation amount. Hence, if the oasis expands towards the desert, the precipitation would increase substantially. However, the total increment of precipitation should be less than the increment precipitation in Kashi Oasis.
3. Kuerle Oasis. As shown in Figure 8 and Table 3, the current width of the oasis is at the first half of the second stage. Hence, it would exhibit similar characteristics to AKesu Oasis.
4. Hetian Oasis. Different from the other three oases, the Hetian Oasis is at the end of the first stage. With the oasis width increase, the precipitation would increase at a slope of approximately 0.65 mm/km. That means the total increment of precipitation is the least if the oasis expanded toward desert in Hetian Oasis because of the much smaller initial condition of precipitation.

5. Discussions

5.1. The Simulated Results for Desertification Prevention

In the Tarim Basin, affected by the substantial increase of local human activities, the destruction of oasis environments is increasing, resulting from the irrational reclamation of land and overuse of natural resources [62]. In fact, the irrigated oasis area has exceeded about 33.6% of the maximum carrying capacity of local precipitation [52]. The increase of the arable land is at the expense of the destruction of vegetation and shrinkage of the water area [53,54]. Even worse, the expansion of artificial oases has led to the degradation of natural oases and the oasis–desert ecotone, which may threaten the security and sustainable development of oases [48]. These problems require understanding the interaction between the oasis and local hydrological cycle, and to selecting the more appropriate areas for desertification prevention and layout of oasis expansion.

According to the simulated results shown in Figure 8 and the current stage of the oasis precipitation, it can be concluded that Kashi Oasis is at the beginning of the fourth stage, which implies that the increment of precipitation would be about 120 mm/year. The increment of precipitation in Akesu and Heian Oasis are the middle level because both oases are at the first half of the second stage. The least promotion would appear in the southwestern basin of Hetian Oasis because it is at the end of the first stage. Therefore, Kashi Oasis of the western basin is the most appropriate area for desertification prevention, while the southwestern basin, i.e., Hetian Oasis, presents a greater challenge.

5.2. The Model for Multisource Data Fusion

The considerable challenge to the data assimilation is the resolution [22]. In this study, the spatial resolution of the assimilated data is 0.1 degrees [23,24], and many researchers have used this dataset to analyze the regional hydrological processes, such as the impact of lake effects on the temporal and spatial distribution of precipitation in the Nam Co basin of the Tibetan Plateau [25], evaporation from the lake [26] and the impact of climatic factors on permafrost in the Tibetan Plateau [27]. The good spatial continuity of the dataset is reflected by gradual changes of the local hydrological processes.

However, the performance of the dataset is still not clear in the desert areas. In this study, the dataset agrees well with simulated results of the model based on information entropy, which is not only demonstrated by the good performance of the model, but also reflected by the good spatial continuity of annual precipitation evolution in the desert–oasis system. Furthermore, the poor performance of the model without the calibration is also an indicator of the different interaction between the oasis and water vapor transportation in the Hetian Oasis. In fact, opposite direction between the water vapor transportation and cold effect of the oasis leads to water vapor from local evapotranspiration accumulating near the desert boundary, where the higher temperature reduces the promotion of oasis on local precipitation substantially [49]. Hence, the model is a useful tool to test the spatial continuity of the dataset obtained from merging multisource data, especially in those areas with extreme natural conditions, for example, desert areas.

6. Conclusions

Based on the established model, parameter inversion, forward simulation, and the discussions on the potential applications of the method, the conclusions are as follows.

1. The model describing precipitation evolution from desert to oasis is based on information entropy theory: the evolution of precipitation is considered as a linear transformed cumulative distribution function, and the probability density function is obtained by solving the Lagrange conditioned extreme value with the objective function of maximum information entropy. The constraints are the constant arithmetic mean and geometric mean of precipitation evolution from desert to oasis when the oasis is wide enough. The general form of the model is a linear transformed cumulative distribution function of a gamma distribution.
2. The model performs well when the oasis is wide enough. The R^2 between the simulated result and the dataset is 0.99, 0.93 and 0.97 for the Kashi Oasis, Akesu Oasis and Kuerle Oasis, respectively. In the Hetian Oasis, however, the model performs poorly with the R^2 of 0.47. The reason is mainly the opposite direction between the oasis cold effect and the water vapor transportation that results in a buffer zone in the first several tens of kilometers. After removing the buffer zone, the model shows a good accuracy with the R^2 of 0.87. However, the model fails to describe the seasonal precipitation evolution because of the non-convergence problem solved by nonlinear fitting and the influence of upper boundary condition solved by the least square method.

3. The simulated results indicate that the evolution of annual precipitation from desert to oasis includes four main stages: slow increase first, then fast increase, followed by a slow increase again and finally a stable stage. The threshold value of the first stage is 30–50 km, implying the promotion of local precipitation is very small if the width of the oasis is less than this value. The second stage can be approximated linearly, with the threshold value of 155–185 km and the slope of 3.32, 2.86, 1.76 and 0.65 mm/10 km for the Kashi Oasis, Akesu Oasis, Kuerle Oasis and Hetian Oasis, respectively. The upper boundary of the third stage is about 220–300 km, after which the precipitation evolution is almost constant.
4. Currently, the four main oases in the Tarim basin are at different stages. The Kashi Oasis has reached the fourth stage, the Akesu and Kuerle Oases are at the first half of the second stage, and the Hetian Oasis is at the end of the first stage. These current stages reflect different local hydrological effect when prevent desertification in the four main oases: maximum promotion on precipitation in the Kashi Oasis, middle promotion in the Akesu Oasis and Kuerle Oasis, and the least promotion in the Hetian Oasis.

Therefore, the theory of the model and simulated results would provide a deeper insight from the perspective of informatics into understanding the precipitation evolution from desert to oasis, which is not only helpful in preventing desertification but also helpful in merging multisource data in the extreme drought desert areas.

Author Contributions: X.Z. as the first author was responsible for establishing the model, collecting the data to calibrate the model, plotting the figures and writing of the main body of the manuscript. Z.N., the second author and corresponding author, searched the background of the hydrological effect of oasis and proposed the idea of this study, and helped to revise this manuscript. W.L., the third author, contributed to the data analysis and manuscript approving. All authors have read and approved the final manuscript.

Acknowledgments: The forcing dataset used in this study was developed by Data Assimilation and Modeling Center for Tibetan Multi-spheres, Institute of Tibetan Plateau Research, Chinese Academy of Sciences.

References

1. Taha, H.; Akbari, H.; Rosenfeld, A. Heat island and oasis effects of vegetative canopies: Micro-meteorological field-measurements. *Theor. Appl. Climatol.* **1991**, *44*, 123–138. [CrossRef]
2. Kai, K.; Matsuda, M.; Sato, R. Oasis Effect Observed at Zhangye Oasis in the Hexi Corridor, China. *J. Meteorol. Soc. Jpn.* **1997**, *75*, 1171–1178. [CrossRef]
3. Chu, P.C.; Lu, S.; Chen, Y. A numerical modeling study on desert oasis self-supporting mechanisms. *J. Hydrol.* **2005**, *312*, 256–276. [CrossRef]
4. Feng, Q.; Si, J.; Zhang, Y.; Yao, J.; Liu, W.; Su, Y.H. Microclimatic characteristics of the Heihe oasis in the hyperarid zone of China. *J. Geogr. Sci.* **2006**, *16*, 34–44. [CrossRef]
5. Potchter, O.; Goldman, D.; Kadish, D.; Iluz, D. The oasis effect in an extremely hot and arid climate: The case of southern Israel. *J. Arid. Environ.* **2008**, *72*, 1721–1733. [CrossRef]
6. Li, X.M.; Yang, J.S.; Liu, M.X.; Liu, G.M.; Yu, M. Spatiotemporal changes of soil salinity in arid areas of south Xinjiang using electromagnetic induction. *J. Integr. Agr.* **2012**, *11*, 1365–1376. [CrossRef]
7. Hao, X.; Li, W. Oasis cold island effect and its influence on air temperature: A case study of Tarim Basin, Northwest China. *J. Arid. Land* **2016**, *8*, 172–183. [CrossRef]
8. Li, X.; Yang, K.; Zhou, Y. Progress in the study of oasis-desert interactions. *Agric. Forest. Meteorol.* **2016**, *230–231*, 1–7. [CrossRef]
9. Li, X.; Cheng, G.D.; Ge, Y.M.; Li, H.Y.; Han, F.; Hu, X.L.; Tian, W.; Tian, Y.; Pan, X.D.; Nian, Y.Y. Hydrological Cycle in the Heihe River Basin and Its Implication for Water Resource Management in Endorheic Basins. *J. Geophys. Res. Atmos.* **2018**, *123*, 890–914. [CrossRef]
10. Zhang, H.; Wu, J.W.; Zheng, Q.H.; Yu, Y.Y. A preliminary study of oasis evolution in the Tarim Basin, Xinjiang, China. *J. Arid. Environ.* **2003**, *55*, 545–553.
11. Qiu, G.Y.; Li, H.Y.; Zhang, Q.T.; Chen, W.; Liang, X.J.; Li, X.Z. Effects of evapotranspiration on mitigation of urban temperature by vegetation and urban agriculture. *J. Integr. Agric.* **2013**, *12*, 1307–1315. [CrossRef]
12. Jin, X.M.; Hu, G.C.; Li, W.M. Hysteresis effect of runoff of the Heihe River on vegetation cover in the Ejina

Oasis in Northwestern China. *Earth Sci. Front.* **2008**, *15*, 198–203. [CrossRef]
13. Kong, Y.; Pang, Z.; Froehlich, K. Quantifying recycled moisture fraction in precipitation of an arid region using deuterium excess. *Tellus B* **2013**, *65*, 19251. [CrossRef]
14. Gat, J.R.; Klein, B.; Kushnir, Y.; Roether, W.; Wernli, H.; Yam, R.; Shemesh, A. Isotope composition of air moisture over the Mediterranean Sea: An index of the air-sea interaction pattern. *Tellus B* **2003**, *55*, 953–965. [CrossRef]
15. Worden, J.; Noone, D.; Bowman, K.; Beer, R.; Eldering, A.; Fisher, B.; Gunson, M.; Goldman, A.; Herman, R.; Kulawik, S.S. Importance of rain evaporation and continental convection in the tropical water cycle. *Nature* **2007**, *445*, 528–532. [CrossRef] [PubMed]
16. Ueta, A.; Sugimoto, A.; Iijima, Y.; Yabuki, H.; Maximov, T.C. Contribution of transpiration to the atmospheric moisture in eastern Siberia estimated with isotopic composition of water vapor. *Ecohydrology* **2014**, *7*, 197–208. [CrossRef]
17. Vallet-Coulomb, C.; Gasse, F.; Sonzogni, C. Seasonal evolution of the isotopic composition of atmospheric water vapour above a tropical lake: Deuterium excess and implication for water recycling. *Geochim. Cosmochim. Acta* **2008**, *72*, 4661–4674. [CrossRef]
18. Li, X.; Liu, S.M.; Xiao, Q.; Ma, M.G.; Jin, R.; Che, T.; Wang, W.Z.; Hu, X.L.; Xu, Z.W.; Wen, J.G. A multiscale dataset for understanding complex eco-hydrological processes in a heterogeneous oasis system. *Sci. Data* **2017**, *4*, 170083. [CrossRef] [PubMed]
19. Robinson, A.R.; Leslie, W.G. Estimation and prediction of oceanic eddy fields. *Prog. Oceanogr.* **1985**, *14*, 385–510. [CrossRef]
20. Moore, A.M.; Cooper, N.S.; Anderson, D.L.T. Initialization and data assimilation in models of the Indian Ocean. *J. Phys. Oceanogr.* **1987**, *17*, 1965–1977. [CrossRef]
21. Balsamo, G.; Mahfouf, J.F.; Belair, S. A land data assimilation system for soil moisture and temperature: An information content study. *J. Hydrometeorol.* **2007**, *8*, 1225–1242. [CrossRef]
22. Moore, A.M. *Some Challenges and Advances in Regional Ocean Data Assimilation. Seminar on Data Assimilation for Atmosphere and Ocean*; European Centre for Medium-Range Weather Forecasts: Reading, UK, 2011.
23. Yang, K.; He, J.; Tang, W.; Qin, J.; Cheng, C.C.K. On downward shortwave and long wave radiations over high altitude regions: Observation and modeling in the Tibetan Plateau. *Agric. For. Meteorol.* **2010**, *150*, 38–46. [CrossRef]
24. Chen, Y.; Yang, K.; He, J.; Qin, J.; Shi, J.C.; Du, J.Y.; He, Q. Improving land surface temperature modeling for dry land of China. *J. Geophys. Res. Atmos.* **2011**, *116*, 132–147. [CrossRef]
25. Dai, Y. *The Impact of Lake Effects on the Temporal and Spatial Distribution of Precipitation at Nam Co, on the Tibetan Plateau*; AGU: Washington, DC, USA, 2015.
26. Ma, N.; Szilagyi, J.; Niu, G.Y.; Zhang, Y.S.; Zhang, T.; Wang, B.B.; Wu, Y.H. Evaporation variability of Nam Co Lake in the Tibetan Plateau and its role in recent rapid lake expansion. *J. Hydrol.* **2016**, *537*, 27–35. [CrossRef]
27. Gao, S.; Wu, Q.; Zhang, Z.; Xu, X. Impact of climatic factors on permafrost of the Qinghai-Xizang Plateau in the time-frequency domain. *Quatern. Int.* **2015**, *374*, 110–117. [CrossRef]
28. Singh, V.P. The use of entropy in hydrology and water resources. *J. Hydrol. Process* **1997**, *11*, 587–626. [CrossRef]
29. Avseth, P.; Mukerji, T.; Mavko, G. Quantitative Seismic Interpretation. *Episodes* **2005**, *3*, 236–237.
30. Ebrahimi, N. Stochastic properties of a cumulative damage threshold crossing model. *J. Appl. Probab.* **1999**, *36*, 720–732. [CrossRef]
31. Crutchfield, J.; Feldman, D.P. Regularities unseen, randomness observed: Levels of entropy convergence. *Chaos* **2003**, *13*, 25–54. [CrossRef] [PubMed]
32. Chapman, T.G. Entropy as a measure of hydrologic data uncertainty and model performance. *J. Hydrol.* **1986**, *85*, 111–126. [CrossRef]
33. Maruyama, T.; Kawachi, T.; Singh, V.P. Entropy-based assessment and clustering of potential water resources availability. *J. Hydrol.* **2005**, *309*, 104–113. [CrossRef]
34. Brunsell, N.A. A multiscale information theory approach to assess spatial-temporal variability of daily precipitation. *J. Hydrol.* **2010**, *385*, 165–172. [CrossRef]
35. Hasan, M.M.; Dunn, P.K. Entropy consistency in rainfall distribution and potential water resource availability in Australia. *J. Hydrol. Process.* **2011**, *25*, 2613–2622. [CrossRef]
36. Yang, Y.; Burn, D.H. An entropy approach to data collection network design. *J. Hydrol.* **1994**, *157*, 307–324.

[CrossRef]

37. Yoo, C.; Jung, K.; Lee, J. Evaluation of rain gauge network using entropy theory: Comparison of mixed and continuous distribution function applications. *J. Hydrol. Eng.* **2008**, *13*, 226–235. [CrossRef]
38. Singh, V.P. Entropy theory for derivation of infiltration equations. *Water Resour. Res.* **2010**, *46*, W03527. [CrossRef]
39. Singh, V.P. Entropy theory for movement of moisture in soils. *Water Resour. Res.* **2010**, *46*, W03516. [CrossRef]
40. Zhou, X.; Lei, W.; Ma, J. Entropy Base Estimation of Moisture Content of the Top 10-m Unsaturated Soil for the Badain Jaran Desert in Northwestern China. *Entropy* **2016**, *18*, 323. [CrossRef]
41. Zanette, D. Generalized Kolmogorov entropy in the dynamics of the multifractal generation. *Phys. A Stat. Mech. Appl.* **1996**, *223*, 87–98. [CrossRef]
42. Guariglia, E. Entropy and Fractal Antennas. *Entropy* **2016**, *18*, 84. [CrossRef]
43. Peng, Y.; Zhou, J.G.; Burrows, R. Modelling the free surface flow in rectangular shallow basins by lattice Boltzmann method. *J. Hydraul. Eng.* **2011**, *137*, 1680–1685. [CrossRef]
44. Peng, Y.; Zhou, J.G.; Burrows, R. Modelling solute transport in shallow water with the lattice Boltzmann method. *Comput. Fluids* **2011**, *50*, 181–188. [CrossRef]
45. Baumer, C. *Southern Silk Road: In the Footsteps of Sir Aurel Stein and Sven Hedin*; Orchid Press: Krung Thep Maha Nakhon, Thailand, 2001.
46. Liu, J.Q.; Qin, X.G. Evolution of the environmental framework and oasis in the Tarim Basin. *J. Quat. Sci.* **2005**, *25*, 533–539.
47. Zhu, Z.D.; Wu, Z.; Liu, S.; Di, X.M. *An Outline of Chinese Deserts*; Science Press: Beijing, China, 1980; Volume 107. (In Chinese)
48. Ren, X.; Mu, G.J.; Xu, L.S.; Lin, Y.C.; Zhao, X. Characteristics of artificial oasis expansion in south Tarim Basin from 2000 to 2013. *Arid Land Res.* **2015**, *38*, 1022–1030. (In Chinese)
49. Zhou, X.; Lei, W.J. Hydrological interaction between oases and water vapor transportation in the Tarim Basin, northwestern China. *Sci. Rep.* **2018**, *8*, 13431. [CrossRef] [PubMed]
50. Li, B.; Chen, Y.; Shi, X.; Chen, Z.; Li, W. Temperature and precipitation changes in different environments in the arid region of northwest China. *Theor. Appl. Climatol.* **2013**, *112*, 589–596. [CrossRef]
51. Halik, W.; Mamat, A.; Dang, J.H.; Deng, B.S.H.; Tiyip, T. Suitability analysis of human settlement environment within the Tarim Basin in Northwestern China. *Quatern. Int.* **2013**, *311*, 175–180. [CrossRef]
52. Zhao, X.F.; Xu, H.L.; Wang, M.; Zhang, P.; Ling, H. Overloading analysis of irrigation area in basins of Tarim River in different years. *Trans. CSAE* **2015**, *31*, 77–81. (In Chinese)
53. Yan, Z.L.; Huang, Q.; Chang, J.X.; Wang, Y. Analysis and dynamic monitoring of land use in the main stream of Tarim river valley using remote sensing. *Trans. CSAE* **2008**, *24*, 119–123. (In Chinese)
54. Zhang, J.S.; Yan, Z.L.; Wang, X.G.; Huang, Q.A.; Gao, F. Remote sensing analysis of spatial-temporal changes of desertification land in Lower Reaches of Tarim River. *Trans. CSAE* **2009**, *25*, 161–165. (In Chinese)
55. Yang, K. User's Guide for China Meteorological Forcing Dataset. 2010. Available online: http://dam.itpcas.ac.cn/data/User_Guide_for_China_Meteorological_Forcing_Dataset.htm (accessed on 14 September 2018).
56. Shannon, C.E. A mathematical theory of communication. *Bell Syst. Technol. J.* **1948**, *27*, 379–423. [CrossRef]
57. Ben-Naim, A. Entropy, Shannon's Measure of Information and Boltzmann's H-Theorem. *Entropy* **2017**, *19*, 48. [CrossRef]
58. Conrad, K. Probability distributions and maximum entropy. *Entropy* **2004**, *6*, 10.
59. Shi, Y.F. Discussion on the present climate change from warm-dry to warm-wet in northwest China. *J. Quat. Sci.* **2003**, *23*, 152–164. (In Chinese)
60. Zhang, J.C. *Monsoon and Precipitation in China*; China Meteorological Press: Beijing, China, 2010; Volume 16. (In Chinese)
61. Shen, L.L.; He, J.H.; Zhou, X.Y.; Chen, L.; Zhu, C.W. The regional variabilities of the summer rainfall in China and its relation with anomalous moisture transport during the recent 50 years. *Acta Meteorol. Sin.* **2010**, *68*, 918–931.
62. Lei, Z.D.; Hu, H.P.; Yang, S.X.; Tian, F.Q. Analysis on water consumption in oases of the Tarim Basin. *J. Hydraul. Eng.* **2006**, *37*, 1470–1475.

12

Time-Averaged Turbulent Velocity Flow Field through the Various Bridge Contractions during Large Flooding

Kwang Seok Yoon [1], Seung Oh Lee [2] and Seung Ho Hong [3,*]

1 Korea Institute of Civil Engineering and Building Technology, Goyang 10223, Korea; ksyoon@kict.re.kr
2 School of Urban and Civil Engineering, Hongik University, 94 Wausan-ro, Mapo-gu, Seoul 04066, Korea; seungoh.lee@hongik.ac.kr
3 Department of Civil and Environmental Engineering, West Virginia University, 1306 Evansdale Drive, Morgantown, WV 26506, USA
* Correspondence: sehong@mail.wvu.edu.

Abstract: Extreme rainfall events, larger than 500-year floods, have produced a large number of flooding events in the land and also close to the shore, and have resulted in massive destruction of hydraulic infrastructures because of scour. In light of climate change, this trend is likely to continue in the future and thus, resilience, security and sustainability of hydraulic infrastructures has become an interesting topic for hydraulic engineering stakeholders. In this study, a physical model experiment with a geometric similarity of the bridge embankments, abutments, and bridge deck as well as river bathymetry was conducted in a laboratory flume. Flow conditions were utilized to get submerged orifice flow and overtopping flow in the bridge section in order to simulate extreme hydrologic flow conditions. Point velocities of the bridge section were measured in sufficient details and the time-averaged velocity flow field were plotted to obtain better understandings of scour and sediment transport under high flow conditions. The laboratory study concluded that existing lateral flow contraction as well as vertical flow contraction resulted in a unique flow field through the bridge and the shape of velocity profile being "fuller", thereby increasing the velocity gradients close to the bed and subsequently resulting in a higher rate of bed sediment transport. The relationships between the velocity gradients measured close to the bed and the degree of flow contraction through the bridge are suggested. Furthermore, based on the location of maximum scour corresponding to the measured velocity flow field, the classification of scour conditions, long setback abutment scour and short setback abutment scour, are also suggested.

Keywords: abutment; overtopping flow; pressure flow; physical hydraulic modeling; scour and velocity field

---

1. Introduction

Usually, bridge foundation failure occurs due to the processes of (1) contraction scour-scouring across the entire channel due to the flow contraction caused by the bridge opening and deflection of floodplain flow into the main channel and (2) local scour at the base of piers and abutments caused by local flow contraction, down flow, and formation of a horseshoe vortex that wraps around the obstructions. Thus, the contraction scour and the local scour (pier scour and abutment scour) have been considered as two separate types of scour caused by different processes.

However, recent studies [1,2] show that abutment scour can be expressed as one type of contraction scour, not as a fundamental mechanism of local scour. When flow area is reduced by the bridge opening, velocity and bed shear stress are increased in order to satisfy continuity and momentum equations. The higher velocity resulted in increased erosive force, so more bed material is removed from the contracted section. In addition to the higher velocity due to the flow contraction, local flow structures associated with the base of the abutment result in additional erosion around the abutment. Due to the local flow structures, the scour depth near the upstream edge or corner of the abutment is usually deeper than that near the center of the channel. Thus, the combined effect of flow contraction as well as the local flow structure around an abutment made the problem of abutment scour more challenging. Although some researchers have suggested the need for more research on the subject of local flow structures and velocity distribution, past studies [3–10] have provided acceptable insight for the mechanisms of local flow structure around single and a group of bridge foundations under free flow conditions. However, for high flow conditions such as in an extremely hydrologic event, there is no widely applicable abutment scour formula, and the term has not been distinctly defined because of difficulties in understanding the complicated flow and scouring mechanism combined with complex geometries of bridge and various flow conditions.

Furthermore, a lot of research on bridge scour has focused only on simple and idealized situations where the bridge is placed in straight rectangular channels, even though many bridges are sited in non-rectangular channels whose geometry and hydraulic characteristics are site-specific in the real world. Thus, in this study, scour experiments were carried out in a compound channel cross-section using various lengths of bridge under free-surface flow (*F*) as well as in submerged orifice flow (*SO*), and overtopping flow (*OT*) cases. To investigate the effect of bridge submergences during large flooding scenarios, a bridge deck model was constructed based on the bridge design and dimensions commonly used in a rural region in the United States (USA). To understand the complex flow physics through the bridge, three components of velocities were measured by acoustic Doppler velocimeters (ADV).

The experimental results show new insights into the velocity flow field through the scour-critical bridges subject to the submerged orifice flow and overtopping flow during extreme flooding events. It is shown that vertical flow contraction by submerged flow resulted in a unique flow field through which the bridge and the shape of vertical velocity profile are "fuller" and thereby push the location of the maximum velocity closer to the bed, increasing the shear stress and erosion. Because the unique shape of velocity profile under the submerged flow causes higher erosion rate on the bed, examination of the relationships between the velocity gradient measured close to the bed and the non-dimensional parameters commonly used for the contraction scour is suggested to understand the effect of local flow variables on the scour. Furthermore, with the locations of the maximum scour corresponding to the measured velocity flow field, the scour conditions, long setback abutment scour and short setback abutment scour were classified for the practitioners to use in bridge design.

2. Methodology

2.1. Experimental Setup

In a previous study [11–13], laboratory experiments were carried out using a 1:60 scaled hydraulic model of the Towaliga River bridge at Macon, Georgia, USA including the full river bathymetry. The field data including measured discharge, bed elevation of cross sections, and gage height from the United State Geology Service (USGS) were reproduced by Froude number similarity. The previous experimental results showed that the hydraulic model can reproduce field scour data.

For the current set of experiments, the cross section shape and river geometry used in the previous experiments were slightly simplified and modified for the experiments to address more general features of the velocity flow field. The shape of the floodplain was horizontal on both sides of the main channel cross-section while preserving the original parabolic shape of the main channel. Also, the channel was constructed to have a straight alignment rather than meandering and all the piers were removed to be able to focus on the abutment scour. Figure 1 shows the modified laboratory model for the experiment in the flume and initial contour before the scouring experiment.

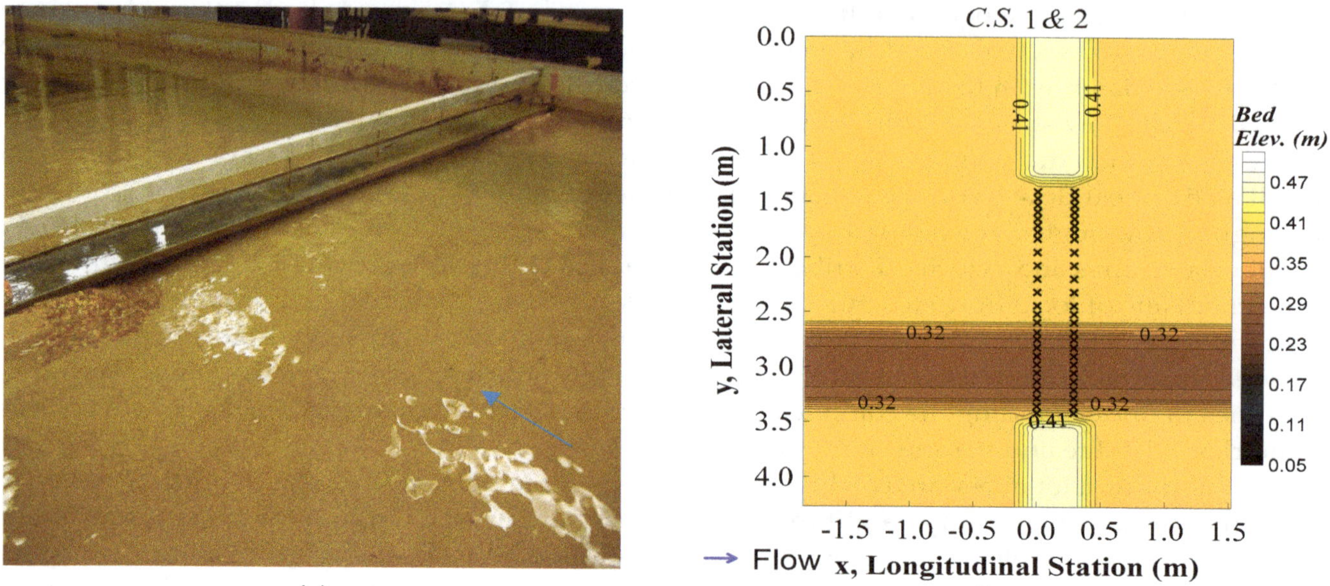

(a) (b)

Figure 1. Laboratory model in this study in (**a**) and initial contour with velocity measurements cross sections in (**b**).

The approach channel upstream of the bridge was 11 m long followed by a working mobile bed section with a length of approximately 5.2 m in which the bridge model was placed. The fixed bed approach section was filled with 3.3 mm of gravel, and the 5.2 m moveable bed working section was filled with sand with d_{50} = 1.1 mm and σ_g = 1.3, where d_{50} and σ_g is the median diameter of sand and the geometric standard deviation often to describe the grain size distribution. To ensure a fully developed boundary layer at the bridge approach section consistent with the moveable bed sediment size, the fixed-bed approach channel consisted of a surface layer of fixed 1.1 mm sand having a depth of 3 cm. Based on the results, the approach channel has enough length to obtain a fully developed turbulent flow before reaching the moveable bed section. The range of approach section length/flow depth is 80 to 150, and the velocity measurements in the approach of the bridge, measured 3 m upstream of the bridge, shows perfect logarithmic velocity profiles for all the experimental cases (R^2 = 0.99) [14]. At the downstream of moveable bed section, there is 1.6 m long sediment trap section; this section trapped the sediment transported out of the working moveable bed section.

The model of the embankment and abutment was constructed as an erodible fill with rock riprap protection [1,2,14]. Three different lengths of erodible-embankment/abutment were modeled to simulate wide range of flow contractions on the left floodplain, but on the right floodplain, the toe of the abutment was maintained at the bankline for all experiments. This arrangement allowed the study of bankline abutment as well as setback abutments in the floodplain simultaneously under realistic geometric conditions similar to in the field. Based on the above modifications, abutment and embankment lengths and river bathymetry to be modeled in the laboratory were constructed as shown in Figure 2a. The ratio between abutment lengths (L_a) to the floodplain width (B_f), L_a/B_f, varied from 0.53 to 0.88 in the left floodplain. One of the main purposes of this study is to find the effect of velocity flow field on scour through the bridge for extreme hydrologic conditions. In those extreme conditions, overtopping or submerged orifice flow is likely to occur at the bridge. To simulate those extreme cases, the following dimensions normally used in a bridge design in rural region were used for the model bridge deck.

(a) Width of bridge deck 12.2 m, in accordance with standard two-lane roads;
(b) Bridge barrier 0.61 m high with 0.46 m top without sidewalks on non-bicycle routes;
(c) Slab depth of 0.46 m including the pavement;
(d) Girders 0.43 m wide and 0.46 m deep with 2.74 m spacing.

These design dimensions are commonly used for rural region two-lane bridges. Based on the prototype dimension, the 1:45 length scale bridge deck was constructed as shown in Figure 2b and a solid bridge deck model was supported and leveled with respect to an upper support beam as shown in Figure 1a.

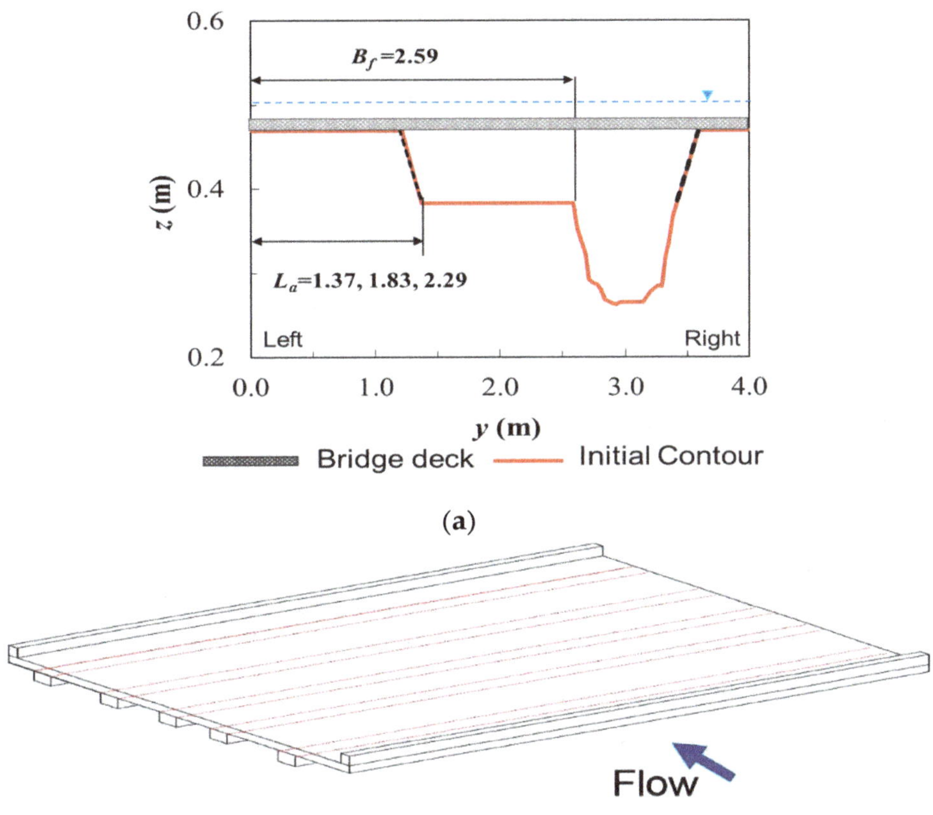

Figure 2. Geometries for (**a**) compound channel and abutment before scouring and (**b**) shape of model bridge deck.

2.2. Experimental Procedure

After completion of laboratory setup, the flume was filled with water slowly enough to saturate sand in working moveable-bed section without changing the original contours. After complete saturation, the initial bed elevations were measured in detail throughout the test section using an ADV and a point gage. The ADV gives distance from the sampling volume to the bed, which can be converted into elevation relative to the fixed datum by reading the point gage. After that, the required discharge and a flow depth larger than the target value were set by using a magnetic flow meter and by adjusting a tailgate, respectively, to prevent un-expected scour (erosion) while the experimental condition was set. Then, the tailgate was lowered slowly enough to achieve the target value of flow depth without abrupt water depth change. During this time, the point gage and/or wave gage mounted on the instrument carriage were used to measure the flow depth. Once the target flowrate and flow depth had been reached, the experiment was continued for five to six days until equilibrium scour was achieved (change in scour depth less than 2% within 24 hour). At the end of experiment, the equilibrium bed elevations were measured in the same way as in the initial bed-elevations measurements using the ADV and the point gage. The end of the experiment was defined when the scour depth reached the equilibrium state at which there were negligible changes in bed elevation with time. Once each experimental study was completed, the original river model was re-constructed, and scour experiment was re-started under same procedure as explained above using different flow variables. Total eighteen experimental conditions were used for the scour experiments to satisfy the purpose of the research.

Then, after the moveable bed experiments, the complete river bathymetry was modeled with a fixed-bed channel by spraying it with polyurethane. In the fixed-bed experiments, water surface profiles and velocities were measured to address the initial hydraulic conditions. Figure 1b shows the velocity measuring cross section (C.S. 1 and C.S. 2) along the bridge. Velocities were taken every 15 cm laterally in both the floodplain and the main channel. A minimum of six measuring points in each vertical profile and as many as 15 points were measured at both C.S. 1 and C.S. 2. With a similar way as in the bridge section, velocities were also measured at 3 m from the upstream face of the bridge for the approach flow velocity. During the velocity measurements, correlation values in these experiments were greater than 80% and the Signal Noise Ratio (SNR) was greater than 15. The sampling frequency of the ADV was chosen to be 25 Hz with a sampling duration of two minutes and perhaps as much as five minutes depending on the turbulence at each measuring location [14–19]. The phase-space despiking algorithm of Goring and Nikora [20] was also employed to remove any spikes in the time record caused by aliasing of the Doppler signal, which sometimes occurs near a boundary.

3. Results and Discussion

Initial hydraulic parameters measured in fixed bed have been summarized in Table 1: Q is the total discharge; V_{f1}/V_{fc1} and V_{m1}/V_{mc1} is approach flow intensity in the floodplain and in the main channel, respectively, where V_{f1} and V_{m1} is approach flow velocity and V_{fc1} and V_{mc1} is approach flow critical velocity calculated by Keulegan's equation; y_{f1} and y_{m1} is the water depth of floodplain and main channel in the approach section, respectively; q_{f2}/q_{f1} and q_{m2}/q_{m1} is unit discharge contraction ratio in the floodplain and main channel, respectively; W is the setback distance; L_m is the traverse distance from the toe of the abutment to the maximum scour hole depth. The definition sketch for the variables are shown in Figure 3.

Table 1. Initial experimental parameters and results for location of maximum scour depth.

Run	Flow Type	Q (m^3/s)	$\frac{L_a}{B_f}$	$\frac{V_{f1}}{V_{fc1}}$	$\frac{V_{m1}}{V_{mc1}}$	Y_{f1} (m)	Y_{m1} (m)	$\frac{q_{f2}}{q_{f1}}$	$\frac{q_{m2}}{q_{m1}}$	$\frac{L_m}{y_{f1}}$	$\frac{W}{y_{f1}}$	Cond-Itions
1	*F*	0.093	0.53	0.61	0.83	0.074	0.152	1.818	1.543	3.70	16.46	A
2	*SO*	0.116		0.60	0.73	0.106	0.184	1.875	1.483	3.75	11.53	A
3	*OT*	0.164		0.61	0.72	0.149	0.227	1.148	1.146	2.05	8.20	A
4	*F*	0.085		0.58	0.77	0.075	0.153	1.755	1.441	4.05	16.19	A
5	*SO*	0.110		0.57	0.68	0.108	0.186	1.781	1.416	3.38	11.27	A
6	*OT*	0.150		0.56	0.64	0.148	0.226	1.25	1.100	1.86	8.26	A
7	*F*	0.085	0.71	0.54	0.74	0.076	0.154	2.236	1.756	5.24	10.08	A
8	*SO*	0.103		0.53	0.71	0.103	0.181	2.257	1.602	5.05	7.42	A
9	*OT*	0.150		0.56	0.66	0.150	0.228	1.176	1.243	1.63	5.09	A
10	*F*	0.074		0.49	0.72	0.073	0.151	2.208	1.539	4.15	10.37	A
11	*SO*	0.091		0.49	0.61	0.105	0.183	2.223	1.647	4.08	7.29	A
12	*OT*	0.130		0.50	0.57	0.147	0.225	1.278	1.262	1.45	5.18	A
13	*F*	0.074	0.88	0.44	0.69	0.076	0.156	*	1.904	5.84	4.03	C
14	*SO*	0.088		0.43	0.63	0.103	0.180	*	1.951	4.76	2.97	C
15	*OT*	0.130		0.45	0.57	0.150	0.227	*	1.422	2.46	2.04	C
16	*F*	0.062		0.38	0.55	0.073	0.155	*	1.976	4.72	4.15	C
17	*SO*	0.074		0.37	0.52	0.105	0.181	*	1.902	4.44	2.92	C
18	*OT*	0.110		0.40	0.50	0.147	0.224	*	1.481	2.09	2.07	C

(Subscript 1 and 2 refers to the approach section and bridge section, respectively; Condition A = long setback and Condition C = short setback abutment; symbol *: discharge per unit width in the bridge section were only measured in the main channel for short setback abutment).

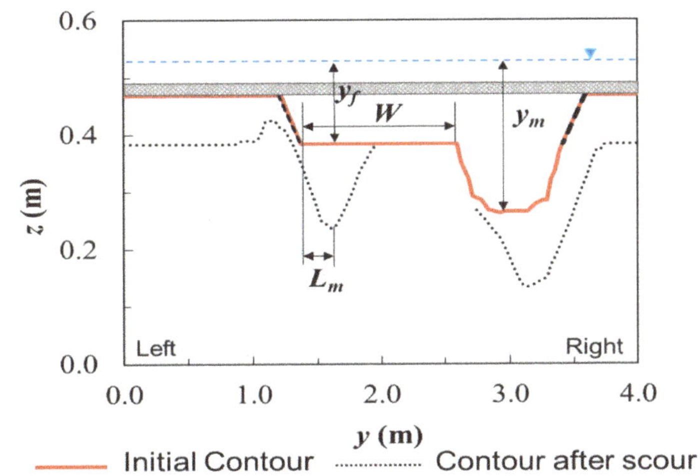

Figure 3. Definition sketch for classification of scour conditions.

As shown in Table 1, in the first three runs, the value of V_{f1}/V_{fc1} (approach flow intensity parameter) was similar. The experimental plan was to reproduce increasing tailwater with increasing discharge according to a tailwater rating curve, just as would occur in the field moving from *F* to *SO* to *OT* flow. The experiments were conducted in such a fashion that V_{f1}/V_{fc1} was held constant in the floodplain for a series of runs encompassing the three flow types for a given abutment length. This arrangement allowed us to find the effect of different flow types in scour because three flow types were encountered for different value of discharge and water depth, but their approach flow intensity parameter remained nearly constant for the three runs. The same conditions were also applied to other sets of the experiment. As explained in Hong et al. [2], the magnitude of maximum scour depth increased as the flow type changed from free flow (*F*) to submerged orifice flow (*SO*) and then decreased again for overtopping flow (*OT*) while holding the flow intensity (V_{f1}/V_{fc1}) constant. This pattern can be found in Figure 4. The scour depth is decreased in *OT* compared to *SO* even if the discharge is higher. This can be explained by the flow relief over the deck in *OT* rather than under the bridge. In most cases, the value of maximum scour depth was greater for *OT* flow than for *F* flow depending on the fraction of the total flow going under the bridge.

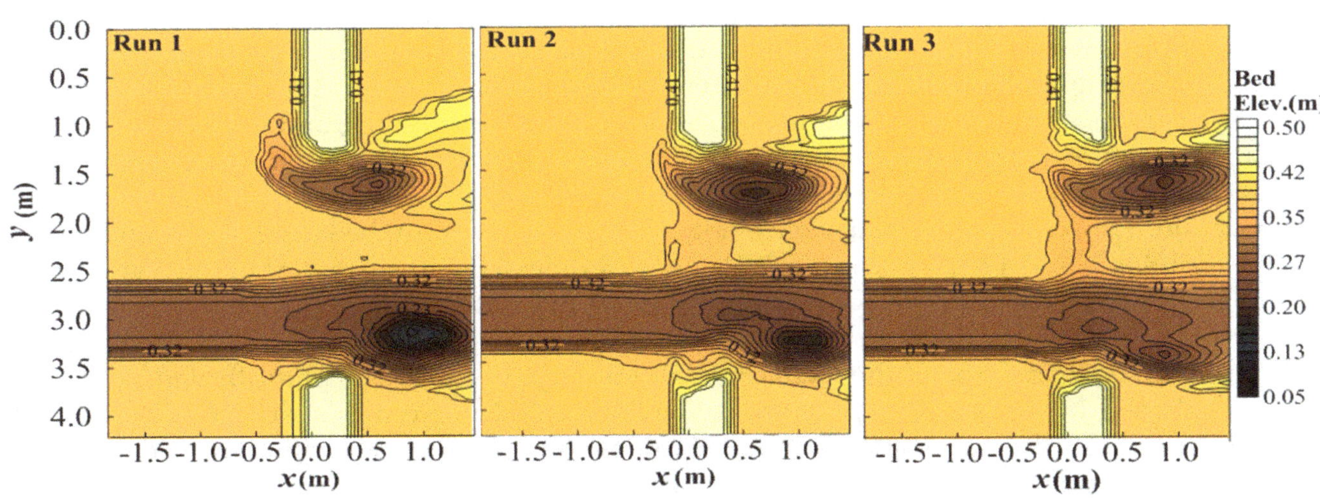

Figure 4. *Cont.*

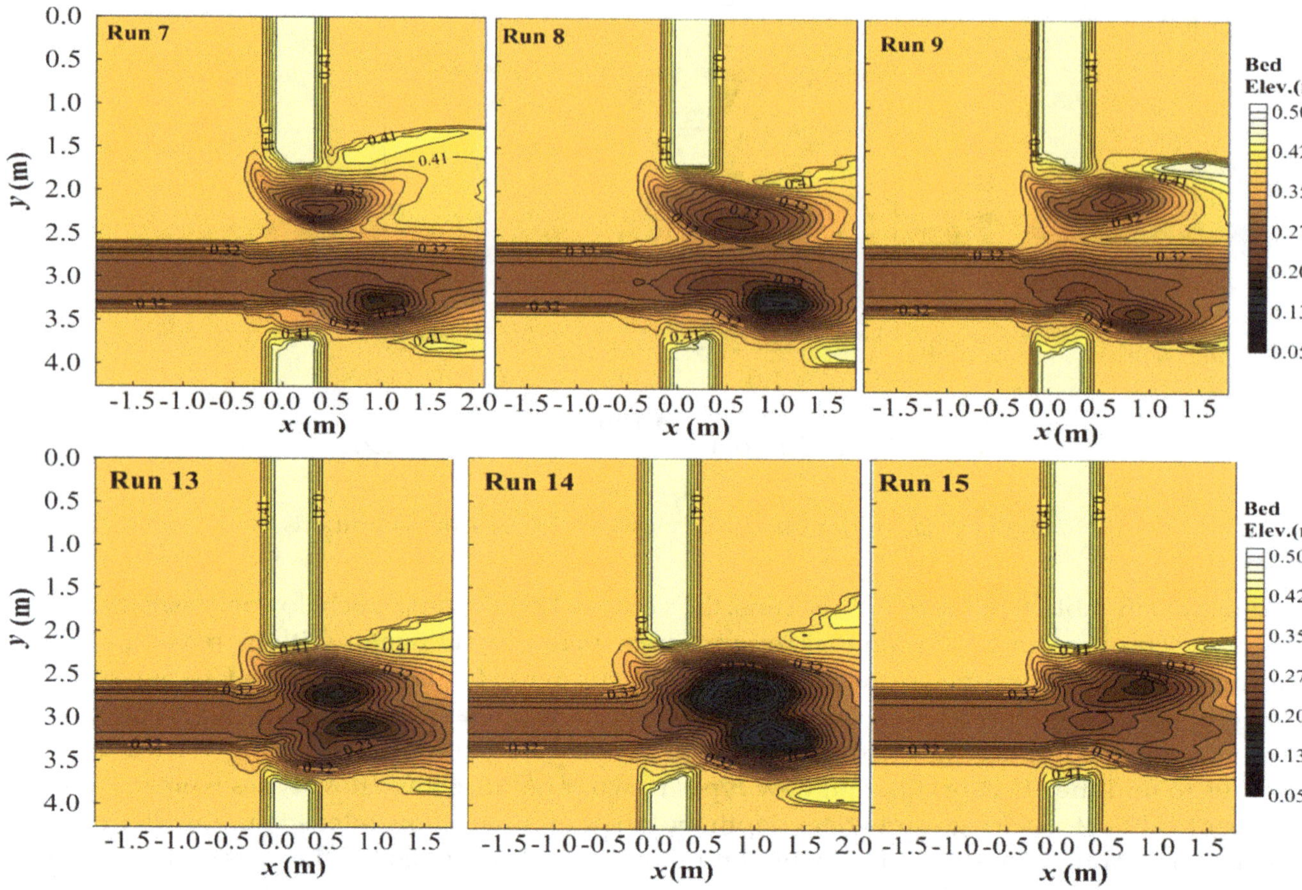

Figure 4. Contours for selected run in equilibrium condition.

3.1. *Classification of Abutment Scour*

Figure 4 shows the selected bed contours after each run. Based on the findings from Hong [14], the scour hole moved downstream from the abutment over time and the region of the deepest scour hole was located near the downstream region of the abutment. Because the scour process differed depending on the flow type over time, the detailed location of the deepest scour depth was case dependent. However, a similar trend can be deduced with respect to the shape of the scour hole and locations of the maximum scour depth in the experiments conducted with the same length of abutment. As shown in Figure 4, for runs 1, 2, and 3, the scour hole shape in the left floodplain showed curvature around the abutment, and the resulting point of the maximum scour was located downstream of the abutment. Similar findings can be applied for runs 7, 8, and 9. However, for runs 13, 14, and 15 with a longer abutment, the scour hole initially developed around the upstream corner of the abutment, the same as in the other runs, and then moved along the toe of the abutment in the left floodplain. However, as the scour hole elongated diagonally from the face of the abutment over time, the resulting point of maximum scour hole depth terminated at the bankline of the main channel, thus, the maximum scour depth in equilibrium was located inside the main channel on the bank side slope. For the bankline abutment in the right floodplain when looking downstream, the maximum scour depth can be found around the main channel side slope downstream of the bridge.

As explained in the previous paragraph, the location of maximum scour depth around the abutment varies in accordance with the length of abutment because the velocity flow field developed by the geometrical characteristics of an abutment is a leading factor to define the location of maximum scour depth. Thus, Melville and Coleman [4] classified the abutment to account for the effect of abutment/embankment length on the scour depth. In their classification, Case A applies to the abutment sited in a rectangular channel, while Case B represents the abutment that is sited on a floodplain and extended into the main channel. Case C is an abutment set well back from the main channel such that all scour takes place on the floodplain only. Case D is the limit of Case B and C where the abutment protrudes out to the edge of the main channel. Also, Chang and Davis [21,22] classified the abutment by the three categories as short, intermediate and long depending on the value of the setback distance and assumed that converging flow under the bridge with the abutment near the channel bank (long abutment) is mixed with the flow in the main channel and distributed uniformly. On the other hand, if the abutment is set well back from the channel bank, it is assumed that the overbank flow and the main channel flow remain separated from each other and do not mix as the flow passes under the bridge. Similar as in the previous research, scour conditions are classified as three cases, Condition A, Condition B, and Condition C in this study, but instead of using the length of the abutment, the scour conditions were classified in accordance with the location of maximum scour hole with respect to geometric ratio between length of abutment and width of the floodplain.

To find whether the maximum scour hole occurs in the floodplain or in the main channel, the non-dimensional value of W/y_{f1} was compared with L_m/y_{f1}. When the value of L_m/y_{f1} is larger than that of W/y_{f1}, the location of maximum scour hole is in the main channel. The maximum value of L_m/y_{f1} in all experiments is 5.84 in Run 13, as shown in Table 1. Thus, if the value of W/y_{f1} is smaller than approximately 6, the location of the maximum scour hole is expected to be outside of the floodplain. Thus, based on the findings in this study, clear water abutment scour conditions can be classified as: Condition A - long setback abutment scour; Condition B - bankline abutment scour; and Condition C - short setback abutment scour. The detailed descriptions and classification in terms of the ratio of setback distance (W) to the approach flow depth in the floodplain, W/y_{f1}, in this study are given below:

- Condition A ($W/y_{f1} > 6$, $L_a/B_f = 0.53$ and 0.77): In a long setback abutment, scour occurs in the floodplain only, well removed from the main channel;
- Condition B ($W/y_{f1} = 0$, $L_a/B_f = 1.0$): For a bankline abutment, maximum scour occurs in the main channel of a compound channel;
- Condition C ($W/y_{f1} < 6$, $L_a/B_f = 0.88$): In a short setback abutment, scour occurs on the floodplain in the initial stage, but maximum scour at equilibrium occurs in the main channel because the setback distance is short.

3.2. *Velocity Flow Field around the Abutment*

The unique shape of the scour hole around an abutment, which results from a longitudinal and diagonal displacement of the deepest scour point relative to the abutment face, can be explained by the velocity flow field around the abutment. Thus, to understand the complex flow physics and resulting sediment transport, velocities were measured at C.S. 1 and C.S. 2. Figure 5 shows the resultant velocity vectors in y direction (v) and z direction (w) in the floodplain and in the main channel for runs 1, 3 (*F*) and 13, 15 (*OT*). As shown in Figure 5, in the cross-sectional velocity plot at upstream face of the bridge (C.S. 1) and the downstream face of the bridge (C.S. 2), higher magnitude velocity vectors are observed around the abutment resulting from local lateral flow contraction where deeper scour can occur as the contracted flow curved around the abutment. For runs 13 and 15, the velocity is even higher than in runs 1 and 3 because of the higher lateral flow contraction in longer abutment. In addition to the existing lateral flow contraction, for the cases with higher discharge (*SO* and *OT* case), the submergence of the upstream face of the bridge produced vertical flow contraction (i.e., the downward component of the velocity vectors in Figure 5c,g). However, at C.S. 2 (downstream of the bridge section in Figure 5d,h), the upward components of velocity are observed. The downward velocity in upstream and upward velocity in downstream flow motion through the bridge induced by the bridge deck resulted in vertical contraction scour because the flow is accelerated, then decelerated. For Run 2 and Run 3, as shown in Figure 4, the lower bed elevation along the location of bridge deck in the floodplain is the result of vertical flow contraction by the unique velocities motions under the bridge. For the cases in other runs, similar observations as in runs 2 and 3 for the contours cannot be discovered because of the extent of abutment scour hole. However, the existing lateral flow contraction as well as vertical flow contraction definitely shows the higher scour depth through the bridge.

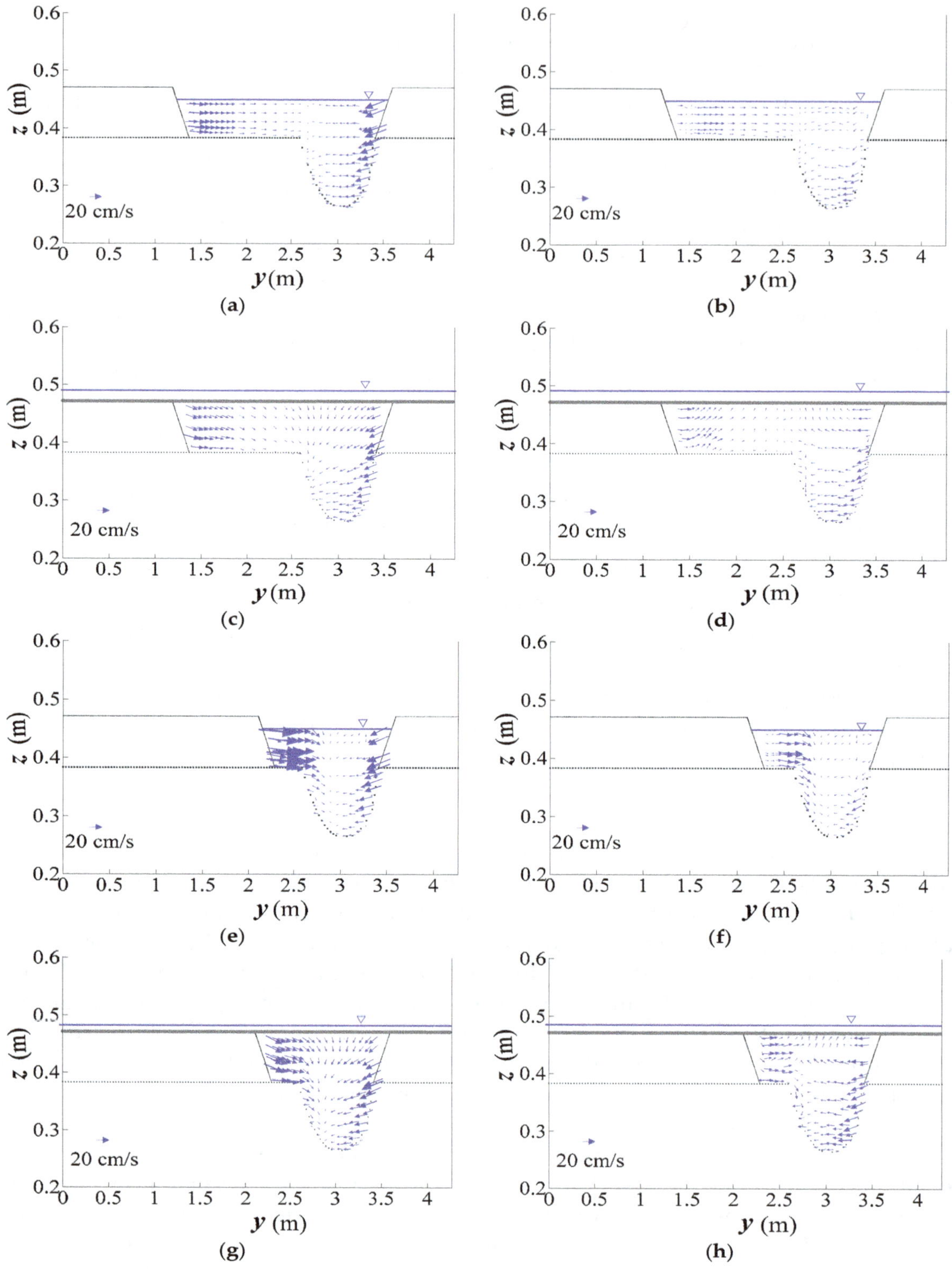

Figure 5. Cross-sectional velocity vectors measured at C.S. 1 and C.S. 2 for each run. (**a**) v-w plot at C.S. 1 (Run 1, F), (**b**) v-w plot at C.S. 2 (Run 1, F) ,(**c**) v-w plot at C.S. 1 (Run 3, OT), (**d**) v-w plot at C.S. 2 (Run 3, OT), (**e**) v-w plot at C.S. 1 (Run 13, F), (**f**) v-w plot at C.S. 2 (Run 13, F), (**g**) v-w plot at C.S. 1 (Run 15, OT), (**h**) v-w plot at C.S. 2 (Run 15, OT).

In the main channel, the v-w velocity plots also show unique flow motions. The narrow bridge opening induced by the abutments forced the water to re-enter through the bridge opening, causing lateral directional velocity. As shown in the Figure 5, for the setback abutment in the left floodplain, the flow moved towards the main channel, but the amount of shifting velocity was small compared to that of the bankline abutment in the right floodplain. Because the direction of lateral velocity vectors from the left side of abutment and the right side of abutment are opposite, the strong momentum transfer occurs when they meet at one point, and the interaction resulted in large counter clock-wise secondary current observed within the main channel. This secondary current initiates another scour hole close to the toe of the left-side slope within the main channel [23]. As shown in Figure 4, however, maximum scour depth around the bankline abutment occurred in the bottom of the main channel near the toe of the right bank. In fact, there appears to be interaction between the two scour holes from the left and right abutments during the initial scour development, but there was one remaining scour hole left in the equilibrium stage. Kara et al. [24] have applied a 3D numerical model to the problem of compound channel flows because the flow and turbulence distributions are so important to the prediction of scour when a bridge abutment is placed in a compound channel. Their results show the important contribution of secondary currents and turbulent stresses to the apparent shear stress at the main channel/floodplain interface when the momentum equation is depth-averaged. Both the secondary current and turbulent stress contributions to the apparent shear stress increase as the relative depth in the floodplain decreases.

Figure 6 shows the vertical velocity profiles for each run measured in the floodplain (Figure 6a) and in main channel (Figure 6b). It is interesting to note that the shape of velocity profiles is different for the case with submerged flow compared to in the free flow case. The vertical velocity measurement location (y_x) on the ordinate axis has been non-dimensionalized by total water depth ($\overline{Y}$) and the point velocity measurements (V_x) was normalized by depth-averaged velocity ($\overline{V}$) in Figure 6. Submergence of the bridge during extreme hydrologic events produced vertical flow contraction leading to more complex flow field through the bridge than in the free flow cases. As shown in Figure 6, the presence of bridge deck resulted in a "fuller" velocity profile than in the free flow cases and tends to shift the higher velocity closer to the bed. The degree of shifting of higher velocity closer to the bed is a key element to understanding the intensity of shear stress that causes erosion.

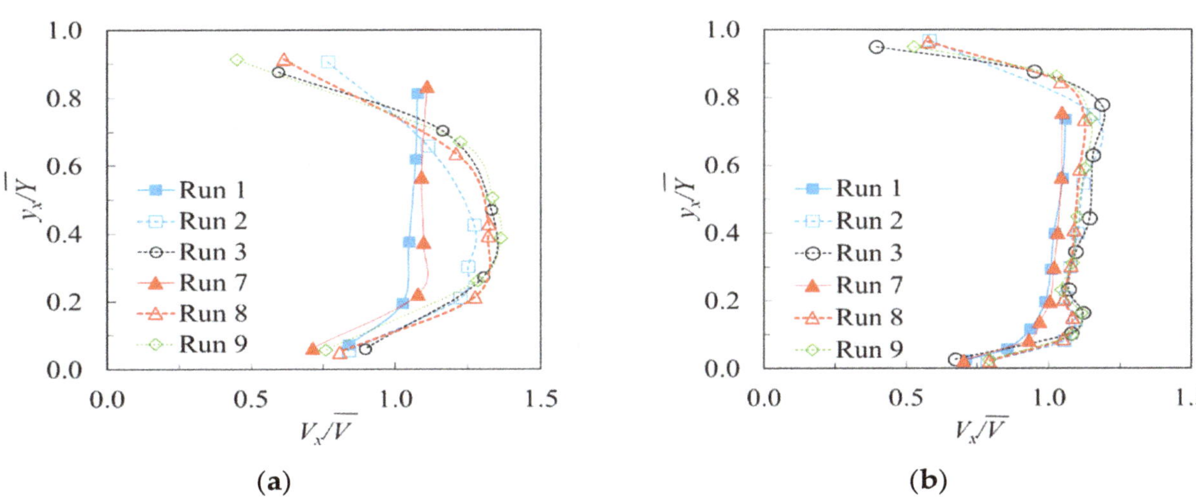

Figure 6. Vertical velocity profiles for (**a**) in the floodplain and (**b**) in the main channel

Thus, the degree of shifting of velocity at the bed is explored with non-dimensional parameters commonly used for the theoretical contraction scour. Hong et al. [2], with their experimental results, confirmed that maximum scour around an abutment can be considered as some amplification of the theoretical contraction scour, $(V_1/V_c)(q_2/q_1)$. Previously, this concept has only applied to the free flow cases. However, they observed that maximum scour depth even in different flow types can be calculated with the product of (V_1/V_c) and (q_2/q_1), as long as the flow contraction ratio(q_2/q_1) is predicted accurately. Thus, the effect of submerged orifice flow and overtopping flow provided by the vertical flow contraction as well as the lateral flow contraction can also be parameterized by the flow contraction ratio. In this study, the velocity gradients (slope) between the origin and a point measured around 20% of water depth was decided using the vertical velocity profiles shown in Figure 6 and are plotted for the floodplain in Figure 7a and for the main channel in Figure 7b, respectively, according to the dimensionless variables, $(V_1/V_c)(q_2/q_1)$, suggested by the theoretical contraction analysis. As shown in Figure 7, as the dimensionless variables, $(V_1/V_c)(q_2/q_1)$, in the x-axis increases, the value of slope decreases because the higher flow contraction increases the velocity close to the bed, leading to the milder slope shown in Figure 6. The measured slope from each profile seems to follow the same trend, even if they have different flow types because the value of q_2/q_1 can be a viable indicator of the combined influence of vertical flow contraction as well as lateral contraction, as explained in the previous paragraph.

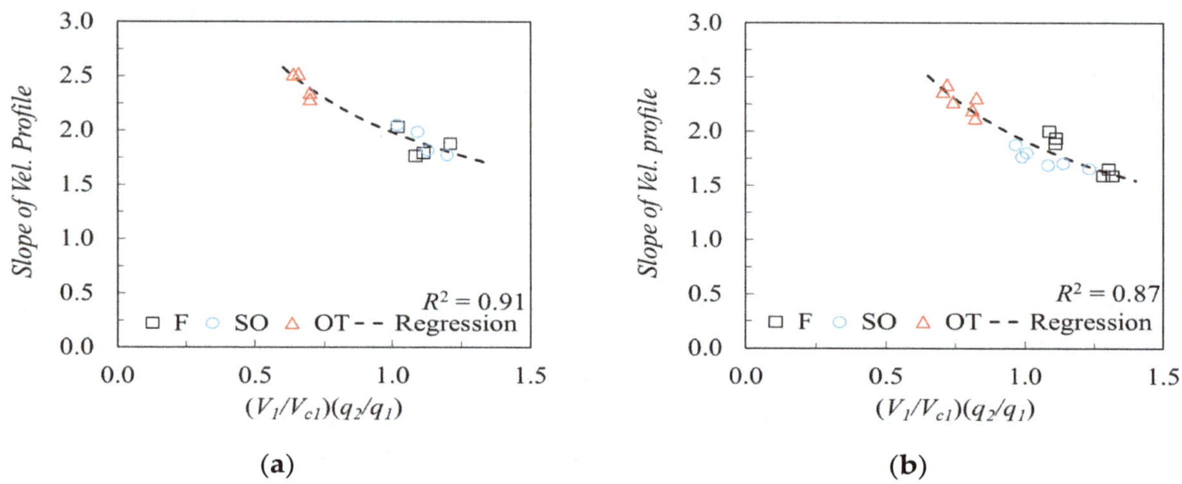

Figure 7. Variation of velocity gradient close to the bed as a function of $(V_1/V_{C1})(q_2/q_1)$ for (**a**) in the floodplain and (**b**) in the main channel

A least-squared regression analysis was conducted on the data given in Figure 7, and the best-fit equation is given by

$$Slope = 1.98 \times \left[(\frac{V_{f1}}{V_{fc1}})(\frac{q_{f2}}{q_{f1}}) \right]^{-0.514}, \text{ for floodplain} \quad (1)$$

$$Slope = 1.91 \times \left[(\frac{V_{m1}}{V_{mc1}})(\frac{q_{m2}}{q_{m1}}) \right]^{-0.635}, \text{ for main channel} \quad (2)$$

with coefficients of determination of 0.91 and 0.87, respectively. Equations (1) and (2) can be used to calculate bed shear stress because shear stress is decided by the velocity gradient (slope). Theoretically, scour is initiated when the bed shear stress induced by the flow is larger than the value of critical shear stress of the bed sediment size and the magnitude of scour depth is directly related to the value of shear stress. Thus, predicting shear stress is key driver to a better understanding of scour around an abutment. Furthermore, the provided equations can be used for bed shear stress closure of numerical modeling around bridge abutments for model developers.

4. Conclusions

A lot of research in past twenty years has focused on a particular type of bridge foundation scour, e.g. pier scour and/or abutment scour using simple experimental set-up. Thus, so far, engineers and researchers have been using SINGLE factors such as mean hydraulic variables, Reynolds stresses, turbulent kinetic energy (TKE), vorticity, and other measures of turbulent structure to understand/calculate the scour depth. However, recent extreme rainfall events associated with climate change can often result in submerged orifice flow with or without overtopping flow, in which the flow field through the bridge is more complex because of the simultaneous existence of local turbulence around the base of the structure as well as vertical flow contraction in addition to the lateral flow contraction. Thus, in this paper, for the investigation of the complex mechanism of scour under extreme hydrologic events, laboratory experiments were conducted using scale down river geometry and the detailed velocity flow field through the bridge were measured. The results show that, in addition to the local turbulence structure being wrapped around the base of abutment, higher cross-sectional velocity around an abutment due to the local flow acceleration cause the maximum scour close to the abutment. During large flooding and bridge submergence, in addition to the higher velocity and the local turbulence structure in the vicinity of the abutment, the unique down-ward and up-ward flow motion lead to the additional scour. Furthermore, the experimental results show that the shape of the velocity profiles is "fuller" compared to the free flow cases, and the unique shape of velocity profiles resulted in higher velocity gradients close to the bed. Based on the measurements, the way of estimating higher velocity gradients close to the bed, which is the key element to exploring shear stress, is suggested with respect to the discharge contraction ratio. For broader impacts, the research is expected to contribute practical design for hydraulic engineers by suggesting the classification of scour conditions with respect to the non-dimensional value of set-back distance because the opening width, location of pier and abutment are important design criteria for the safety of bridges vulnerable to scour.

Even if this study provides new insights into the velocity flow field through the scour-critical bridges subject to the submerged orifice flow and submerged flow with overtopping, the entire erosion development process should be simulated numerically to find the changes in turbulent structure, flow contraction and their interactions with the bed in both time and space. The qualitative understanding of the flow pattern and the resulting sediment transport shown in this study can serve to stimulate and guide quantitative experiments and numerical simulation. Furthermore, based on the research for scouring under ice cover, very rough ice can push the maximum velocity even further towards the bed [25,26]. These findings alarmed us and suggest that additional studies should be conducted with different types of bridge deck because the value of roughness under the model bridge deck varies depending on the size and shape of the girders.

Author Contributions: K.S.Y., S.O.L., and S.H.H. provided descriptions and motivation. K.S.Y. and S.H.H. designed experiments and measured the results; S.O.L., S.H.H., and K.S.Y. analyzed the data and interpretation of results; K.S.Y. and S.H.H. prepared the manuscript; and all authors show their efforts in reviewing and editing of the manuscript.

Acknowledgments: Authors would like to show our sincere appreciation to Terry W. Sturm in Georgia Tech for his support. The detailed velocity measurement data for additional research using numerical simulation are available upon request from an interested party.

References

1. Ettema, R.; Nakato, T.; Muste, M. *Estimation of Scour Depth at Bridge Abutments*; Final Report, NCHRP 24-20; TRB: Washington, DC, USA, 2010.
2. Hong, S.H.; Sturm, T.W.; Stoesser, T. Clear water abutment scour in a compound channel for extreme hydrologic events. *J. Hydraul. Eng.* **2015**, *141*, 04015005. [CrossRef]
3. Kwan, T.F.; Melville, B.W. Local scour and flow measurements at bridge abutments. *J. Hydraul. Res.* **1994**, *32*, 661–673. [CrossRef]
4. Melville, B.W.; Coleman, S.E. *Bridge Scour*; Water Resources Publications: Highlands Ranch, CO, USA, 2000.
5. Chrisohoides, A.; Sotiropoulos, F.; Sturm, T.W. Coherent structures in flat-bed abutment flow: Computational fluid dynamics simulations and experiments. *J. Hydraul. Eng.* **2003**, *129*, 177–186. [CrossRef]

6. Kim, H.; Roh, M.; Nabi, M. Computational modeling of flow and scour around two cylinders in staggered array. *Water* **2017**, *9*, 654. [CrossRef]
7. Zang, L.; Wang, P.; Yang, W.; Zuo, W.; Gu, X.; Yang, X. Geometric characteristics of spur dike scour under clear-water scour conditions. *Water* **2018**, *10*, 680. [CrossRef]
8. Chen, S.; Tfwala, S.; Wu, T.; Chan, H.; Chou, H. A hooked-collar for bridge piers protection: Flow fields and scour. *Water* **2018**, *10*, 1254. [CrossRef]
9. Zang, L.; Wang, H.; Zang, X.; Wang, B.; Chen, J. The 3-D morphology evolution of spur dike scour under clear-water scour conditions. *Water* **2018**, *10*, 1583. [CrossRef]
10. Yang, Y.; Qi, M.; Li, J.; Ma, X. Evolution of hydrodynamics characteristics with scour hole developing around a pile group. *Water* **2018**, *10*, 1632. [CrossRef]
11. Hong, S.H.; Sturm, T.W. Physical Model Study of Bridge Abutment and Contraction Scour under Submerged Orifice Flow Conditions. In Proceedings of the 33rd IAHR Congress: Water Engineering for a Sustainable Environment, Vancouver, BC, Canada, 9–14 August 2009.
12. Hong, S.H.; Sturm, T.W. Physical Modeling of Abutment Scour for Overtopping, Submerged Orifice and Free Surface Flows. In *Scour and Erosion, Proceedings of the Fifth International Conference on Scour and Erosion, San Francisco, CA, USA, 7–10 November 2010*; American Society of Civil Engineers: Reston, VA, USA, 2010.
13. Saha, R.; Lee, S.; Hong, S.H. A comprehensive method of calculating maximum bridge scour depth. *Water* **2018**, *10*, 1572. [CrossRef]
14. Hong, S.H. Prediction of Clear Water Abutment Scour Depth in Compound Channel for Extreme Hydrologic Events. Ph.D. Thesis, School of Civil and Environmental Engineering, Georgia Institute of Technology, Atlanta, GA, USA, 2013.
15. Ge, L.; Lee, S.; Sotiropoulos, F.; Sturm, T.W. 3D unsteady RANS modeling of complex hydraulic engineering flows. Part II: Model validation and flow physics. *J. Hydraul. Eng.* **2005**, *131*, 809–820. [CrossRef]
16. Garcia, C.M.; Cantero, M.I.; Nino, Y.; Garcia, M.H. Closure to "Turbulence measurements with acoustic doppler velocimeters". *J. Hydraul. Eng.* **2007**, *131*, 1062–1073. [CrossRef]
17. Hong, S.H.; Abid, I. Physical model study of bridge contraction scour. *KSCE J. Civ. Eng.* **2016**, *20*, 2578–2585. [CrossRef]
18. Hong, S.H.; Lee, S.O. Insight of Bridge Scour during Extreme Hydrologic Events by Laboratory Model Studies. *KSCE J. Civ. Eng.* **2018**, *22*, 12871–12879. [CrossRef]
19. Lee, S.O.; Hong, S.H. Reproducing field measurements using scaled-down hydraulic model studies in a laboratory. *Adv. Civ. Eng.* **2018**, *2018*, 9091506. [CrossRef]
20. Goring, D.; Nikora, V. Despiking acoustic Doppler velocimeter data. *J. Hydraul. Eng.* **2002**, *128*, 117–126. [CrossRef]
21. Chang, F.; Davis, S. Maryland SHA Procedure for Estimating Scour at Bridge Abutments, Part 2—Clear Water Scour. In *Proceedings of Water Resources Engineering '98*; ASCE: Memphis, TN, USA, 1998; pp. 169–173.
22. Chang, F.; Davis, S. Maryland SHA Procedure for Estimating Scour at Bridge Waterways, Part 1—Live Bed Scour. In *Stream Stability and Scour at Highway Bridges*; Richardson, E., Lagasse, P., Eds.; American Society of Civil Engineers: Reston, VA, USA, 1999; pp. 4001–4011.
23. Hong, S.H.; Abid, I. Scour around an erodible abutment with riprap apron over time. *J. Hydraul. Eng.* **2019**, in press.
24. Kara, S.; Stoesser, T.; Sturm, T.W. Flow dynamics through a submerged bridge opening with overtopping. *J. Hydraul. Res.* **2015**, *53*, 186–195. [CrossRef]
25. Zabilansky, L.J.; Hains, D.B.; Remus, J.I. Increased Bed Erosion Due to Ice. In *Cold Regions Engineering 2006: Current Practices in Cold Regions Engineering*; American Society for Civil Engineers: Reston, VA, USA, 2006; pp. 1–12.
26. Hains, D.; Zabilansky, L. *Laboratory Test of Scour under Ice: Data and Preliminary Results*; ERDC/CRREL TR-04-9; U.S. Army Engineer Research and Development Center: Hanover, NH, USA, 2004.

13

An Experimental Study on Mechanisms for Sediment Transformation due to Riverbank Collapse

Anping Shu [1,*], Guosheng Duan [1], Matteo Rubinato [2], Lu Tian [1], Mengyao Wang [1] and Shu Wang [1]

1 School of Environment, Key Laboratory of Water and Sediment Sciences of MOE, Beijing Normal University, Beijing 100875, China; duanguosheng@mail.bnu.edu.cn (G.D.); tianlu2011@mail.bnu.edu.cn (L.T.); miawang@mail.bnu.edu.cn (M.W.); wangshu861217@gmail.com (S.W.)

2 Department of Civil and Structural Engineering, The University of Sheffield, Sir Frederick Mappin Building, Mappin Street, Sheffield S1 3JD, UK; m.rubinato@sheffield.ac.uk

* Correspondence: shuap@bnu.edu.cn.

Abstract: Riverbank erosion is a natural process in rivers that can become exacerbated by direct and indirect human impacts. Unfortunately, riverbank degradation can cause societal impacts such as property loss and sedimentation of in-stream structures, as well as environmental impacts such as water quality impact. The frequency, magnitude, and impact of riverbank collapse events in China and worldwide are forecasted to increase under climate change. To understand and mitigate the risk of riverbank collapse, experimental/field data in real conditions are required to provide robust calibration and validation of hydraulic and mathematical models. This paper presents an experimental set of tests conducted to characterize riverbank erosion and sediment transport for banks with slopes of 45°, 60°, 75°, and 90° and quantify the amount of sediments transported by the river, deposited within the bank toe or settled in the riverbed after having been removed due to erosion. The results showed interesting comprehension about the characterization of riverbank erosion and sediment transport along the river. These insights can be used for calibration and validation of new and existing numerical models.

Keywords: sediment; transforming mechanism; riverbank collapse

---

1. Introduction

Rivers and streams are dynamic systems and are continuously changing their structure due to different flow conditions. Riverbank erosion [1] commonly refers to the removal of bank material by flowing water or carried sediment. This is a phenomenon that derives from two categories of factors [2]: natural (e.g., climate parameters such as precipitation type, intensity, and variability, or soil properties like water content and shear strength and type of vegetation) and human actions [3] (e.g., construction of dams, logging, and intensive grazing).

The phenomenon of riverbank collapse incorporates a variety of bank and riverbed deformations in the affected sections (e.g., the longitudinal erosion and deposition of material on the riverbed or the transverse deformation of riverbed channels) [4]. These deformations assume different shapes and typically alter the cross-sectional morphology of the river affected. This phenomenon plays a very important role in the evolution of rivers, and despite the fact that some stable rivers have a healthy amount of erosion from which they benefit, unstable rivers and the erosion taking place on their banks are a cause for economic, environmental, and social concern: for example, (i) people are forced to migrate due to land erosion; (ii) riverbank collapse causes the loss of large areas of farmland; (iii) riverbank collapse can change the original boundary conditions of the river as well as the water and sediment conditions when large amounts of sediments enter the river channel and siltation occurs; (iv) the sediments eroded due to riverbank collapse are one of the main sources of river sediments and make the river muddy, originating environmental and ecological problems. Therefore, this phenomenon is a great concern for the society and researchers should accurately characterize and assess the several causes that typically accelerate this phenomenon that leads to major impacts, and they should identify feasible solutions for mitigation and adaptation strategies to be implemented.

For non-cohesive riverbanks, the sediment particles on the bank slope are mainly affected by the thrust on the bank, the uplift force, and the gravity effect generated by the water flowing in the river [5]. When the slope of the riverbank is larger than the underwater angle formed by the deposition of eroded sediment, the soil within the upper layer typically collapses along a sliding surface, usually in the form of "shallow collapse" [6]. Particle entrainment can be quantified using the magnitude of the shear stress and the particle size [7,8] for each soil type. On the other hand, cohesive riverbanks, more commonly found worldwide within river streams, are not only subject to these forces but also to the inter-particle cohesion magnitude. Lohnes et al. [9] completed a study to analyze the stability of cohesive riverbanks by calculating the ratio of the driving forces to provide resistance on the collapse surface, and proposed a hypothetical collapse model to be associated with cohesive riverbanks. However, the model that Lohnes et al. [9] developed does not take into consideration the effects of tensile cracks [10], pore water pressure [11], and hydrostatic pressure, and only assumes that the collapse surface can pass through the slope foot and therefore this method can only be applied to relatively steep cohesive riverbanks [12,13]. On this basis, Osman et al. [10] established an additional collapse model for cohesive riverbanks, which takes into account the effects of tensile cracks and assumes that the collapse surface passes through the bank slope foot. The study conducted by Darby et al. [14] also considered the actual topography of the riverbank, plus the pore water pressure and the hydrostatic pressure. Few years later, Rinaldi and Casagli [15] introduced the suction component of saturated and unsaturated parts of the riverbank into the identified collapse mode. Over the last decade, the bank stability and toe erosion model (BSTEM) proposed by the US National Sediment Laboratory has been widely used to numerically quantify the erosion within riverbanks. In this model, the process of riverbank collapse is divided into two parts [16]: (a) the slope foot erosion and (b) the riverbank stability analysis. However, other studies have been completed to provide alternative numerical and mathematical solutions. Huang et al. [17] also established a mathematical model for the collapse, quantifying the factors affecting the riverbank stability. Wang and Kuang [18] derived an equation for calculating the critical height of the initial and secondary riverbank collapse while Xia et al. [19] established a secondary collapse model to be used for cohesive riverbanks that analyzes the forces applied on the bank using the soil typical of the lower Yellow River.

Furthermore, the composite riverbank of an alluvial river generally exhibits a dual structure of two single layers (with different thickness and same distribution) or a tiered structure (with different thickness and distributions). For this last configuration, cantilever collapse [20] and piping collapse [21] are more likely to occur. Xia et al. [22] studied the dual structure experienced within the lower Jingjiang River and specified the three stages of dual structure cantilever collapse, quantifying the influencing factors and analyzing the collapse process. Once the riverbank collapse has taken place, the eroded material is transported within the river and this process is composed of three steps: (i) riverbank collapse and movement of eroded particles; (ii) sediment deposition and its interaction within the riverbed; and (iii) sediment transport along the river. Nagata et al. [23] combined the riverbank collapse model with a two-dimensional mathematical model to simulate the deformation process of the river channel. Darby and Delbono [24] combined a two-dimensional model with the collapse mode of cohesive riverbanks to calculate the deformation process of curved channels. Simon et al. [25] comprehensively considered the impact of sediment accumulation from eroded banks and collapsed banks on the riverbed, using the BSTEM model to estimate the riverbank collapse and material sedimentation under different flood conditions. Darby et al. [26] predicted the evolution of river channels composed of fine sand by coupling an infiltration model and the collapse model. Jia et al. [27] combined the dual structure collapse mode with the three-dimensional model to simulate the evolution of Shishou Bay in the lower Jingjiang River. Nardi et al. [28] simulated the evolution of rivers where bed is composed by medium grained sand. Xiao et al. [29] simulated the river evolution process under the influence of vegetation by combining non-cohesive riverbank collapse models with the shallow water equation. Xia et al. [30] established a mixed model of two-dimensional riverbed deformation in the orthogonal curvilinear coordinate system, and simulated the evolution process of the wandering sections of the lower Yellow River. Jia et al. [31], based on the Osman model, built a three-dimensional water sediment model with bank deformation considered and effectively simulated the horizontal oscillations of river channels caused by cohesive riverbank collapse. These studies demonstrate that the riverbed erosion is a phenomenon clearly observed but they do not consider the characterization of the motion of the sediment eroded and the load in the riverbed. Without considering these aspects, it is very difficult to replicate the riverbed erosion phenomenon with numerical models due to the paucity of existing datasets useful for calibration and validation of numerical tools. Mathematical models developed to date can be used to obtain riverbed and riverbank deformations but are applicable only under limited boundary conditions. Furthermore, it is still very challenging to numerically replicate multiple conditions associated with various slopes and flow conditions because of the paucity of high resolution localized data for this type of phenomenon. Hence, more field or experimental studies are needed to calibrate and validate the dynamic features associated with riverbank erosion under numerous conditions.

Focusing on previous studies based on physical models, Yu et al. [32] used a flume test to qualitatively analyze the interactions between material eroded due to hydraulic effect and its deposition in the river. Yu and Guo [33] studied the coupling relationship between material eroded and its transportation and deposition in the riverbed using a curved channel flume. Wu and Yu [34] revealed new insights on this natural phenomenon for cohesive riverbank collapse, non-cohesive riverbank collapse, and riverbed deposition via experimental tests. Deng et al. [35] simulated the collapse process of the upper Jingjiang Riverbank by combining the longitudinal deposition of the riverbed surface with the secondary collapse mode of cohesive riverbanks. Yu et al. [36] studied the interactions between bank collapse and riverbed deposition for different near-shore riverbed compositions by using a curved channel flume to complete the experimental tests.

Physical experiments can qualitatively measure the amount of material eroded and the amount of material deposited within the close riverbed sections to fulfill the gaps previously described. This paper presents the results obtained with an experimental flume constructed at the Key Laboratory of Water and Sediment located within the School of Environment of Beijing Normal University, investigated the phenomenon of riverbank collapse utilizing a variety of slopes (45°, 60°, 75°, and 90°) for cohesive banks, and characterized the transport of riverbank eroded material within rivers under dissimilar flow conditions (45 and 60 L/s). The datasets provided led to new insights for riverbank erosion under novel specific physical and hydraulic conditions and can be used by numerical modelers to validate relationships between variables associated with this natural phenomenon.

2. Methodology

2.1. Experiments

The experiments were conducted in the multi-function flume (0.8 m wide, 0.8 m deep, and total length of 25 m), located at the School of Environment, Key Laboratory of Water and Sediment of Beijing Normal University, China. By using a set of rigorous procedures authors guaranteed the continuous accurate re-construction of the bank slopes tested (four configurations) within this facility, ensuring the regularity of the experiments.

2.1.1. Experimental Setup

Gravel material was located upstream and downstream of the bank re-created to enable constant boundary conditions. Authors constructed in the laboratory a container with the dimensions of the banks to be replicated and fit it (once filled with material) within the flume for each experiment, guaranteeing the same position each time (±2 mm) by controlling reference points identified at the edge of the investigation area. Six sections (1#, 2#, 3#, 4#, 5#, and 6# as displayed in Figure 1) were identified within the bank to monitor the relevant parameters for this study. Four water level gauges were fitted in the flume to monitor the water level, and a pore water pressure gauge was embedded to monitor changes in the pore water pressure inside the bank as shown in Figure 2b. An example of constructed bank slope fitted within the experimental flume is shown in Figure 2a. For each experiment, velocities associated to each flow rate were measured at multiple locations within sections by using a propeller (accuracy 0.01 m/s), as represented in Figure 3, for the six sections illustrated in Figure 4. Water samples containing material were taken every three minutes at three monitoring section (1#, 3#, 5#) plus at the tailgate to measure the correspondent sediment concentration. A camera was set up on the side of the flume to record the whole process of bank slope collapse.

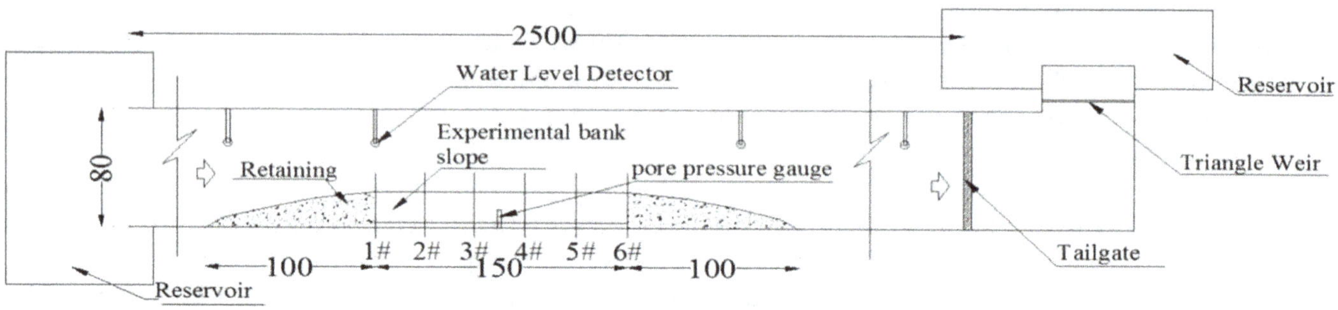

Figure 1. Schematic diagram of test flume and its equipment (not to scale) (Unit: cm).

Figure 2. Scheme of the experimental model. (**a**) Constructed bank slope fitted within the experimental flume; (**b**) Position of the pore water pressure gauge embedded inside the bank.

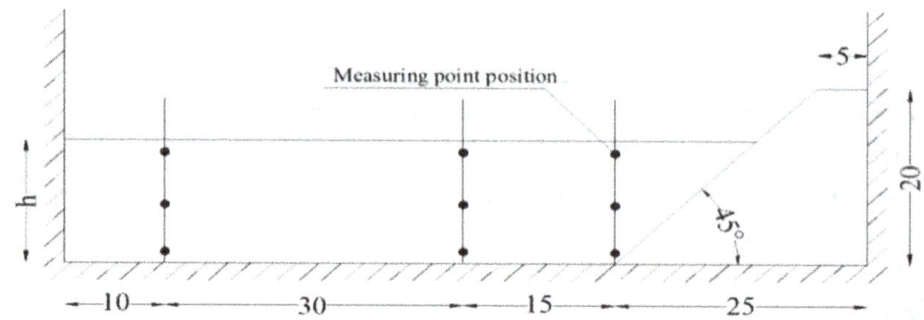

Figure 3. Example of one cross section and correspondent monitoring point for the velocity (Unit: cm).

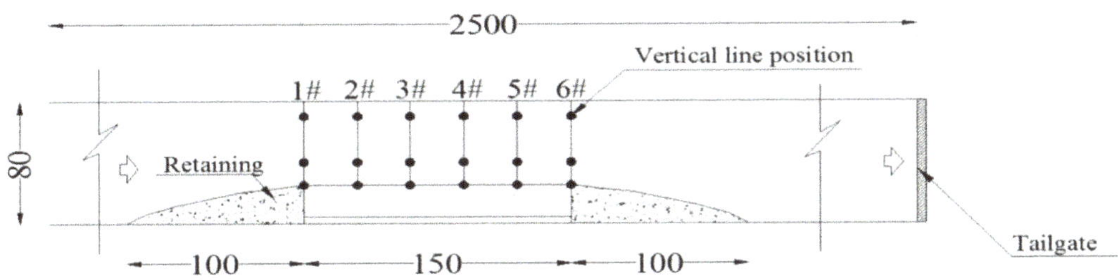

Figure 4. Side-view of the experimental setup and location of the sections for data collection (Unit: cm).

2.1.2. Experimental Testing Conditions

Based on field observations [37], four bank slope configurations (No.1, No.2, No.3, and No.4 as shown in Table 1) were considered and replicated with the experimental facility. Material utilized for this study was collected in the natural riverbank located in the upstream section of the Yellow River, from Dengkou County of the Ningmeng reach (Figure 5a) and the particle size distribution is shown in Figure 5b. All the physical properties of the tested materials are listed in Table 2.

Table 1. Configurations and bank morphology details.

Group	Slope Degree (°)	Bank Morphology			Flux (L/s)	Water Discharge Time (min)	Water Level (cm)
		Bank Top Width (cm)	Bank Height (cm)	Bank Toe Width (cm)			
No.1	45	5	20	25	45 60	40 30	11.2 13.2
No.2	60	13.45	20	25	45 60	40 30	11.5 12.8
No.3	75	19.64	20	25	45 60	40 30	11.8 13.9
No.4	90	25	20	25	45 60	40 30	11.25 13

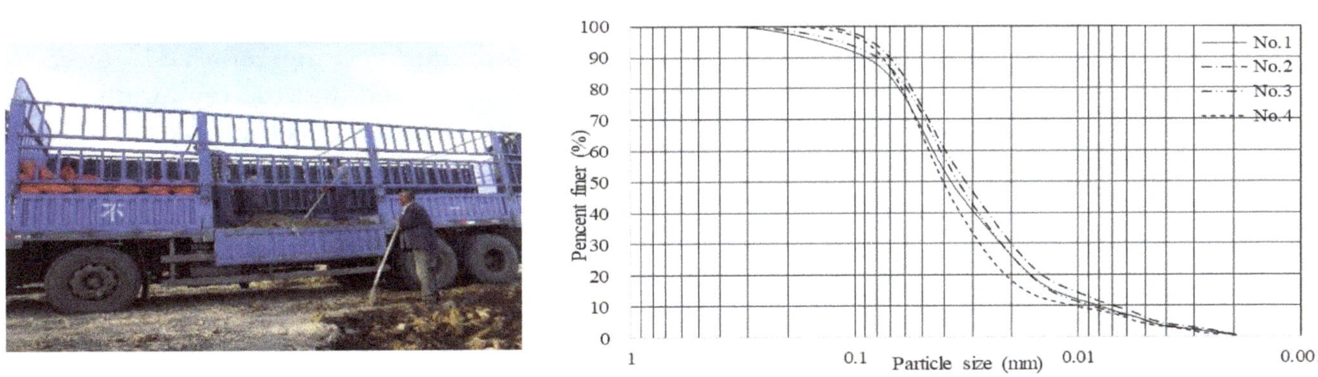

(a) (b)

Figure 5. Real site and characteristics of the material. (**a**) Collection of the material on site; (**b**) Particle size distribution of the material collected.

Table 2. Physical properties of the material tested for each configuration.

Group	Soil Position	Moisture Content (%)	Unit Weight (g/cm^3)	Cohesion (kPa)	Friction Angle (°)
No.1	Bank toe	15.50	1.83	14.77	19.46
	Bank top	15.33	1.82	14.45	19.19
No.2	Bank toe	15.50	2.01	14.77	19.46
	Bank top	15.33	1.76	14.45	19.19
No.3	Bank toe	14.92	1.66	13.66	18.54
	Bank top	14.54	1.58	12.93	17.92
No.4	Bank toe	14.98	1.80	13.77	18.63
	Bank top	15.80	1.61	15.37	19.96

2.1.3. Experimental Procedure

Each experiment was conducted following these steps:

(1) Before each group of experiments, soil samples were taken from the monitored sections on the bank slope to obtain parameters such as unit weight, water content, and size distribution associated to initial conditions and hence parameter of comparison with data collected successively.

(2) In order to avoid bank toe erosion caused by a drastic change of water level during the initial water discharge released, the tailgate was kept close and water was entering the flume very slowly. When the water reached the designed level, the release was interrupted to let the part of the bank under slope to be soaked as in the natural scenario. When the soil was saturated, water was then discharged according to the designed flow, and the tailgate was opened as designed. This moment was considered the start of the experiment.

(3) After each experiment began, the monitoring devices in the flume were turned on to monitor the water level in real time, as well as the camera to record the collapse process.

(4) Samples were taken regularly at the monitoring section (1#, 3#, 5#) plus at the tailgate to measure the sediment concentration. A ladle was used to take samples of sandy water from the flow, then the samples collected were put into a beaker. The weight of the beaker and sandy water could be obtained by using a scale. Successively, the beaker was inserted in the oven, and once it was dry, the weight of sediment remained by the evaporation of the water could be obtained by using the weight scale. Finally, the concentration could be obtained dividing the weight of sediment by the weight of sandy water. The accuracy of scale is 0.01 g.

(5) For each configuration group, 45 L/s flowed over a pre-constructed bank for 40 min and 60 L/s flowed over a water worked bank for additional 30 min.

(6) When there was no more observation of riverbank material being removed or eroded, the erosion process was considered completed, the water was stopped and the remaining terrain (and its shape) of the riverbank was measured by regularly removing 20 cm of section each time and create profiles at each removal step.

2.2. Data Processing

2.2.1. Estimation of Riverbank Collapse Volume

After each experiment, the amount of material collapsed was obtained by combining the volume associated with the topography of the riverbank before and after the entire process and calculating the difference. As previously stated, multiple measuring sections were selected every 20 cm in the direction of water flow to characterize the final riverbank shape, and for each step the full topography profile was measured by using a system of gridlines as shown in Figure 6. Finally, the coordinates obtained were entered into AutoCAD© [38] to recreate the profiles acquired and an example (first configuration, 45° slope) is shown in Figure 7.

Figure 6. Schematic diagram of a topographical measurement conducted after the collapse section using a system of gridlines.

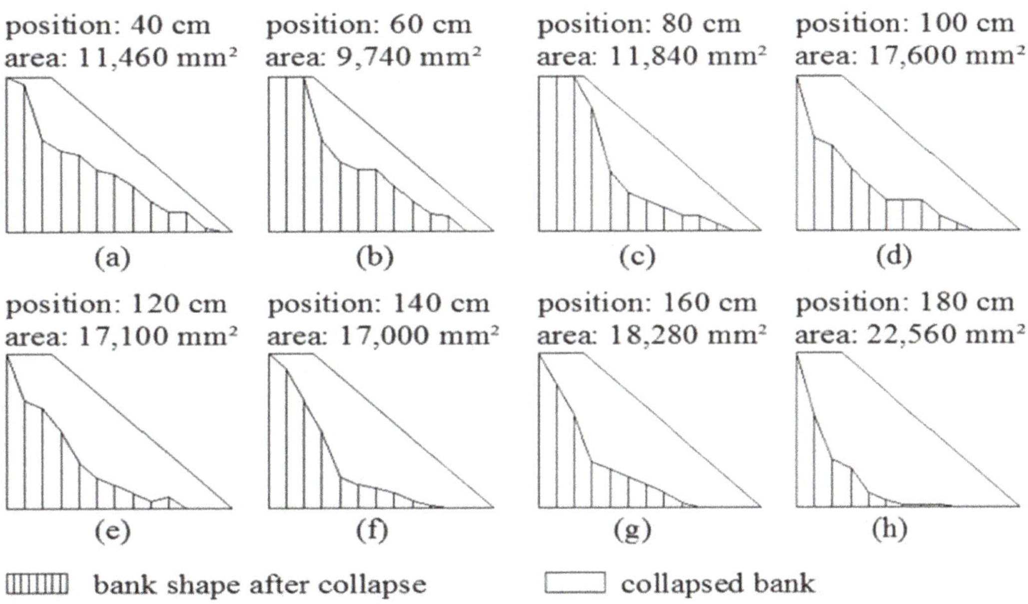

Figure 7. Example of riverbank profiles reconstructed via AutoCAD for the 45° riverbank configuration. (**a**) Position 40 cm; (**b**) Position 60 cm; (**c**) Position 80 cm; (**d**) Position 100 cm; (**e**) Position 120 cm; (**f**) Position 140 cm; (**g**) Position 160 cm; (**h**) Position 180 cm.

2.2.2. Estimation of Velocity for Incipient Sediment Motion

When the sediments enter the channel due to riverbank collapse, part of them was carried downstream by the water flowing, and the remaining part deposited at the foot of the existing bank. The deposited sediment, under the action of water flow, was then further activated and transformed into bed load and suspended load, which was transported towards the downstream section of the channel [39]. Therefore, to study the sediment transport, it was necessary to analyze the characteristics of the sediment deposited at the bank toe. The particle size of the experimental soil tested was between 0.0016 to 0.352 mm, and these values are typical of cohesive particles. It is generally believed that the newly deposited cohesive soil is compacted sediment that deposited rapidly. It can still be treated as single-particle sediment, but there is an additional bonding force between particles applied to it. Based on this view, the particle size was calculated by using the equation provided by Qian and Wan [40]. This equation enabled to compute D (sediment particle size) under the experimental conditions as showed in Table 3.

$$\frac{U_i}{\sqrt{gD}} = \sqrt{\frac{\gamma_s - \gamma}{\gamma}\left(6.25 + 41.6\frac{h}{Ha}\right) + \left(111 + 740\frac{h}{Ha}\right)\frac{Ha\delta_0}{D^2}} \quad (1)$$

where Ha is the atmospheric pressure expressed in terms of water column height, and Ha = 10 m; δ_0 is thickness of a water molecule, and $\delta_0 = 3.0 \times 10^{-8}$ cm; γ_s is unit weight of sediment, and γ_s = 17,542 N/m^3; γ is unit weight of water, and γ = 9800 N/m^3; g is the gravitational acceleration, and g = 9.8 m/s^2, D is sediment particle size, m; U_i is velocity for incipient sediment motion and U is the velocity, m/s (for this study $U_i = U$); h is water depth, m.

Table 3. Characterization of percentage of transported sediment under different experimental conditions.

Group	Slope Gradient (°)	Flow Rate (L/s)	Average Water Level (cm)	Average Flow Rate (m/s)	Incipience Motion Particle Size (Lower Limits) (μm)	Incipience Motion Particle Size (Upper Limits) (mm)	Incipience Motion Percentage (%)
No.1	45	45	12.40	0.62	9.25	7.29	89.85
		60	13.25	0.70	7.23	7.84	91.46
No.2	60	45	11.50	0.69	7.46	8.95	91.06
		60	12.80	0.74	6.38	10.42	92.92
No.3	75	45	11.80	0.65	8.48	8.88	90.56
		60	13.90	0.72	6.88	10.63	92.06
No.4	90	45	11.25	0.65	8.20	9.52	90.81
		60	13.00	0.77	5.98	11.10	93.60

2.2.3. Estimation of Sediment Load Ratio

In a sediment gravity flow, the coarse sand is mainly bed load movement, and the energy is primarily provided by rapid whirlpools revolving the current in the river [41]. The fine sand is mainly suspended load movement, and its energy is provided by smooth change in water direction. In spite of the mutual transformation of the two in the process of movement, under specific water flow conditions, the amount of bed load and suspended load carried by the water flow would barely change in a certain period of time. Therefore, from the point of view of quantity, it is reasonable to divide the sediment in water into coarse particles and fine particles moving with the water flow. Then, it is particularly important to introduce a boundary particle size between the bed load and the suspended load. The following procedure was applied to estimate the amount of collapsed bank sediment converted to bed load and suspended sediment after entering the river channel.

Different from the clear water flow, the movement of sediment gravity flow needs to overcome the internal cohesive resistance of water flow carrying suspended load and the friction of the bed surface during the movement of the bed load, in addition to the boundary resistance. Therefore, in the presence of the sediment gravity flow, total energy slope loss (J) is equal to the sum of the energy slope loss of water flow carrying suspended sand (J_l) and the energy slope loss of bed load motion (J_s). The automatic adjustment of alluvial riverbed is accompanied by energy loss, transformation and modification, that is, various elements (such as bed sand, section, and longitudinal slope) tend to regulate toward minimum energy consumption of the water flow. In the process of sediment transport, the adjustment tends to minimize the total energy consumption of the suspension movement and the layer movement, which can be expressed in the form of energy slope [42].

$$J_l + J_s = J_{\min} \quad (2)$$

The water flow with suspended sediment is composed of water and fine particles in suspension motion and its energy slope loss can be expressed by the Darcy–Weisbach equation. After considering the unit weight correction of the suspended sediment flow, its energy slope loss can be calculated as follows:

$$J_l = \frac{f}{8} \times \frac{U^2}{gR} \times \frac{\gamma_f}{\gamma_m} \tag{3}$$

where γ_f and γ_m are the unit weight of the suspended sediment flow and the sediment gravity flow, respectively, N/m^3; R is the hydraulic radius, m. The resistance coefficient of the suspended sediment flow f is expressed as [43,44] provided:

$$f = 0.11a\left(\frac{d_0}{4R} + \frac{68}{Re}\right)^{0.25} \tag{4}$$

$$a = 1 - 0.41g(\mu_r) + 0.08(\lg\mu_r)^2\left(\frac{4R}{d_{90}}\right)^{\frac{1}{6}} \tag{5}$$

where d_0 is the boundary particle size of the suspended sediment and bed load, m; a is resistance-reducing coefficient; Re is the Reynolds number of the sediment gravity flow.

$$Re = \frac{4RU_c\gamma_m}{g\mu} \tag{6}$$

where μ is dynamic viscosity of the sediment gravity flow (water flowing with both coarse sand and fine sand), Pa·s; U_c is the non-silting critical velocity of the suspended sediment flow (water flowing with fine sand), m/s; μ_r is the relative viscosity of the sediment gravity flow, it can be obtained based on Equation (7).

$$\mu_r = \mu/\mu_0 = 1 + 2.5S_v \tag{7}$$

where μ_0 is dynamic viscosity of water in the same temperature as the sediment gravity flow, Pa·s; S_v is the sand content of the sediment gravity flow.

According to the movement characteristics of sediment gravity flows, suspended sediment and bed load are continuously exchanged in the process of moving with the water flow, and the mechanism is very complicated. Based on previous research results, it is believed that the fine particles are carried away by the water flow in the form of suspended sediment, while the coarse particles move in the form of bed load. Then, the non-silting critical flow rate of the fine particles in sediment gravity flow can be obtained.

$$U_c = 27.8\sqrt{\frac{8}{f}\omega_0 \cdot S_v^{\frac{2}{3}}}\left(\frac{4R}{d_{90}}\right)^{\frac{1}{9}} \tag{8}$$

where ω_0 stands for the corresponding sedimentation rate of d_0. Within the range of Stokes, ω_0 can be calculated using Equation (9):

$$\omega_0 = (\gamma_m - \gamma_w)gd_0^2/25.6\gamma_w\nu \tag{9}$$

where γ_w is the unit weight of water, N/m^3; ν is the kinematic viscosity of water, m^2/s.

If the upper particle size is replaced by the boundary particle size d_0, then the weight percentage of bed load in all solids is X. The suspended sediment concentration and the bed load concentration are S_{vf} and S_{vc}, which can be expressed as follows:

$$S_{vc} = X \cdot S_v \tag{10}$$

$$S_{vf} = \frac{S_v(1-X)}{1-XS_v} \tag{11}$$

If the sediment concentration (S_v) in Equation (8) is expressed by the suspended sediment concentration (S_{vf}), the non-silting critical flow rate of the fine particles in sediment gravity flow can be obtained.

$$U_c = 27.8\sqrt{\frac{8}{f}\omega_0 \cdot S_{vf}^{\frac{2}{3}}}\left(\frac{4R}{d_0}\right)^{\frac{1}{9}} \tag{12}$$

For this study, $U = U_c$, and by including Equation (12) into Equation (3), it is possible to calculate the energy slope loss of the water flowing with the suspended sediment.

$$J_l = \frac{1}{gR}\left[27.8\omega_0 \cdot S_{vf}^{\frac{2}{3}}\left(\frac{4R}{d_0}\right)^{\frac{1}{9}}\right]^2 \frac{\gamma_f}{\gamma_m} \tag{13}$$

As can be seen, the energy slope loss J_l of the suspended sediment water flow is closely related to the boundary particle size d_0.

According to Bagnold's research, the energy slope consumed by the movement of solid-phase particles in solid–liquid flow to overcome internal resistance is related to the coarse particle concentration (S_{vc}) and the macroscopic coefficient of friction between particle interactions (tanα). After considering the buoyancy of suspended sediment water unit weight on bed load particles, their movement resistance can be expressed in the form of energy slope.

$$J_s = XS_{vc}\left(\frac{\gamma_s - \gamma_f}{\gamma_m}\right)\tan\alpha \tag{14}$$

where $\gamma_f = \gamma + S_{vf}(\gamma_s - \gamma)$ is the unit weight of suspended sediment flow in the sediment gravity flow, and $\tan\alpha$ is the macroscopic coefficient of friction between bed load particles, α is friction angle between bed load particles, °. For this study, the Begnold parameter $\tan\alpha$ was selected to be 0.63.

Based on the principle of Equation (2), the critical particle size of different groups of simulation experiments is obtained by using the sediment gradation in the flume experiment. According to the experiment soil gradation of each group in Figure 5b and characterization of incipience motion particle size under different experimental conditions, the amount of collapsed bank sediment converted to bed load and suspended sediment after entering the river channel can be obtained.

The calculating process can be summarized as follows:

- Firstly, it is essential to assume a series of d_0 values (The value of d_0 is from 0.050 mm to 0.070 mm, and d_0 should be taken every 0.002 mm, for example, 0.050 mm, 0.052 mm, ..., 0.070 mm). For each value of d_0, the corresponding X can be obtained through Figure 5b;
- Consequently, because the parameters (S_v, γ_f, γ_m, γ_s, γ_w, g, R, and ν) are the basic data and can be obtained through experimental conditions, S_{vc} and S_{vf} can be obtained through Equations (10) and (11);
- Then ω_0 can be obtained by applying Equation (9);
- J_l can be obtained from Equation (13), J_s can be obtained from Equation (14);
- $J = J_l + J_s$ can be obtained from Equation (2);

A series of J values can then be obtained: $J_{0.050}$, $J_{0.052}$, ... , $J_{0.070}$. It is important to choose the smallest value each time for each range selected, which corresponds to the critical particle size.

3. Results

3.1. Dynamic Characteristics of the Bank Collapse Process

3.1.1. Collapse Process of Cohesive Riverbanks

The various collapse processes of cohesive riverbank associated with different hydraulic conditions were recorded (some examples are shown in Figure 8) and we have characterized the collapse process via exhaustive and reiterated observations of the experiments completed. After releasing the water, the bank toe was initially scoured by the water flowing in the channel. Consequently, small fragments of material begun to fall down along the bank slope, occasionally, and a groove was gradually formed on the bank toe, as shown in Figure 8a.

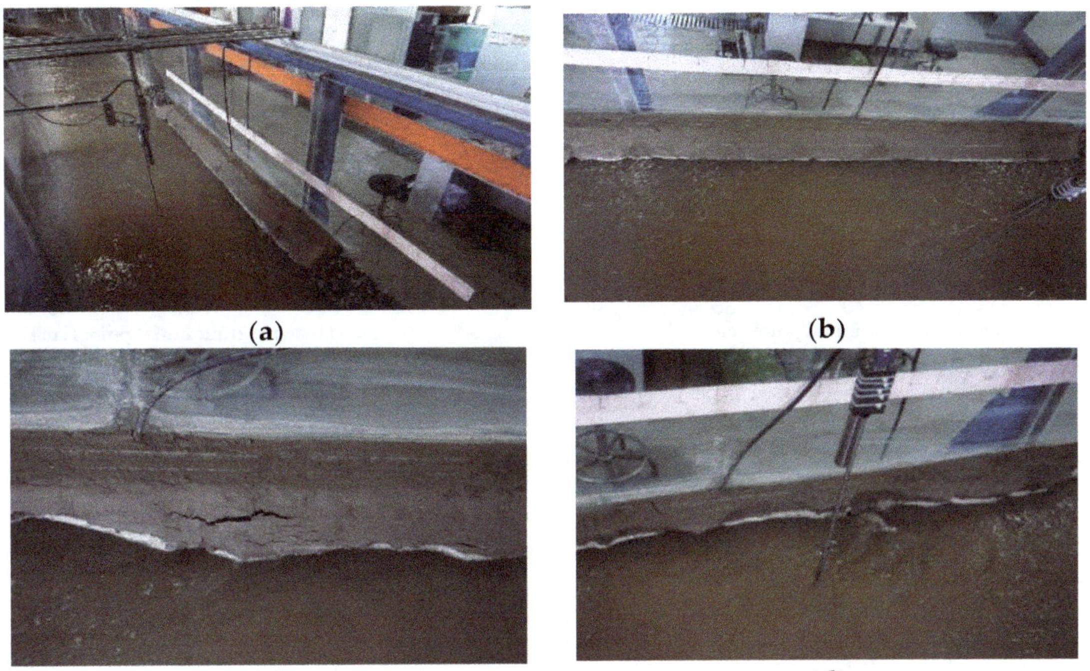

(a) (b) (c) (d)

Figure 8. Example of collapse processes recorded for cohesive riverbank. (**a**) Upstream bank toe erosion; (**b**) Lateral view of the bank; (**c**) Tension crack; (**d**) Complete bank collapse.

As the water continued to flow washing the bank, the groove became deeper and deeper but due to the strong bonding effect typical of the cohesive soil, the upper part of the bank affected by the groove was maintained suspended. When the gravity of the suspended soil was then greater than its anti-sliding force and exceeded its stable slope, multiple cracks appeared on the bank, as shown in Figure 8b. The magnitude of these cracks deepened with the time and eventually the collapse occurred on the bank causing the removal of partial blocks as shown in Figure 8c. Finally, in the case in which the bank slope was steep, its collapse surface was recorded to be almost vertical, as shown in Figure 8d, which is in agreement with the natural phenomenon observed in the Dengkou river section [37].

3.1.2. Velocity Distribution

Flow velocity is a major factor influencing the phenomenon of riverbank collapse, hence it was important to obtain the localized velocity estimation for the experiments conducted under different flow conditions. Figure 9 shows an example of velocity distribution recorded for section #3 (displayed in Figure 4). These velocity measurements were taken before the bank collapse and among the collapse process. The main reason is due to the fact that the exact location and timing of the collapse were difficult to predict. The greater riverbank slope, the higher the nearshore velocity measured under the same hydraulic conditions. Typically, a larger slope of the bank corresponds to a larger obstructed area since the bottom width of the bank is imposed, and as a consequence, velocities are larger when the contraction ratio is larger.

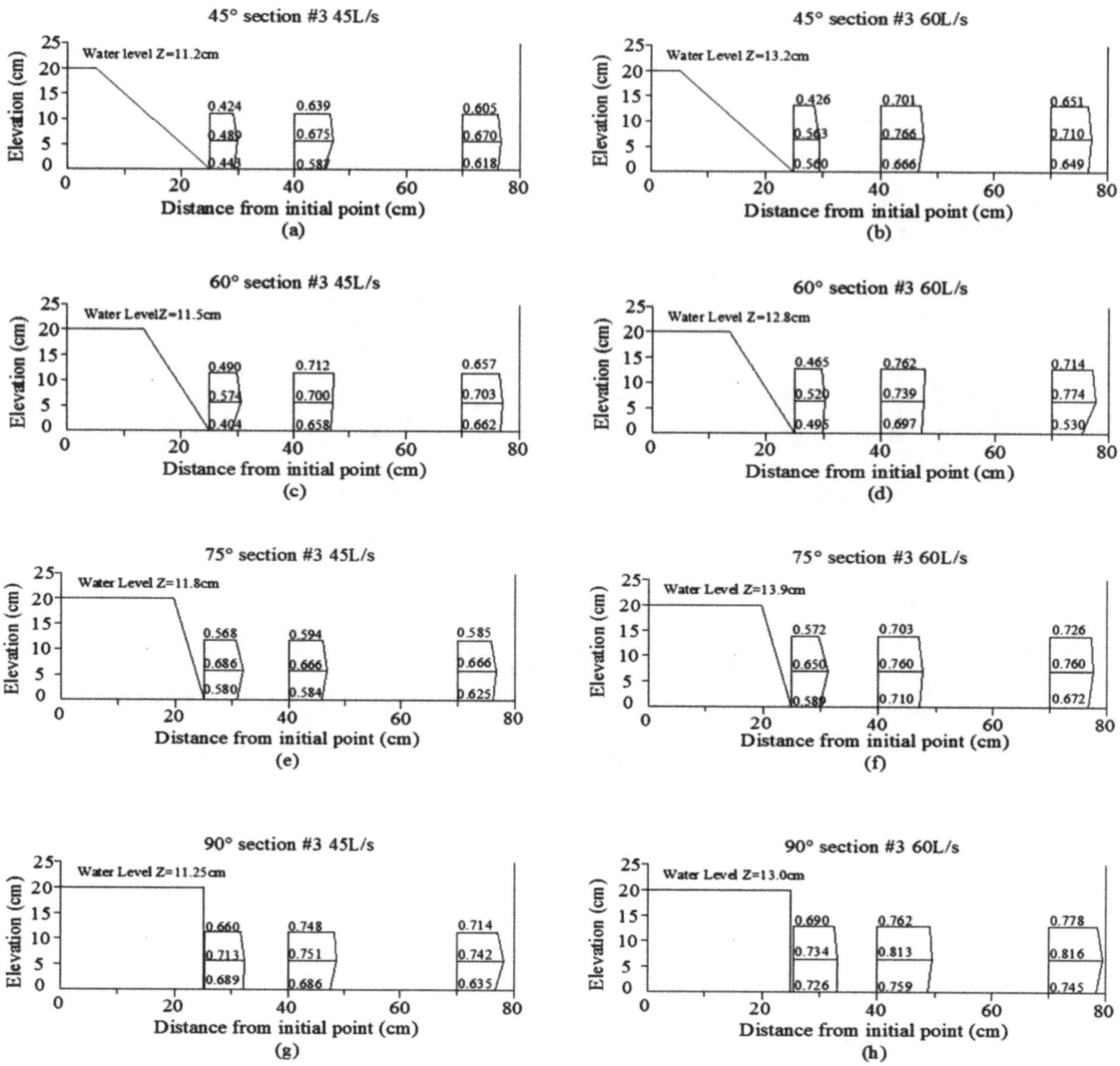

Figure 9. Velocity distribution of section #3 under different flow conditions (45 L/s and 60 L/s) and slope angles (45°, 60°, 75°, and 90°) (Unit: m/s). (**a**) 45° section #3 45 L/s; (**b**) 45° section #3 60 L/s; (**c**) 60° section #3 45 L/s; (**d**) 60° section #3 60 L/s; (**e**) 75° section #3 45 L/s; (**f**) 75° section #3 60 L/s; (**g**) 90° section #3 45 L/s; (**h**) 90° section #3 60 L/s.

3.1.3. Pore Water Pressure

Water content within the riverbank is another important factor to consider when replicating riverbank collapse within experimental facilities. An example showing the changes of pore water pressure for each slope tested (45°, 60°, 75°, and 90°) and the flow rate 45 L/s is shown in Figure 10. By an analysis completed by the authors assessing all the tests conducted including scenarios with all the slopes tested and flow rate 60 L/s, it can be detected that the steeper is the slope of the riverbank, faster is the initial collapse. For gentle slopes, the internal pore water pressure of the soil fluctuates for a longer time, particles seem to interact more frequently, and the shear resistance is stronger than the one measured for steep slopes.

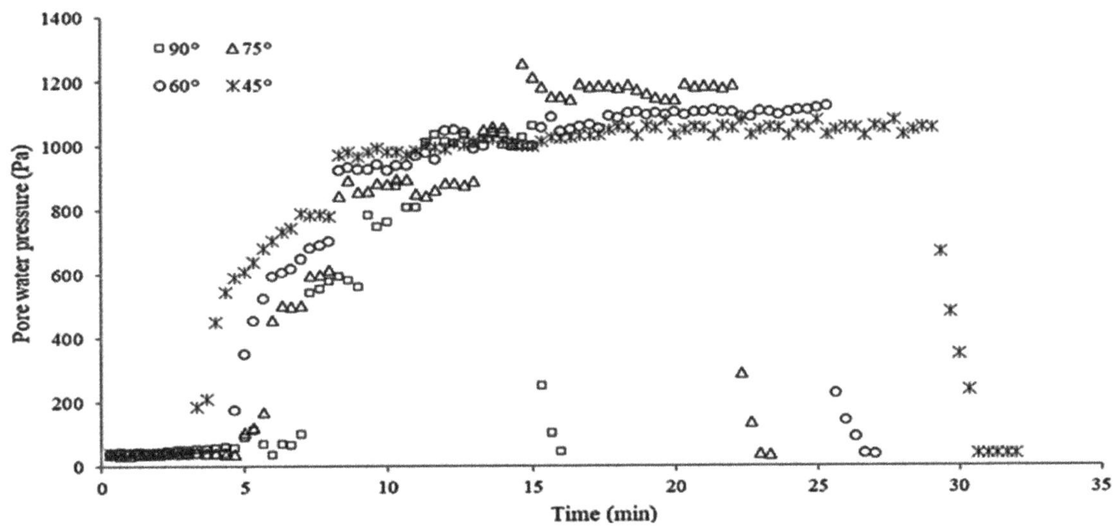

Figure 10. Pore water pressure measured under the flow rate of 45 L/s for each slope angle (45°, 60°, 75°, and 90°).

As shown in Figure 10, pore water pressure is very small at the beginning of the experiment because there is a low quantity of water in the riverbank. Gradually the pore water pressure increases as the water penetrates the bank, reaching an almost constant value prior the bank collapse and consequently drastically decreases because due to the removal of the bank material, the pore water pressure gauge used to measure this parameter is directly exposed to the water in the flume.

3.2. *Quantitative Analysis of the Sediments Transformation*

3.2.1. Sediment Concentration Variation

In order to analyze the variation in time of sediment concentration during the collapse process, this parameter was monitored in real time for the sections selected for this study during each experiment (see Methodology section) and an example is provided in Figure 3 where results collected at Section #3 (Figure 11a) and the tailgate (Figure 11b) are displayed.

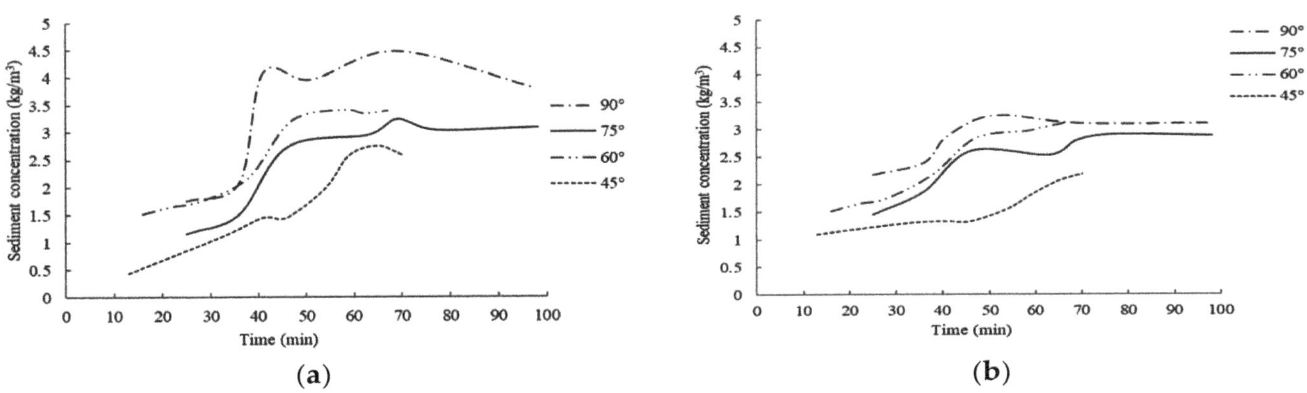

Figure 11. The diagram of sediment concentration change in typical sections. (**a**) Sediment concentration measured in Section #3; (**b**) Sediment concentration measured at the tailgate section.

By comparing the sediment concentration recorded in Section #3 for each slope configuration of 45°, 60°, 75°, and 90°, it is possible to notice a consistent trend regardless of the different initial slope (Figure 11a). At the beginning, for each configuration, the sediment concentration increases slowly, and then rises dramatically. The rapid increase in sediment concentration corresponds with the moment when the bank slope collapses. When this phenomenon happens, a large amount of debris enters the river, mixes with the water, increasing its sediment concentration that reaches its maximum level within the stream. When the water continues to flow towards the downstream section, the sediment concentration recorded gradually decreases and becomes stable (Figure 11b). For the tests recorded, the time measured to reach the collapse and hence the highest sediment concentration in the river is between 35 and 45 min. Furthermore, it has been noticed that highest is the slope and faster is this process.

3.2.2. Sediment Transformation from Bank Collapsed Materials to River Sediment

The duration to achieve the collapse phenomenon (min), the amount of material collapsed from the riverbank (kg) and the average collapse magnitude (g/s) for each configuration tested (45°, 60°, 75°, and 90°) is displayed in Table 4. The steeper the slope, the less time it takes the collapse to occur and higher the average collapse magnitude. The larger the near bank velocity, the larger is the bank collapse magnitude (Figure 9 and Table 4).

Table 4. Experimental parameters measured within the experimental tests.

Group	Slope Gradient (°)	Collapse Time (min)	Collapse Amount (kg)	Average Collapse Magnitude (g/s)
No.1	45	29	38.87	9.25
No.2	60	27	41.58	9.90
No.3	75	23	62.45	14.87
No.4	90	16	82.43	19.63

As displayed in Figure 5b the sediment particle sizes used in the experiment range 2 μm–0.32 mm. As shown in Table 3, all the incipience motion particle sizes have lower and upper limits and these can be calculated using Equation (1). For example, considering 45° and 45 L/s, when the sediment particle distribution is between 9.25 μm and 7.29 mm, material will start to move as observed under the experimental conditions. When the size of the sediment is smaller than 9.25 μm, the force between particles is greater than the flow shear force, hence the sediments tend to deposit at the bank toe. When the size of the sediment particle is bigger than 7.29 mm, it requires a greater flow shear force and this means that only the sediments smaller than 9.25 μm (10.15%) deposit at the bank toe, while the remaining part (89.85%) moves. For all the experimental groups, about 89.85–93.6% of the collapse amount typically moves, and 6.4–10.15% deposits at the bank toe.

3.2.3. Sediment Transformation from Bed Load to Suspended Load

The critical particle size distribution associated with different experiments was obtained by using the sediment gradation in the flume experiment, as shown in Table 5. It can be seen that the critical particle size of the bed load and suspended sediment obtained by this method is about 0.06–0.068 mm, which is close to the critical particle size (0.05 mm) between coarse and fine sands in the upper reaches of the Yellow River.

Table 5. The critical particle size of collapsed riverbank.

Group	Slope Degree (°)	Flow Rate (L/s)	Critical Particle Size (mm)
No.1	45	45	0.06
		60	0.062
No.2	60	45	0.064
		60	0.062
No.3	75	45	0.068
		60	0.066
No.4	90	45	0.068
		60	0.068

It can be seen from Table 6 that after the collapsed bank sediment enters the river, it is mainly converted into three parts. About 7–10% of the sediment typically deposits locally, and the rest continues to move downstream with the water flowing, part of which (70–77%) forms suspended sediment and another part (15–20%) forms bed load.

Table 6. Mass percentage of sediment fractions.

Group	Bank Slope	Flow Rate (L/s)	Deposited Sediment (%)	Suspended Sediment Percentage (%)	Bed Load Percentage (%)
No.1	45°	45	10.15	70.35	19.50
		60	8.54	71.61	19.85
No.2	60°	45	8.94	75.76	15.30
		60	7.08	77.31	15.61
No.3	75°	45	9.44	70.91	19.65
		60	7.94	72.08	19.98
No.4	90°	45	8.39	73.38	18.23
		60	9.06	74.16	16.78

4. Conclusions and Discussion

The conclusions can be summarized as follows:

(1) Experimental tests conducted have enabled the characterization of the riverbank collapse process within four different slopes simulated. This phenomenon can be subdivided into multiple steps: (i) the foot of the bank is frequently affected by the washing effect of the water flowing within the river; (ii) small fragments start to fall down along the bank and a groove is usually gradually formed on the bank toe. The groove becomes deeper and deeper as the water flow continues to wash the bank. Due to the strong bonding effect, the cohesive soil in the upper part of the groove is suspended. When the gravity of the suspended soil is greater than its anti-sliding force and exceeds its stable slope, cracks form along the bank slope. The cracks deepen until the collapse occurs and part of the bank falls in blocks inside the river. When the bank slope is steep, its collapse surface is almost flat, which is consistent with the collapse observed at Dengkou River section [37].

(2) Topographic surveys were completed to characterize the riverbank collapse under different slope scenarios and the bank collapse magnitude is directly correlated with the water flow rate; in fact, the greater is the flow rate, the faster the collapse occurs and faster is the time for the collapse to initiate. Moreover, the pore water pressure monitored during the occurrence of the riverbank collapse, is an indicator of how quickly this phenomenon arises (e.g., for gentle slopes, the internal pore water pressure of soil fluctuates for a longer duration, particles interact more frequently, and the shear resistance seems to be stronger than the shear resistance of steep slopes).

(3) The process of sediment transformation from bank collapsed materials to river sediment was analyzed through monitoring the sediment concentration at typical sections and the tailgate. It was found that the change of sediment concentration with time is basically consistent regardless of the initial bank slope. At the beginning, the sediment concentration increases slowly, and then rises dramatically. The rapid rise of the curve corresponds with the occurrence of the collapse along the riverbank. When the high amount of debris enters the river, it mixes with the water, increasing the sediment concentration along the stream and reaching the maximum value recorded. Consequently, it gradually decreases and becomes stable. The rapid increase of sediment concentration due to collapse is between 35 and 45 min after the start of the experiment, but for the tailgate section, the maximum sediment concentration is between 40 and 50 min. The maximum sediment concentration of the tailgate section arrives later than that of section #3. This is due to the fact that when collapsed soil enters into the water, the water requires time to

carry it downstream. Additionally, it was observed that the maximum sediment concentration of the tailgate is smaller than that of section #3. That is due to the typical sediments transport process within river.

(4) When the sediment enters the river, part of it is carried downstream by the current, and the remaining part deposits at the foot of the bank. The deposited sediment, under the action of water flowing, is further activated and then transformed into bed load and suspended load, which is transported downstream as well within the water. In terms of quantity, about 7–10% of the sediment typically deposits locally, and the rest continues to move downstream with the water flowing, part of which (70–77%) forms suspended sediment and another part (15–20%) forms bed load.

Despite the insights provided with this study, it is fundamental to state its limitations considering the complexity of the riverbank collapse phenomenon. To study the mechanisms of riverbank collapse, it is necessary to consider not only the longitudinal erosion and the siltation effect of the riverbed due to water and sand, but also the transverse deformation of the river channel (or the collapse of riverbank). In natural rivers, the longitudinal erosion and siltation of the riverbed, riverbank collapse, and sediment transportation are simultaneous, hence it has been very complex to simulate all these phases. Additionally, since the mechanism of bank collapse and the interaction between sediments and silt bed are not completely explored yet, it is difficult to accurately simulate the process of sediment transportation in the bank collapse section by using either a mathematical or a physical model. However, we were able to provide new insights regarding the longitudinal deformation and the transverse deformation of the river, involving the degree of deformation in a certain period of time under various flow conditions, furnishing details about the amount of riverbed erosion, the amount of siltation, and the amount of riverbank collapse. Hence, this study can be a considered as a preliminary phase to investigate bank erosion and sediment transportation mechanisms due bank collapse. Future research should also address other limitations. Firstly, bank shapes in natural rivers are irregular, so more bank angles should be considered. Secondly, the bank is typically eroded by sediment gravity flow in natural rivers, so the effect of sediment concentration on bank collapse should be added when simulating the water flowing, which is never completely clear. However, results obtained with this study should be used to calibrate existing and new numerical models and such recommendations can support current best practices for river management and environmentally sustainable restoration, preventing riverbank destabilization.

Author Contributions: All the authors jointly contributed to this research. A.S. was responsible for the proposition and design of experiment and the calculation method; G.D. and M.R. analyzed the experimental datasets and wrote the paper; L.T., M.W., and S.W. participated in the experiments.

References

1. Yao, Z.Y.; Ta, W.Q.; Jia, X.P.; Xiao, J.H. Bank erosion and accretion along the Ningxia-Inner Mongolia reaches of the Yellow River from 1958 to 2008. *Geomorphology* **2011**, *1–2*, 99–106. [CrossRef]
2. Shu, A.; Gao, J.; Duan, G.S.; Zhang, X. Cluster analysis for factor classification and riverbank collapse along the desert wide valley reach of the upper Yellow River. *J. Tsinghua Univ. (Sci. Tech.)* **2014**, *8*, 1044–1048. (In Chinese) [CrossRef]
3. Ta, W.Q.; Xiao, H.L.; Dong, Z.B. Long-term morphodynamic changes of a desert reach of the Yellow River following upstream large reservoirs' operation. *Geomorphology* **2008**, *97*, 249–259. [CrossRef]
4. Yu, M.H.; Wei, H.Y.; Liang, Y.J.; Hu, C.W. Study on the stability of non-cohesive riverbank. *Int. J. Sediment. Res.* **2010**, *4*, 391–398. [CrossRef]
5. Xia, J.Q.; Wang, G.Q.; Zhang, H.W.; Fang, H.W. A review of the research on lateral widenning mechanisms of alluvial rivers and their simulation approaches. *J. Sediment. Res.* **2001**, *6*, 71–78. (In Chinese) [CrossRef]
6. Thorne, C.R. ASCE task committee on hydraulic, bank mechanics, and modeling of river width adjustment. River width adjustment. I: Processes and mechanisms. *J. Hydraul. Eng.* **1998**, *9*, 881–902. [CrossRef]
7. Buffington, J.M.; Montgomery, D.R. A systematic analysis of eight decades of incipient motion studies,

with special reference to gravel-bedded rivers. *J. Water Resour. Res.* **1997**, *8*, 1993–2029. [CrossRef]
8. Hooke, J.M. An analysis of the processes of riverbank erosion. *J. Hydrol.* **1979**, *1–2*, 39–62. [CrossRef]
9. Lohnes, R.A.; Handy, R.L. Slope Angles in Friable Loess. *J. Geol.* **1968**, *3*, 247–258. [CrossRef]
10. Osman, A.M.; Thorne, C.R. Riverbank Stability Analysis. I: Theory. *J. Hydraul. Eng.* **1988**, *2*, 134–150. [CrossRef]
11. Simon, A.; Wolfe, W.J.; Molinas, A. Mass wasting algorithms in an alluvial channel model. In Proceedings of the 5th Federal Interagency Sedimentation Conference, Las Vegas, NV, USA, 18–21 March 1991; pp. 22–29.
12. Taylor, D.W. Fundamentals of Soil Mechanics. *Soil Sci.* **1948**, *2*, 161. [CrossRef]
13. Millar, R.G.; Quick, M.C. Discussion: Development and Testing of Riverbank-Stability Analysis. *J. Hydraul. Eng.* **1997**, *11*, 1051–1053. [CrossRef]
14. Darby, S.E.; Thorne, C.R. Development and Testing of Riverbank-Stability Analysis. *J. Hydraul. Eng.* **1996**, *8*, 443–454. [CrossRef]
15. Rinaldi, M.; Casagli, N. Stability of streambanks formed in partially saturated soils and effects of negative pore water pressures: The Sieve River (Italy). *Geomorphology* **1999**, *4*, 253–277. [CrossRef]
16. Simon, A.; Curini, A.; Darby, S.E.; Langendoen, E.J. Bank and near-bank processes in an incised channel. *Geomorphology* **2000**, *3–4*, 193–217. [CrossRef]
17. Huang, B.S.; Bai, Y.C.; Wan, Y.C. Model for dilapidation mechanism of riverbank. *J. Hydraul. Eng.* **2002**, *9*, 49–54. (In Chinese) [CrossRef]
18. Wang, Y.G.; Kuang, S.F. Critical height of bank collapse. *J. Hydraul. Eng.* **2007**, *10*, 1158–1165. (In Chinese) [CrossRef]
19. Xia, J.Q.; Wu, B.S.; Wang, Y.P.; Zhao, S.G. An analysis of soil composition and mechanical properties of riverbanks in a braided reach of the Lower Yellow River. *Chin. Sci. Bull.* **2008**, *15*, 2400–2409. [CrossRef]
20. Thorne, C.R.; Tovey, N.K. Stability of composite riverbanks. *Earth Surf. Proc. Land.* **1981**, *5*, 469–484. [CrossRef]
21. Hagerty, D.J. Piping/Sapping Erosion. I: Basic Considerations. *J. Hydraul. Eng.* **1991**, *8*, 991–1008. [CrossRef]
22. Xia, J.Q.; Zong, Q.L.; Xu, Q.X.; Deng, C.Y. Soil properties and erosion mechanisms of composite riverbanks in Lower Jingjiang Reach. *Adv. Water Sci.* **2013**, *6*, 810–820. (In Chinese) [CrossRef]
23. Nagata, N.; Hosoda, T.; Muramoto, Y. Numerical Analysis of River Channel Processes with Bank Erosion. *J. Hydraul. Eng.* **2000**, *4*, 243–252. [CrossRef]
24. Darby, S.E.; Delbono, I. A model of equilibrium bed topography for meander bends with erodible banks. *Earth Surf. Proc. Landf.* **2002**, *10*, 1057–1085. [CrossRef]
25. Simon, A.; Pollenbankhead, N.; Mahacek, V.; Langendoen, E. Quantifying reductions of mass-failure frequency and sediment loadings from streambanks using toe protection and other means: Lake Tahoe, United States. *J. Am. Water Resour. Assoc.* **2009**, *1*, 170–186. [CrossRef]
26. Darby, S.E.; Trieu, H.Q.; Carling, P.A.; Sarkkula, J.; Koponen, J.; Kummu, M.; Conlan, I.; Leyland, J. A physically based model to predict hydraulic erosion of fine-grained riverbanks: The role of form roughness in limiting erosion. *J. Geophys. Res. Earth Surf.* **2010**, *115*, F04003. [CrossRef]
27. Jia, D.D.; Shao, X.J.; Wang, H.; Zhou, G. Three-dimensional modeling of bank erosion and morphological changes in the Shishou bend of the middle Yangtze River. *Adv. Water Res.* **2010**, *3*, 348–360. [CrossRef]
28. Nardi, L.; Rinaldi, M.; Solari, L. An experimental investigation on mass failures occurring in a riverbank composed of sandy gravel. *Geomorphology* **2012**, *9*, 56–59. [CrossRef]
29. Xiao, Y.; Shao, X.J.; Zhou, G.; Zhou, J.Y. 2D mathematical modeling of fluvial processes considering influences of vegetation and bank erosion. *J. Hydraul. Eng.* **2012**, *6*, 149–153. (In Chinese)
30. Xia, J.Q.; Wang, G.Q.; Wu, B.S. *Evolution of Wandering Rivers and Its Numerical Simulation*; Deng, Q., Zhang, J., Li, L., Eds.; China Water Conservancy and Hydropower Press: Beijing, China, 2005; pp. 182–188, ISBN 9787508426259.
31. Jia, D.D.; Shao, X.J.; Wang, H.; Zhou, G. 3D mathematical modeling for fluvial process considering bank erosion. *Adv. Water Sci.* **2009**, *3*, 311–317. (In Chinese) [CrossRef]
32. Yu, M.H.; Shen, K.; Wu, S.B.; Wei, H.Y. An experimental study of interaction between bank collapse and riverbed evolution. *Adv. Water Sci.* **2013**, *5*, 675–682. (In Chinese) [CrossRef]
33. Yu, M.H.; Guo, X. Experimental study on the interaction between the hydraulic transport of failed bank soil and near-bank bed evolution. *Adv. Water Sci.* **2014**, *5*, 677–683. (In Chinese) [CrossRef]

34. Wu, S.B.; Yu, M.H. Experimental study on bank failure process and interaction with riverbed deformation due to fluvial hydraulic force. *J. Hydraul. Eng.* **2014**, *6*, 649–657. (In Chinese) [CrossRef]
35. Deng, S.S.; Xia, J.Q.; Zhou, M.R.; Li, J. Conceptual model of bank retreat processes in the Upper Jingjiang Reach. *Chin. Sci. Bull.* **2016**, *33*, 3606–3615. (In Chinese) [CrossRef]
36. Yu, M.H.; Chen, X.; Wei, H.Y.; Hu, C.W.; Wu, S.B. Experiment of the influence of different near-bank riverbed compositions on bank failure. *Adv. Water Sci.* **2016**, *2*, 176–185. (In Chinese) [CrossRef]
37. Duan, G.S.; Shu, A.; Matteo, R.; Wang, S.; Zhu, F.Y. Collapsing mechanisms of the typical cohesive riverbank along the Ningxia-Inner Mongolia catchment. *Water* **2018**, *9*, 1272. [CrossRef]
38. Xue, Y. *Basic Course of AutoCAD 2010*; Tsinghua University Press: Beijing, China, 2009; pp. 113–135, ISBN 9787302207504. (In Chinese)
39. Dong, Y.H. Review on 10 groups of demarcating grain sizes of river sediment. *J. Yangtze Sci. Res. Inst.* **2018**, 2, 1–7. (In Chinese) [CrossRef]
40. Qian, N.; Wan, Z.H. *Sediment Motion Mechanics*; Yang, J., Ed.; Science Press: Beijing, China, 1983; pp. 265–268, ISBN 9787030112606.
41. Qian, N.; Zhang, R.J.; Zhou, Z.D. *Riverbed Evolution*; Zhu, S., Ed.; Science Press: Beijing, China, 1987; pp. 258–269, ISBN 13031-3498.
42. Shu, A.; Wang, L.; Yang, K.; Fei, X.J. Investigation on movement characteristics for non-homogeneous and solid-liquid two-phase debris flow. *Chin. Sci. Bull.* **2010**, *31*, 3006–3012. (In Chinese) [CrossRef]
43. Shu, A.; Zhang, Z.D.; Wang, L.; Fei, X.J. Method for determining the critical grain size of viscous debris flow based on energy dissipation principle. *J. Hydraul. Eng.* **2008**, *3*, 257–263. (In Chinese) [CrossRef]
44. Fei, X.J. A model for calculating viscosity of sediment carrying flow in the middle and lower Yellow River. *J. Sediment. Res.* **1991**, 2, 1–13. (In Chinese) [CrossRef]

14

Study on the Best Depth of Stilling Basin with Shallow-Water Cushion

Qiulin Li [1], Lianxia Li [1,*] and Huasheng Liao [2]

1 State Key Laboratory of Hydraulics and Mountain River Development and Protection, College of Water Resource and Hydropower, Sichuan University, Chengdu 610065, Sichuan, China; leeqlin@stu.scu.edu.cn

2 Department Civil and Environmental Engineering, Michigan State University, East Lansing, MI 48824, USA; liao@egr.msu.edu

* Correspondence: lianxiali@scu.edu.cn.

Abstract: The depth of the stilling basin with shallow-water cushion (SBSWC) is a key factor that affects the flow regime of hydraulic jump in the basin. However, the specific depth at which the water cushion is considered as 'shallow' has not been stated clearly by far, and only conceptual description is provided. Therefore, in order to define the best depth of SBSWC and its relationship between the Froude number at the inlet of the stilling basin, a large number of experiments were carried out to investigate SBSWC. First of all, 30 cases including five different Froude numbers and six depths were selected for which large eddy simulation (LES) was firstly verified by the experiments and then adopted to calculate the hydraulic characteristics in the stilling basin. Finally, three standards, based on the flow regime of hydraulic jump, the location of the main stream and the energy dissipation rate, were proposed to define the best depth of SBSWC. The three criteria are as follows: (1) a complete hydraulic jump occurs in the basin (2) the water cushion is about 1/10–1/3 deep of the stilling basin, and (3) the energy dissipation rate is more than 70% and the unit volume energy dissipation rate is as high as possible. It showed that the best depth ratio of SBSWC (depth to length ratio) was between 0.1 and 0.3 and it also indicated the best depth increased with the increase in Froude number. The results of the work are of significance to the design and optimizing of SBSWC.

Keywords: energy dissipation; hydraulic jump; Froude number; stilling basin with shallow-water cushion (SBSWC); large eddy simulation; best depth

---

1. Introduction

At present, the energy dissipation by hydraulic jump is used for downstream in many hydraulic projects [1], of which the stilling basin plays an important role in energy dissipation. However, the traditional stilling basin has disadvantages such as high underflow speed near the bottom [2–4], apparent damages by erosion and cavitation as well as insufficient energy dissipation rate [5]. Many researchers have made studies on such problems. The concept of a stilling basin with shallow-water cushion (SBSWC) was introduced [6]; namely, a shallow-water cushion is added to the ordinary stilling basin. In the structure of the new type stilling basin, the water cushion formed at the basin bottom can be used as the 'flexible bottom plate', which applies a flexible counterforce to the water stream in the steep slope section and 'absorbs' impact force of high speed flow on the floor of the stilling basin, so as to protect the bottom of the basin. After that, a series of studies on the impacts caused by inlet type, inflow angle, and low Froude number of the stilling basin with shallow-water cushion on its hydraulic characteristics were conducted [7]. For example, the concept of a stilling basin with double shallow-water cushions is an extension of SBSWC, which has the same advantages as SBSWC, but with better hydraulic characteristics [8], lower underflow speed, and better distribution of dynamic pressures of bottom plate of the stilling basin. The stilling basin with a drop sill is similar to SBSWC in shape [9–14]. However, the former one is focused on the connection type of a drop sill at the stilling basin inlet, while the latter one is focused on the energy dissipation mechanism of the stilling basin and the effect of the shallow-water cushion.

The depth of SBSWC is of critical importance to the energy dissipation effect of the stilling basin; however, no systematic study has been made on it for now. Besides, whether the water cushion is considered as deep or shallow has not been defined clearly since the concept of SBSWC was proposed. This paper attempt to propose an accurate and practical definition of the best depth of SBSWC in terms of three aspects; namely, flow regime, main stream location, and energy dissipation rate. If the water cushion depth is higher than the best depth, it is not necessary; if the water cushion depth whereas is lower than the best depth, the water cushion cannot play the role of the buffer action fully. Therefore, it is expected that in a stilling basin with the best depth, a complete hydraulic jump (critical hydraulic jump and submerged hydraulic jump are considered complete hydraulic jumps) occurs and the main stream is not located too close or far from the bottom of the basin, and the energy dissipation rate is as high as possible.

This study was not about the thickness of a shallow water cushion, but to study the depths of a stilling basin with shallow-water cushion required at different Froude numbers. In order to obtain the best stilling basin depth (represented by dimensionless depth-to-length ratio, which was convenient for application) corresponding to the specified Froude number; 30 cases including five different Froude numbers and six depths were selected firstly in the study. Then the large eddy simulation (LES [15]) was adopted to calculate hydraulic characteristics in the stilling basin, and the reliability and accuracy of which were verified by the results of physical model. Large eddy simulation has a proportional advantage, such as (1) prediction of turbulent flow from laminar flow, (2) prediction of high-speed turbulence. Therefore, according to the simulation results of the flow regime of hydraulic jump, the location of the main stream and the energy dissipation rate, the definition or the standard of the best depth of SBSWC was proposed; finally the relationship between Froude number and the best depth of the basin was revealed.

2. Materials and Methods

2.1. Experimental Setup

Experiments were conducted in a smoothened glass flume. It was composed of several sections including a 363 cm-length inflow discharge chute, baffle, 120 cm-length stilling basin with shallow-water cushion, and 300 cm-length tail water section (Longitudinal cross-sectional view as shown in Figure 1). The discharge chute section and all downstream parts were with rectangular cross section, with a width of 30 cm. The inlet control is a gate that is 10 cm wide, which controls the head of water to the inflow discharge chute. The slope of the inflow discharge chute used in this study is fixed to 17°. At downstream, the water level was controlled, and outflow condition was applied.

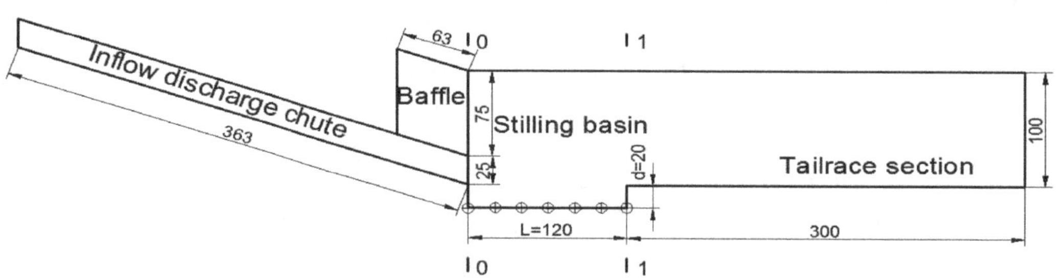

Figure 1. Schematic design of the model test (in cm).

2.2. Mathematical Model

In the present study, ANSYS 16.0 was utilized to investigate the flow characteristics over the stilling basin with shallow-water cushion with different Froude numbers. An iterative format selected an implicit time-marching scheme. The volume of fluid (VOF) [16,17] method was adopted to track and simulate the free surface and two-phase flow of air and water. The mathematical model of 3D large eddy simulation (LES) was used in the calculation, which is introduced below.

(1) Governing equations

In LES [18–22], the large-scale eddy was simulated directly by solving the momentum equation, and the small-scale eddy was expressed in the sub-grid scale model. The governing equation was,

$$\frac{\partial \rho}{\partial t} + \frac{\partial (\rho \overline{u_i})}{\partial x_i} = 0 \tag{1}$$

$$\frac{\partial (\rho \overline{u_i})}{\partial t} + \frac{\partial (\rho \overline{u_i u_j})}{\partial x_j} = -\frac{\partial \overline{p}}{\partial x_i} + \frac{\partial}{\partial x_j}\left(\mu \frac{\partial \overline{u_i}}{\partial x_j}\right) - \frac{\partial \tau_{ij}^{sgs}}{\partial x_j} + \rho g_i \tag{2}$$

where, the value with "–" represents the large-scale value obtained after filtering; ρ represents the density; U represents the velocity; t represents the time; p represents the pressure; g represents the acceleration of gravity; x represents the coordinate; i, j represents the coordinate orientation; and τ_{ij}^{sgs} represents the sub-grid stress, which indicates the impact caused by small-eddy movement on the large-eddy movement. Generally, the sub-grid stress is calculated by the eddy viscidity model,

$$\tau_{ij}^{sgs} - \frac{1}{3}\tau_{kk}^{sgs}\delta_{ij} = -2\mu_{sgs}\overline{S_{ij}} \tag{3}$$

where, μ_{sgs} represents the sub-grid turbulent viscosity coefficient, $\delta_{ij} = \begin{cases} 1, i = j \\ 0, i \neq j \end{cases}$; $\overline{S_{ij}}$ represents the strain rate tensor under the scale to be solved, which is defined as;

$$\overline{S_{ij}} = \frac{1}{2}\left(\frac{\partial \overline{u_i}}{\partial x_j} + \frac{\partial \overline{u_j}}{\partial x_i}\right) \tag{4}$$

The Smagorinsky–Lilly model is used to calculate the sub-grid turbulent viscosity coefficient: $\mu_{sgs} = \rho L_s^2 \sqrt{2\overline{S_{ij}} \times \overline{S_{ij}}}$, $L_s = \min\left(\kappa d, C_s V^{1/3}\right)$; where, represents the mixed length of the sub-grid scale; κ represents the Karman constant; d represents the distance to the nearest wall face; V represents the volume of computing control body; and C_s represents the Smagorinsky constant, which is 0.1 in this study [23].

(2) Discretization and algorithm

In this test, the finite volume method [24] was used to discrete of the governing equations, and second-order implicit scheme was used in the discretization of time item, and the PISO algorithm [25–27] was used to solve and control the coupling of speed and pressure in the equation. The VOF [16,17] method was adopted to track and simulate the free surface and two-phase flow of air and water.

(3) Computational domain and boundary conditions

(1) Computational domain

The computational domain and grid are shown in Figure 2 with the same dimensions as Figure 1. The longitudinal cross-section in Figure 2a is located at the center of Figure 2b. The grids in the stilling basin and tail water section were transited from coarse to fine from top to bottom, of which size is 2–3 cm, and the total grids of the model is 272,000.

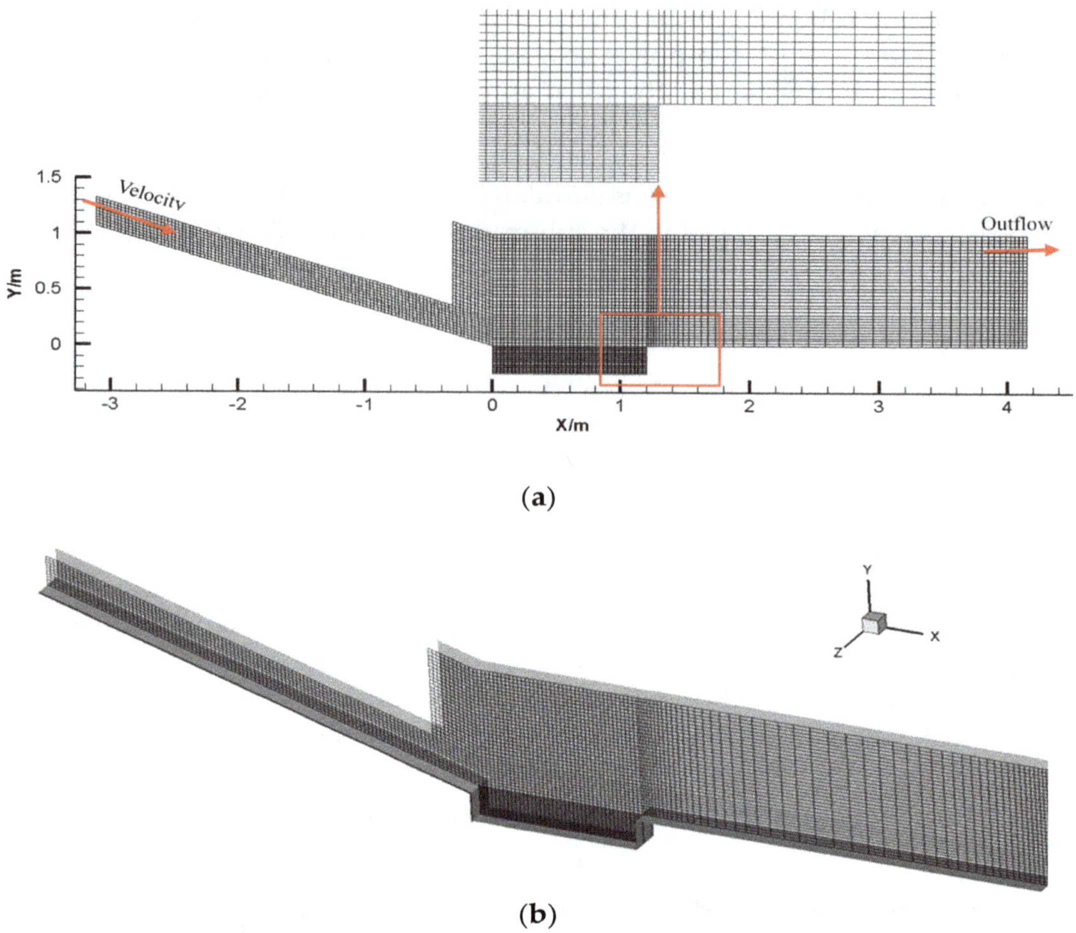

Figure 2. Schematic diagram for computational domain and grids of mathematical model. (**a**) Axial cross-sectional view; (**b**) The entire calculation domain.

(2) Boundary conditions

The boundary condition of the inlet of the chute was set as velocity inlet condition which could be determined according to the flow and the area of inlet of the chute. The outflow condition [28] was applied at the outlet boundary. The viscous sub-layer of the near wall was treated by the wall function method [29]. Non-slip condition was applied on the fixed wall.

2.3. Verification of Mathematical Model

The calculation results of the large eddy simulation were compared with the test results of three cases based on the Froude number (Fr) at the inlet of stilling basin and the depth of SBSWC, which are (a) Fr = 6.92, d = 20 cm, (b) Fr = 7.64, d = 15 cm, and (c) Fr = 9.25, d = 25 cm, respectively.

The reliability and accuracy of LES for this study were verified by the computing and experimental results. Similar hydraulic jumps in the stilling basin were observed from the regime of both the test and computation (Figure 3) in all 3 cases, and the numerical model captured the flow profiles [30] accurately compared with test data (Figure 4) and the differences of the average pressure between test results and calculated results (Figure 5) were very small except for individual locations in case 3.

It can be seen from Figures 4 and 5 that, at most locations, the calculated results were fairly consistent with that of laboratory tests. The maximums of the relative error for flow profile and average pressure were 7.1% and 6.5%, respectively, indicating that the numerical simulations produced reliable and acceptable results.

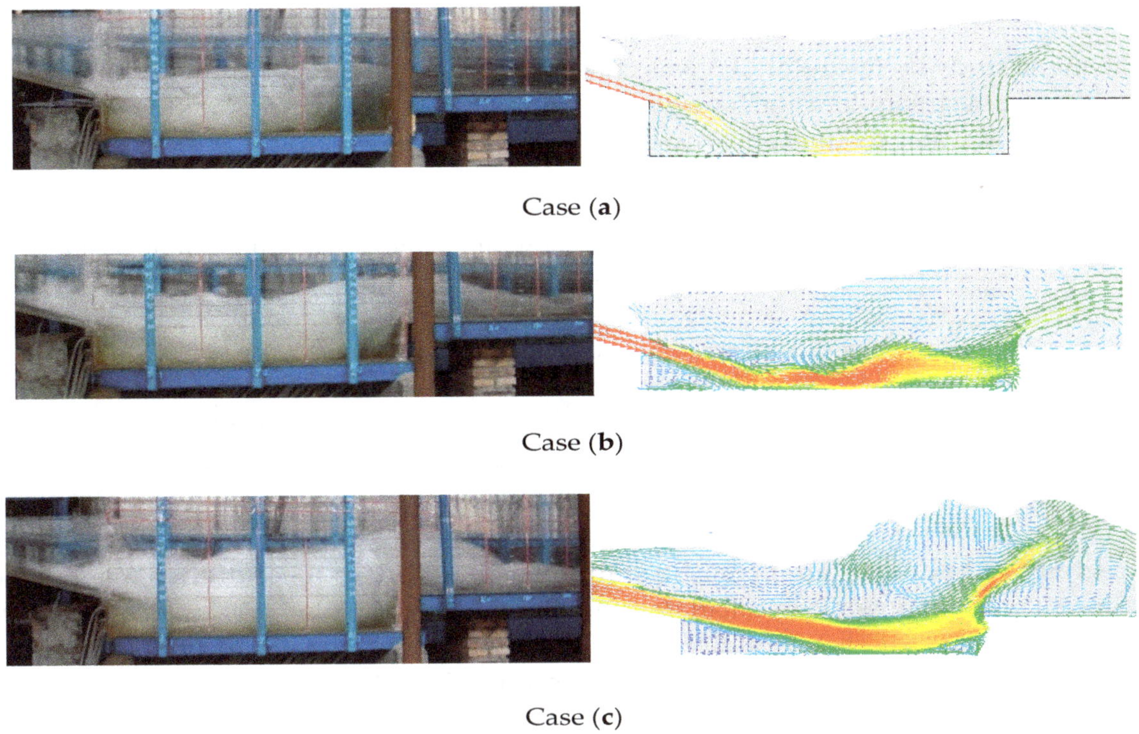

Figure 3. Schematic diagram for flow regimes under three cases.

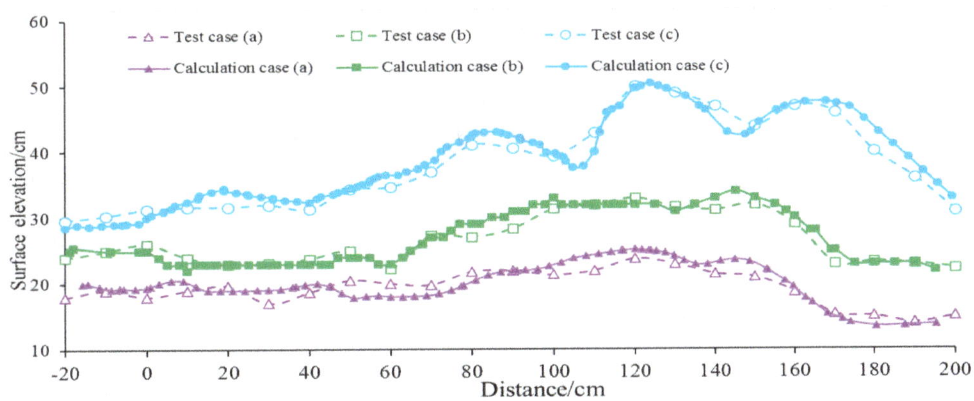

Figure 4. Comparison of flow profile.

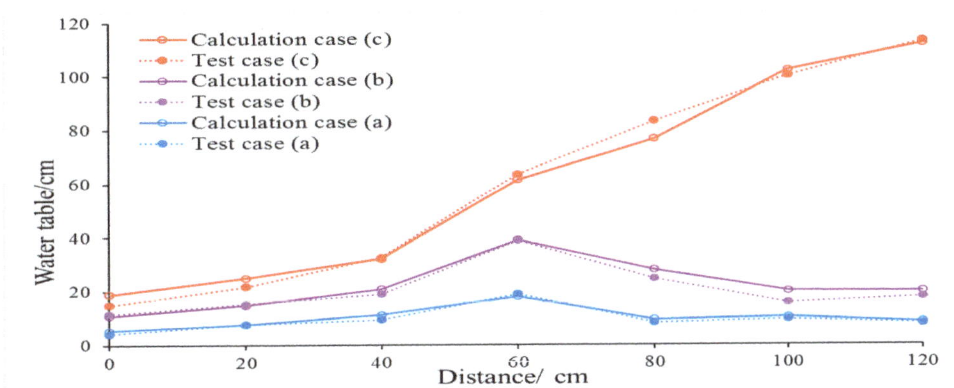

Figure 5. Comparison of average pressure.

3. Results and Discussion

A systematic study by means of LES was conducted to simulate the hydraulic characteristics such as flow regime, main stream location, and unit volume energy dissipation rate to determine the best depth of the shallow-water cushion on 30 cases including five typical Froude numbers and six different depths of shallow-water cushion shown in Table 1.

Table 1. The case of Froude number corresponding to the depth of the stilling basin.

Fr	d (cm)	d/L	Fr	d (cm)	d/L	Fr	d (cm)	d/L	Fr	d (cm)	d/L	Fr	d (cm)	d/L
7.64	0	0.000	6.92	0	0.000	8.05	0	0.000	9.25	0	0.000	11.24	0	0.000
	5	0.042		5	0.042		5	0.042		5	0.042		5	0.042
	10	0.083		10	0.083		10	0.083		10	0.083		10	0.083
	15	0.125		15	0.125		15	0.125		15	0.125		15	0.125
	20	0.167		20	0.167		20	0.167		20	0.167		20	0.167
	25	0.208		25	0.208		25	0.208		25	0.208		25	0.208

3.1. Flow Regime

The regimes of hydraulic jumps in the stilling basin at different conditions were presented in Figure 6. Since it was not possible to calculate every depth as a calculated working condition, the same six depths were selected for each Froude number, which were 0, 5, 10, 15, 20, and 25 cm respectively. Therefore, the calculated best depth in each corresponding Froude has an error of ± 2.5 cm.

It is seen that the regimes and velocity distribution are different in the stilling basins for different depths with the same Froude number and with different Froude number but same depth. For the same Froude number, if the depth is not enough, no hydraulic jump takes place in the basin, and the regime of jump gets more complete as the depth increases; for example, when the depth of stilling basin is 0 cm or 5 cm, there is no hydraulic jump in the stilling basin; when the depth of stilling basin is 10 cm, 15 cm, 20 cm, or 25 cm, there is a hydraulic jump in the stilling basin. This pattern holds for all the groups (a)–(e) with different Froude numbers. For different Froude numbers, the depth required to form a complete hydraulic jump is different. For example, a complete hydraulic jump occurred in the depth of 15 cm, 20 cm, and 25 cm in the stilling basin when the Froude numbers are 7.64 and 6.92. In case that the Froude number is 8.05 and 9.25, the complete hydraulic jump occurred in the depth of 20 and 25 cm in the stilling basin. While in the case of Froude number 11.24, the complete hydraulic jump occurs in the depth of 25 cm in the stilling basin.

From the prospective of flow regime, the minimum depth of stilling basin with shallow water cushion (SBSWC) with different Froude numbers were depicted in Figure 7. It is indicated that the higher the Froude number, the deeper the basin required to form a complete hydraulic jump. In detail, in cases that the Froude number are 6.92, 7.64, 8.05, 9.25, and 11.24, the minimum depths of stilling basin with shallow water cushion are 15 cm, 15 cm, 20 cm, 20 cm and 25 cm.

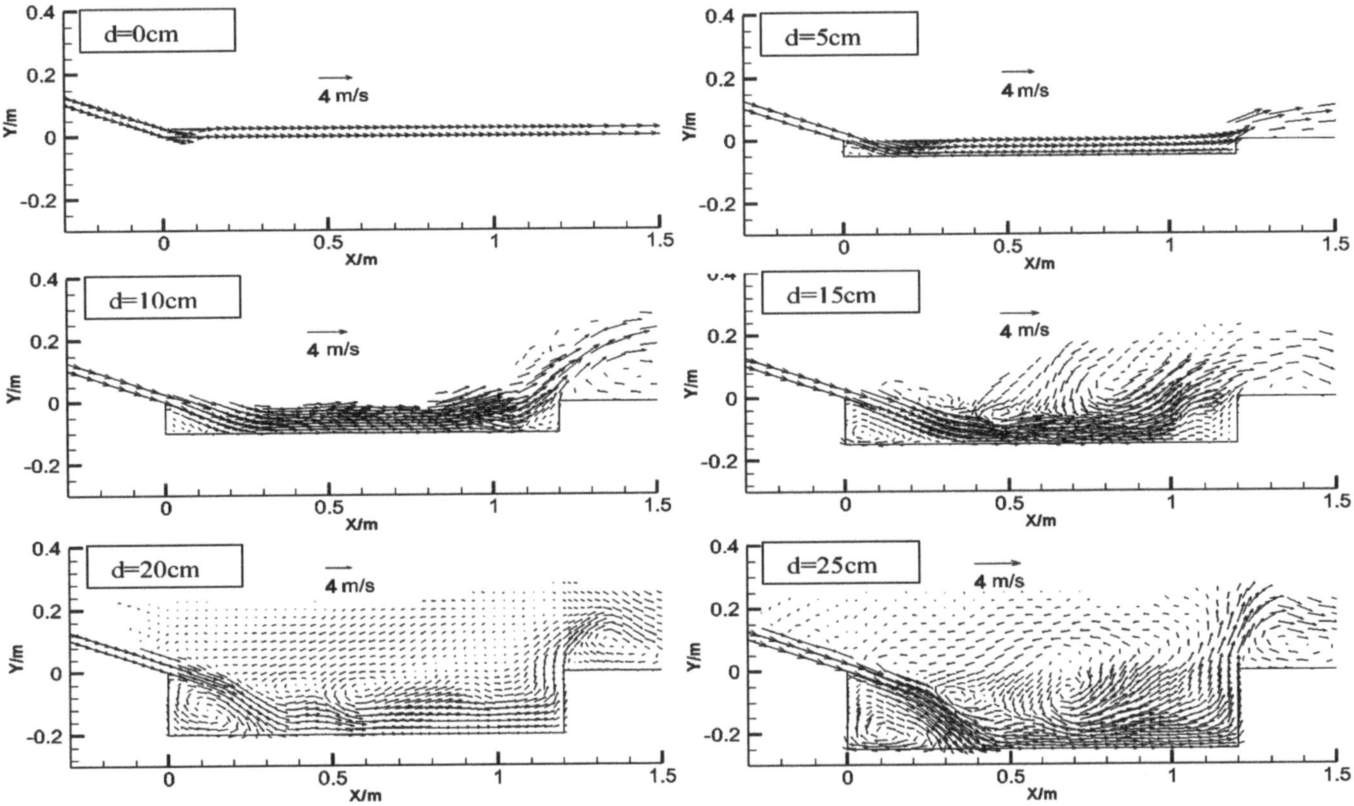

(**a**)

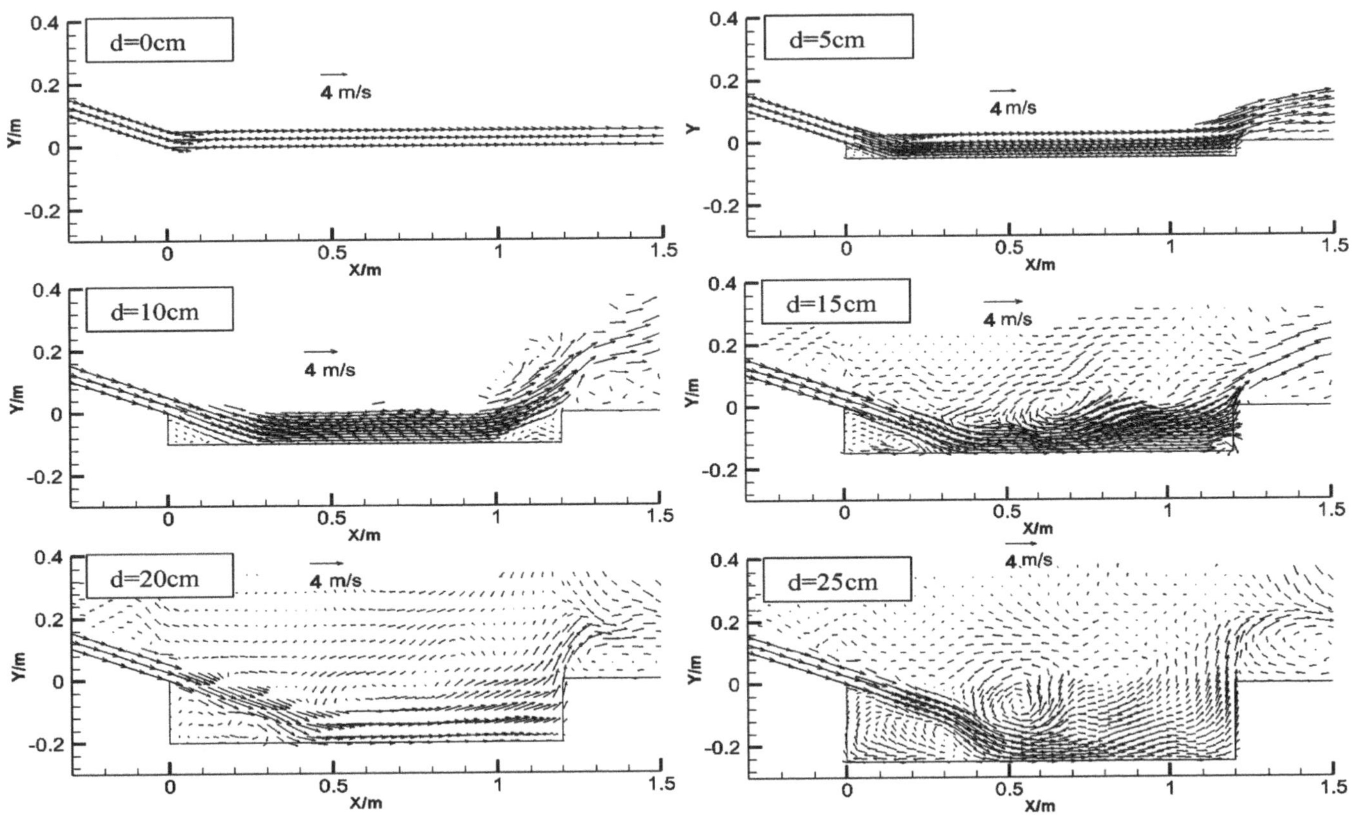

(**b**)

Figure 6. *Cont.*

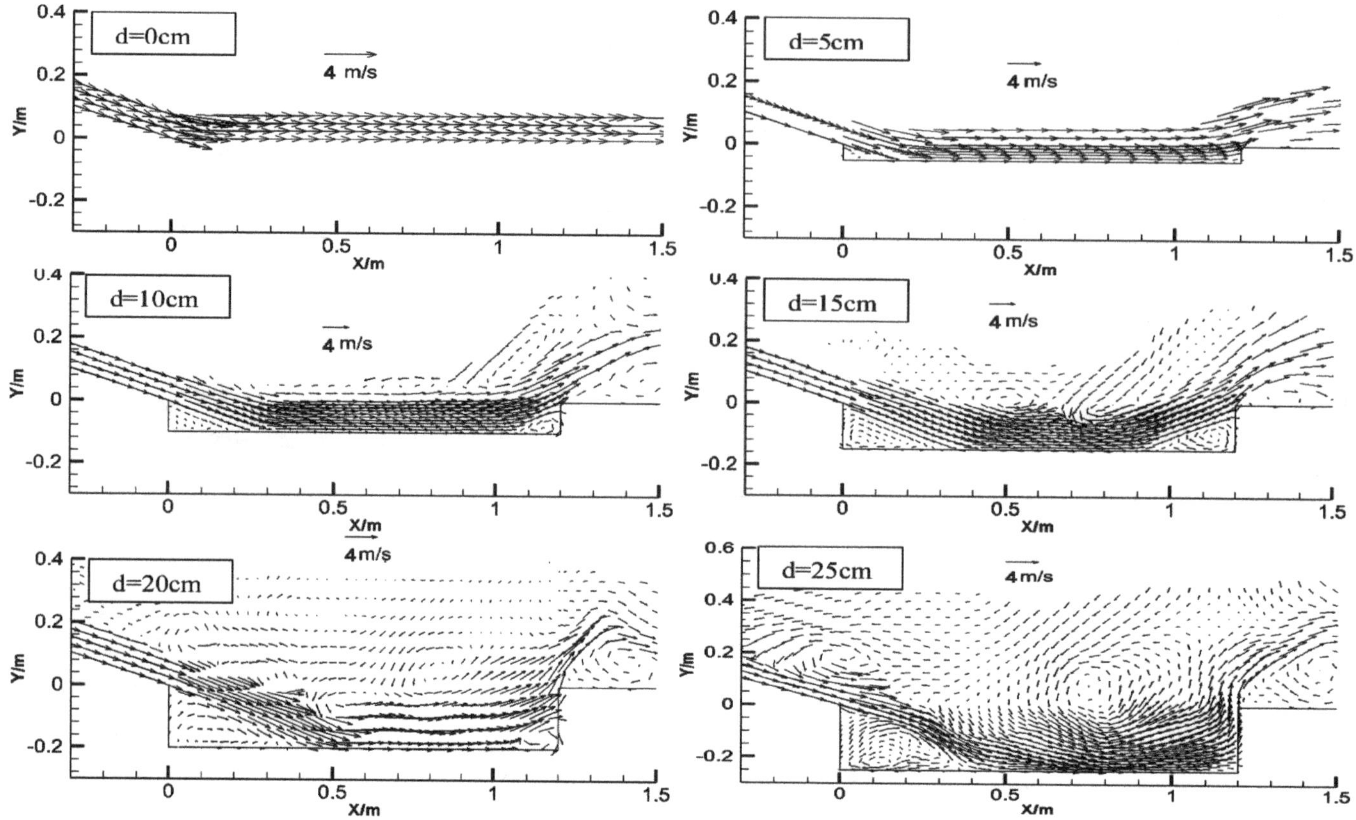

(c)

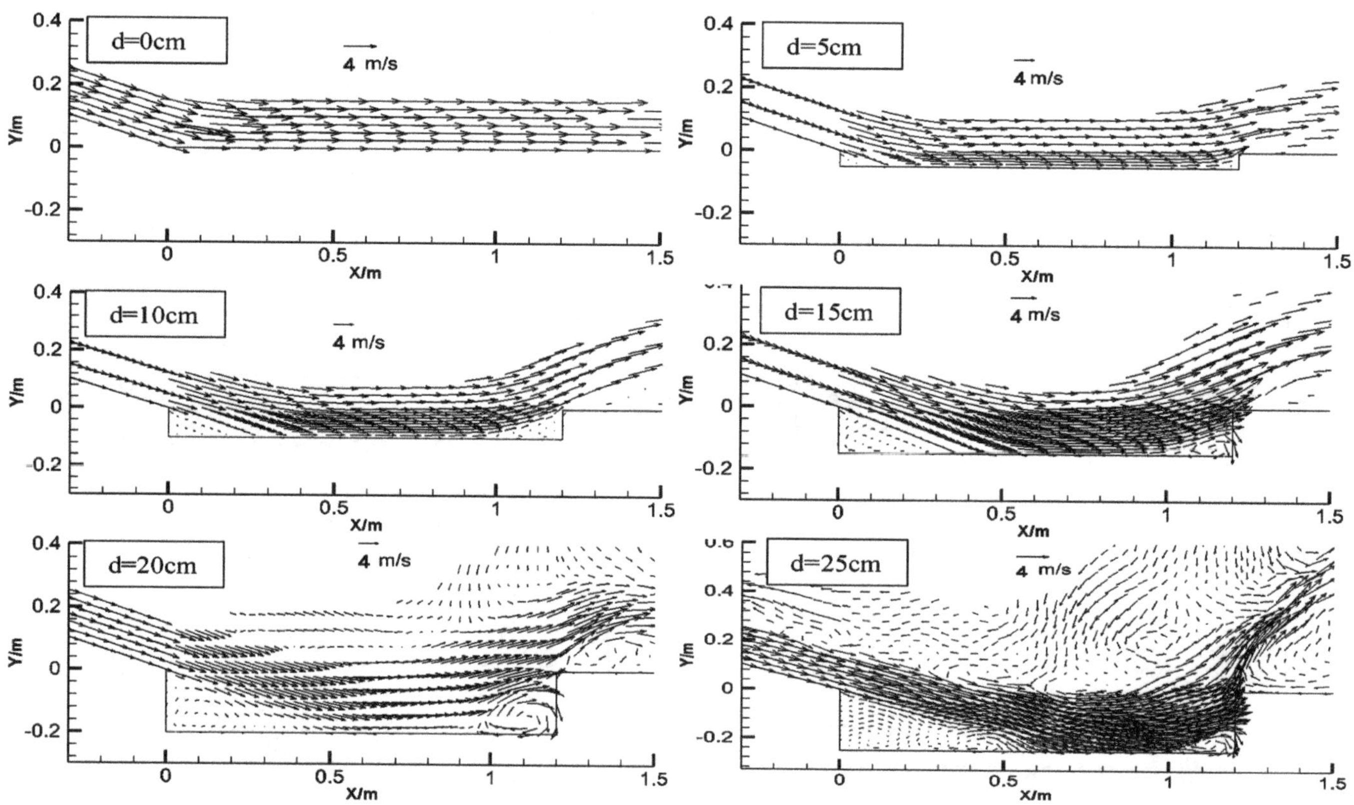

(d)

Figure 6. *Cont.*

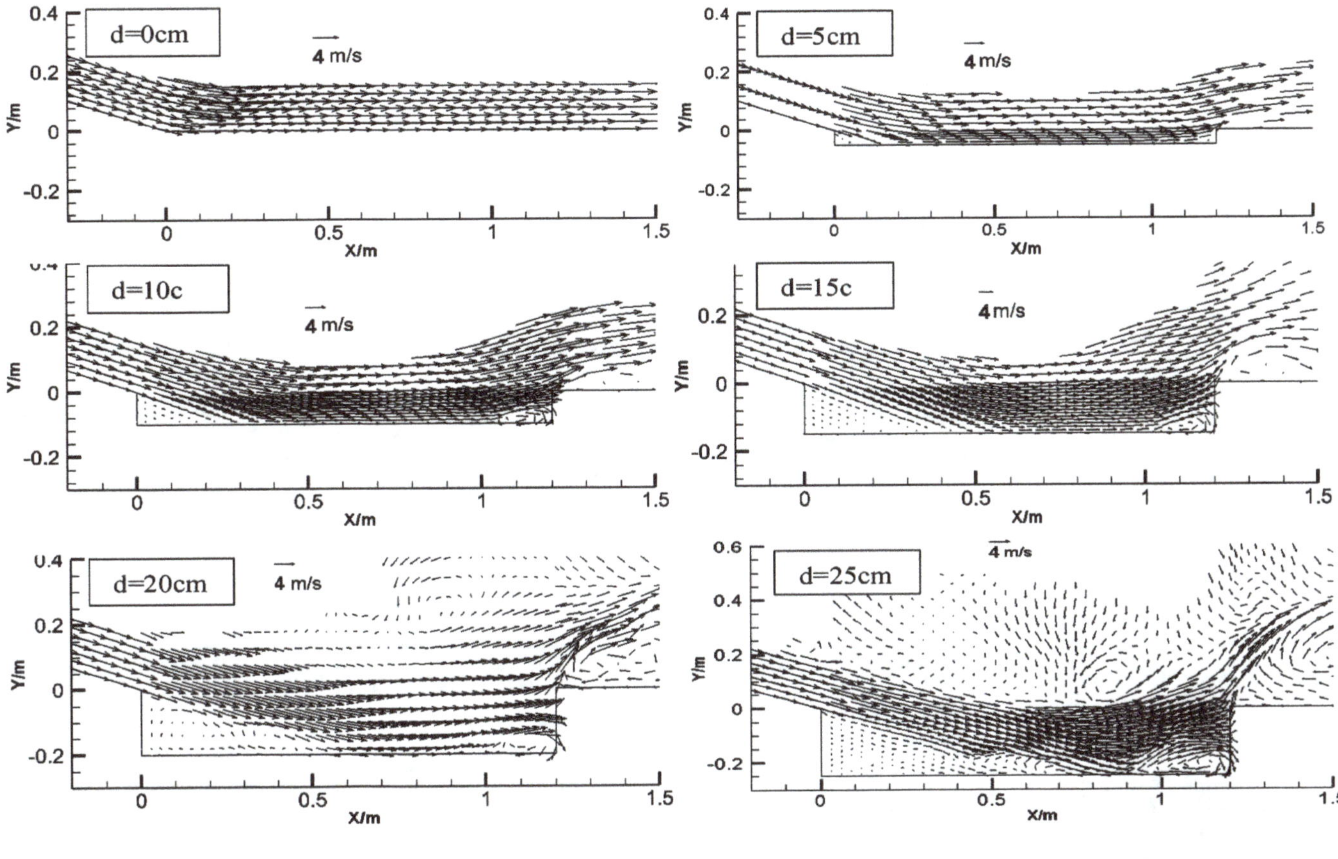

Figure 6. Flow regimes and velocity distribution at axial plane. (**a**) Fr = 6.92; (**b**) Fr = 7.64; (**c**) Fr = 8.05; (**d**) Fr = 9.25; (**e**) Fr = 11.24. Fr: Froude number.

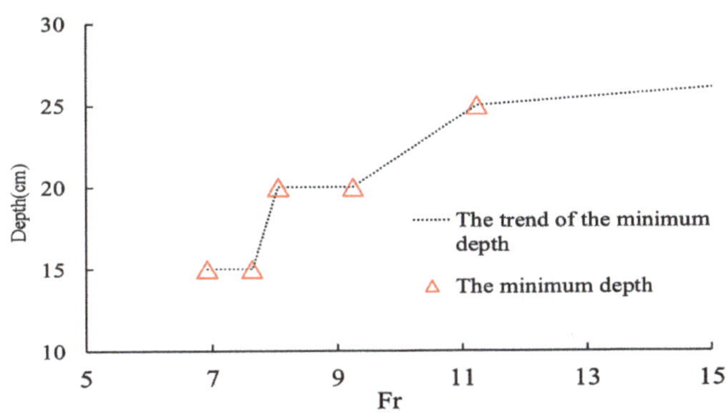

Figure 7. The minimum depth of the stilling basin for the complete hydraulic jump.

3.2. Main Stream Location

The main stream zone was defined as the area that the flow speed was larger than the average speed of cross section in this study. The water cushion was then defined as the water body under the lower boundary of the main stream and the dimensionless thickness of water cushion was defined as the thickness of water cushion referred to the depth of the basin.

The main stream location of the 30 cases were grouped and shown in Figure 8 in which the main stream was represented out by velocity iso-surface in some cases. It shows that the center of the main stream closest to the bottom of the basin is located at about 1/3 depth of the basin from the bottom in cases that hydraulic jumps occur which agrees with the results of previous study [7]. In cases that the Froude numbers are 6.92, 7.64, and 8.05, it can be clearly observed that the water cushions are formed between the main stream and the cushion bottom in the cases of the stilling basin depth of 15 cm, 20 cm and 25 cm. While in the case of Froude number 9.25 and 11.24 the water cushion locates between the main stream and the cushion bottom when the stilling basin depth is 20 cm or 25 cm.

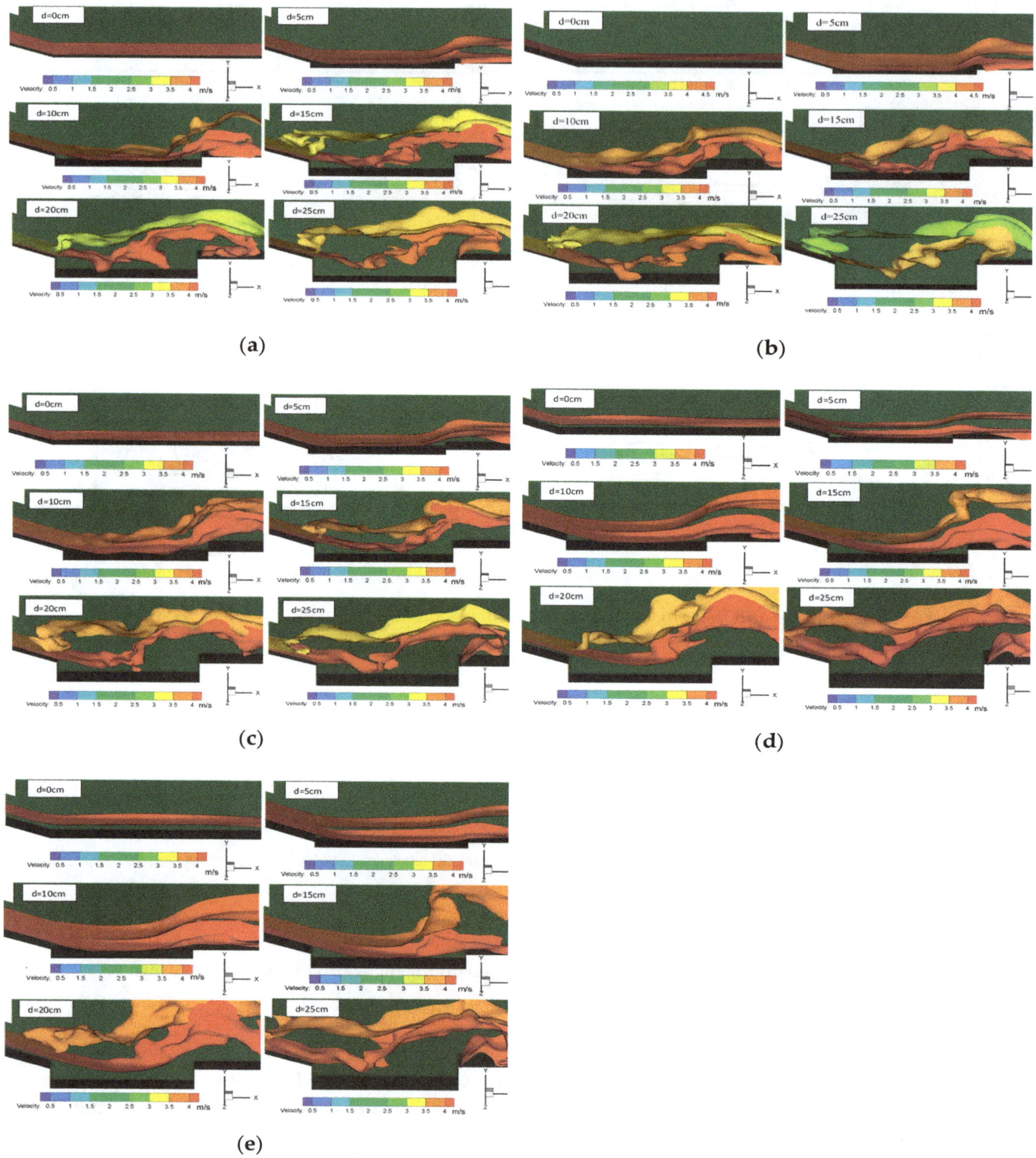

Figure 8. Main stream position in stilling basin. (**a**) Fr = 6.92; (**b**) Fr = 7.64; (**c**) Fr = 8.05; (**d**) Fr = 9.25; (**e**) Fr = 11.24.

Among the 30 cases, for each Froude number, the minimum dimensionless thickness (T) of water cushion (depth of water cushion over depth of stilling basin) varies with the depth of the stilling basin

(Table 2). It is seen that the thickness are different in the stilling basins with different depths with the same Froude number and with different Froude number but same depth. For the same Froude number, the deeper the stilling basin, the thicker the thickness of the water cushion. For the same depth, the bigger the Froude number, the thinner the thickness of the water cushion.

Table 2. Minimum of the dimensionless thickness of the water cushion.

Group	Fr	d = 0	d = 5 cm	d = 10 cm	d = 15 cm	d = 20 cm	d = 25 cm
1	6.92	0	0.08	0.09	0.13	0.15	0.16
2	7.64	0	0.08	0.08	0.13	0.14	0.14
3	8.05	0	0.07	0.09	0.12	0.13	0.13
4	9.25	0	0.06	0.07	0.09	0.11	0.12
5	11.24	0	0.06	0.04	0.08	0.10	0.11

3.3. Energy Dissipation Rate

The energy dissipation [31] rate of the stilling basin could be expressed as Equation (5),

$$e = \frac{E_0 - E_1}{E_0} \tag{5}$$

In which, E_0 and E_1 stand for the energy of the control section at the inlet and outlet of the stilling basin and taken the elevation at the tailrace section as the reference elevation in Figure 1.

Then the unit volume energy dissipation rate could be defined by Equation (6),

$$\omega = \frac{e}{\eta} \tag{6}$$

where, ω is the unit volume (dimensionless) energy dissipation rate; $\eta = \Omega/\Omega_0$ is dimensionless volume; $\Omega = LdB$ is the volume of the stilling basin, L, d and B are the length, the depth and the width of the stilling basin respectively, Ω_0 is the volume of the water between inlet and outlet of the stilling basin.

The energy dissipation rates of the 30 cases were calculated and listed in Table 3, in which, column 1 is the Froude number at inlet of the stilling basin; column 2 is the depth of the stilling basin; column 3 is the ratio of the depth to length of the stilling basin (the total length is 120 cm); columns 4 to 10 are the mean speed of different cross sections (the heads of 0, 20, 40, 60, 80, 100, and 120 indicate the distance of the cross sections away from the inlet of the stilling basin entrance with the unit of cm); column 11 is the energy dissipation rate; column 12 is the dimensionless volume of the stilling basin; and column 13 is the unit volume energy dissipation rate.

Using columns 2, 11, and 13 in Table 3, the energy dissipation rate and the unit volume energy dissipation rate related with the depth of the stilling basin were calculated and depicted in Figure 9 with the different Froude number as a parameter.

Generally speaking, the total energy dissipation (e) increases rapidly with the depth of the basin and changes little after reaching a certain value with the depth (Figure 9a); whereas, when the Froude number is smaller, the unit volume energy dissipation rate goes high with the increase of the depth first, and then drops down after reaching the maximum value, which means some of the water body does not involve the energy dissipation (Figure 9b); when the Froude number is larger, it goes higher with the increase of the depth. It also shows that the total energy dissipation reaches the maximum but the unit volume energy dissipation rate may not reach its maximum value, which indicates the depth of the basin is not good enough for the hydraulic jump; for example, the unit volume energy dissipation rate keep increasing with Froude number in the cases of Froude numbers equal to 9.25 and 11.24 which means the depth of 25 cm is sufficient for them getting to the maximum unit volume energy dissipation rates in the study.

Table 3. The unit volume energy dissipation rate of the stilling basin.

Fr (1)	d (cm) (2)	d/L (3)	*V* (m/s) at Different Sections 0 (4)	20 (5)	40 (6)	60 (7)	80 (8)	100 (9)	120 (10)	e (11)	η (12)	ω (13)
7.64	0.00	0.00	4.25	4.25	4.28	4.18	4.13	4.10	4.01	10.98%	0.00	/
	5.00	0.04	4.19	3.87	4.30	4.20	4.20	4.05	3.80	17.75%	0.76	0.30
	10.00	0.08	3.60	3.30	3.50	3.70	3.81	3.40	2.90	35.11%	0.72	0.55
	15.00	0.13	2.84	3.15	2.30	1.48	1.36	1.25	1.20	82.15%	0.56	1.46
	20.00	0.02	2.60	1.55	1.33	0.96	1.09	0.99	0.96	86.37%	0.43	2.10
	25.00	0.21	1.66	1.75	1.23	0.76	0.77	0.68	0.56	88.62%	0.45	2.05
6.92	0.00	0.00	4.37	4.38	4.37	4.29	4.22	4.20	4.14	10.25%	0.00	/
	5.00	0.04	4.32	4.17	4.11	4.08	3.90	3.90	3.83	21.40%	0.54	0.25
	10.00	0.08	4.07	3.89	3.85	3.88	3.80	3.50	3.03	44.58%	0.62	0.50
	15.00	0.13	4.03	3.85	3.13	2.77	2.15	1.53	1.36	88.61%	0.48	1.85
	20.00	0.02	3.30	1.73	1.55	1.13	0.91	0.73	0.98	91.18%	0.44	2.20
	25.00	0.21	2.30	1.83	1.44	1.16	0.88	0.72	0.66	91.77%	0.46	1.99
8.05	0.00	0.00	7.78	7.63	7.54	7.43	7.51	7.46	7.40	9.53%	0.00	/
	5.00	0.04	7.90	7.55	7.49	7.43	7.35	7.30	7.26	15.55%	0.42	0.37
	10.00	0.08	7.55	7.45	7.29	6.98	6.85	6.33	6.25	31.47%	0.42	0.60
	15.00	0.13	7.53	7.20	6.70	6.50	6.29	5.94	5.09	54.31%	0.46	1.19
	20.00	0.02	5.81	5.01	4.53	4.01	3.82	3.15	2.85	75.94%	0.36	2.01
	25.00	0.21	4.73	4.16	3.86	3.38	3.19	2.65	2.23	77.77%	0.37	2.10
9.25	0.00	0.00	10.39	10.14	10.01	10.02	9.97	9.91	9.88	9.58%	0.00	/
	5.00	0.04	10.36	9.89	9.93	9.76	9.71	9.64	9.63	13.60%	0.27	0.46
	10.00	0.08	10.20	9.77	9.53	9.52	9.44	9.29	8.61	28.75%	0.45	0.65
	15.00	0.13	10.42	9.56	9.44	9.36	9.20	8.09	7.43	49.16%	0.53	1.05
	20.00	0.02	9.44	9.49	9.26	8.79	7.85	6.25	5.30	68.48%	0.43	1.80
	25.00	0.21	7.89	6.18	6.03	5.82	5.33	4.85	3.88	75.82%	0.35	2.27
11.24	0.00	0.00	13.36	13.11	13.12	13	12.92	12.88	12.76	8.78%	0.00	/
	5.00	0.04	13.23	12.96	12.88	12.69	12.43	12.42	12.42	11.87%	0.26	0.50
	10.00	0.08	13.2	12.91	12.8	12.56	12.21	11.52	11.33	26.33%	0.43	0.70
	15.00	0.13	13.19	12.88	12.62	12.12	11.63	11.15	10.07	41.71%	0.31	0.92
	20.00	0.02	13.26	12.56	11.45	10.62	9.53	8.05	7.79	65.49%	0.32	1.60
	25.00	0.21	12.21	11.64	10.88	9.48	8.38	7.23	6.11	74.96%	0.32	2.34

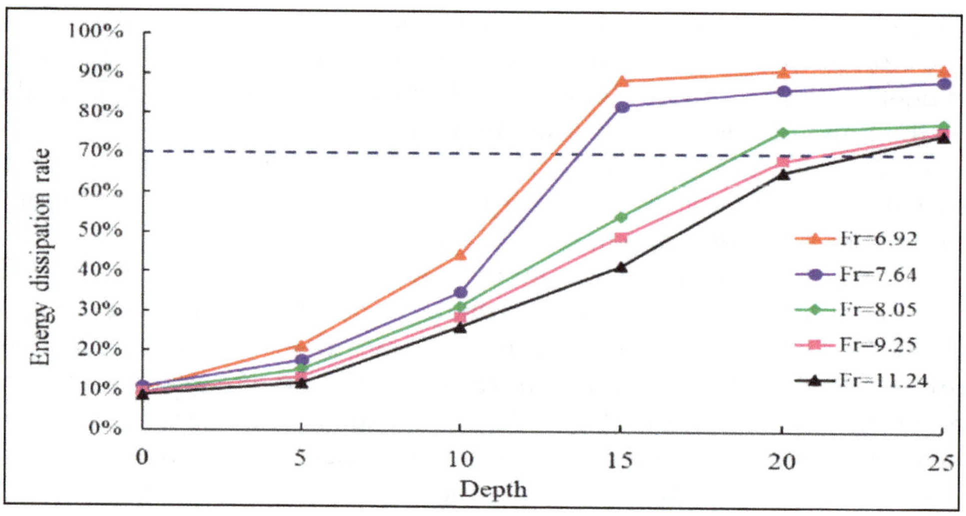

(a)

Figure 9. *Cont.*

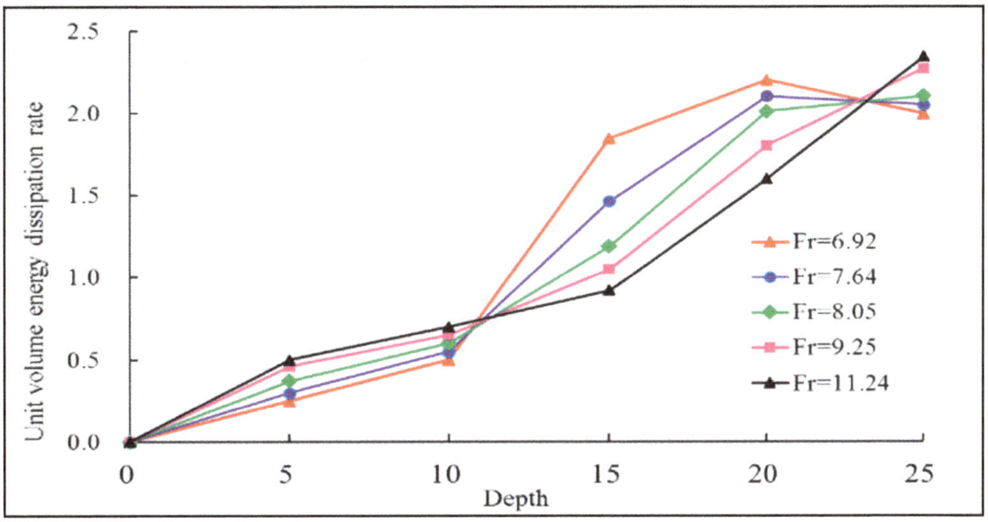

(b)

Figure 9. Energy dissipation rate. (**a**) Energy dissipation rate; (**b**) Unit volume energy dissipation rate.

3.4. The Best Depth of the Stilling Basin

For a given Froude number, it is expected that a complete hydraulic jump occurs in the stilling basin and the velocity near the bottom is relatively low and with a high energy dissipation rate in total and per volume. From the perspectives of flow regime, total energy dissipation rate and bottom velocity, the deeper a stilling basin is, more easily a complete hydraulic jump forms, higher the total energy dissipation rate and farther the main stream away from the bottom of the basin; however, from the perspective of the unit volume energy dissipation rate, the depth of the basin should not be too large, it otherwise is a waste as part of the water body in the basin is not involved in the energy dissipation. In addition, the thickness of the water cushion may be too large to regard the stilling basin as a 'shallow-water cushion' stilling basin and may be turned into a plunge pool which performs very different than the stilling basin with shallow water cushion in energy dissipation. Therefore, according to the results above of flow regime, flow profile, main stream location, energy dissipation, etc., the standards for defining the best depth of the stilling basin with shallow-water cushion can be proposed as (1) a complete hydraulic jump must occur in the basin (2) the thickness of water cushion is about 1/10–1/3 deep of the basin, and (3) the energy dissipation rate is more than 70% and the unit volume energy dissipation rate is as high as possible.

According to the definition for the best depth of SBSWC, three ranges of the depth were depicted in the 5 different Froude numbers respectively.in Figure 10. The blue line stands for the depth range for the standard (1) of flow regime according to the minimum depth for a complete hydraulic jump from Figure 6, and the green line stands for the depth range for the standard (2) of the relationship between the main stream and the thickness of the water cushion in Figure 8 and Table 2, and the red line stands for the depth range for the standard (3) of energy dissipation according to Table 3 and Figure 9. It is shown that all the 3 standards are necessary to obtain the range as neither the minimums nor the maximums of the range can be determined by the same standard alone for different Froude numbers. For example, for cases of Froude numbers equal to 6.92 and 9.25, the minimums are determined by the standards of flow regime and energy dissipation, respectively, while the maximums by energy dissipation and main stream location, respectively.

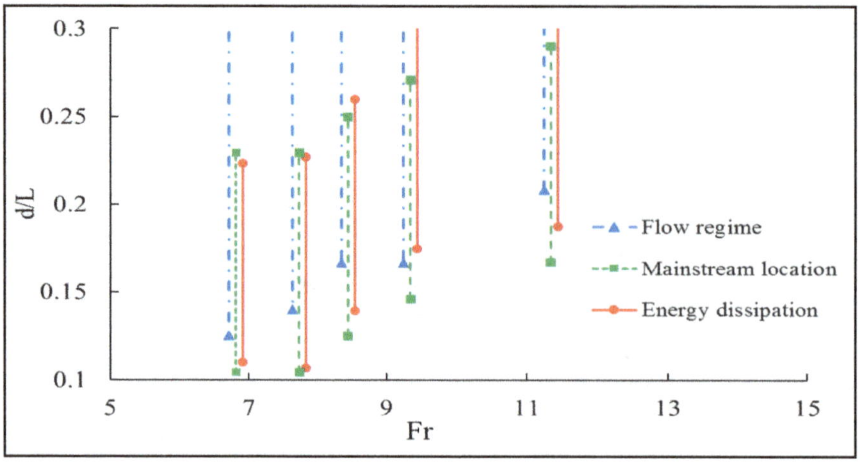

Figure 10. The range of the best depth of the stilling basin under three judgment conditions.

The overlap of all the three ranges for the same Froude number represented that the three standards were met in this integral (the integral between the maximum of the 3 minimums and the minimum of the maximums) and its mean is taken as the best depth of the stilling basin with shallow-water cushion shown in Figure 11 where the dimensionless depth (d/L, the depth over the length of the stilling basin) was adopted. It indicates that the ranges of the best depth are large for some cases while small for some other cases and that the best depth of the stilling basin increases with the increase of the Froude number.

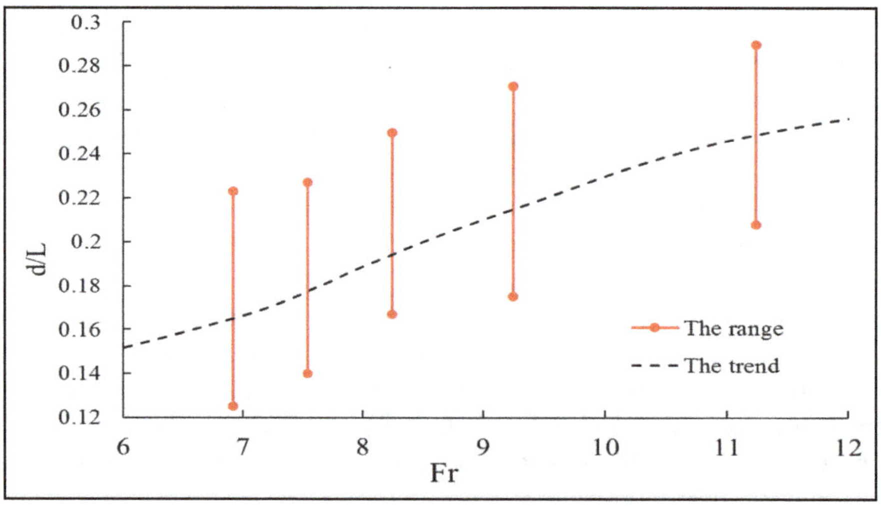

Figure 11. The range of the best depth of the stilling basin.

4. Conclusions

In this study, 30 cases consisting of five different Froude numbers and six depths were selected for which large eddy simulation (LES) was adopted to simulate the hydraulic characteristics in the stilling basin, including the flow regime of hydraulic jump, the location of the main stream and the unit volume energy dissipation rate. The following conclusions can be drawn:

1. After comparing large eddy simulation calculation results and model tests, the computation of the flow profile and hydraulic jumps were similar to the test, and the differences of the average pressure between test results and calculated results were very small except for individual locations. The numerical data and experimental results were in good agreement, indicating the reliability and accuracy of LES in the present work.

2. Based on the experimental results of flow regime of hydraulic jump, the location of the main stream and the energy dissipation rate, for the first time three standards of the best depth of stilling basin with shallow water cushion (SBSWC) are proposed in this paper; namely, (1) a complete hydraulic jump must occur in the basin (2) the thickness of water cushion is about 1/10–1/3 deep of the basin, and (3) the energy dissipation rate is more than 70% and the unit volume energy dissipation rate is as high as possible.

3. According to the definition for the best depth of SBSWC, three ranges of the depth were depicted in the 5 different Froude numbers respectively. Therefore, the overlapping sections are the range of the best depth of stilling basin with shallow water cushion for this study.

4. It showed that the best depth radio of stilling basin with shallow-water cushion (depth to length ratio) was between 0.1 and 0.3 and it also indicated that the best depth of the stilling basin with shallow water cushion increased with the increase of Froude number. The results in this paper are of significance for design and optimizing of SBSWC.

Author Contributions: Q.L. carried out the model simulations and result analysis as well as writing of the manuscript. L.L. provided the prototype field data and relevant engineering documents. H.L. took the leadership of whole project and participated in the discussion and decision-making process. All the authors participated and contributed to the final manuscript.

References

1. Chanson, H. Current knowledge in hydraulic jumps and related phenomena. A survey of experimental results. *Eur. J. Mech. B Fluids* **2009**, *28*, 191–210. [CrossRef]
2. Alikhani, A.; Behrozi-Rad, R.; Fathi-Moghadam, M. Hydraulic jump in stilling basin with vertical end sill. *Int. J. Phys. Sci.* **2010**, *5*, 25–29. Retrieved from http://www.academicjournals.org/IJPS.
3. Chanson, H. Hydraulic condition for undular-jump formations. *J. Hydraul. Res.* **2001**, *39*, 203–209. [CrossRef]
4. Zare, H.K.; Doering, J.C. Forced Hydraulic Jumps below Abrupt Expansions. *J. Hydraul. Eng.* **2011**, *137*, 825–835. [CrossRef]
5. Champagne, T.M.; Barlock, R.R.; Ghimire, S.R.; Barkdoll, B.D.; Gonzalez-Castro, J.A.; Deaton, L. Scour Reduction by Air Injection Downstream of Stilling Basins: Optimal Configuration Determination by Experimentation. *J. Irrig. Drain. Eng.* **2016**, *142*. [CrossRef]
6. Tianxiang, L.I. Investigation on Hydraulic Performances of Stilling Basin with Shallow-Water Cushion. Master's Thesis, State Key Laboratory of Hydraulics and Mountain River R & D in Sichuan University, Chengdu, China, 2006.
7. Li, L.X.; Liao, H.S.; Liu, D.; Jiang, S.Y. Experimental investigation of the optimization of stilling basin with shallow water cushion used for low Froude number energy dissipation. *J. Hydrodyn. B Ser.* **2015**, *27*, 522–529. [CrossRef]
8. Liu, D.; Li, L.X.; Huang, B.S.; Liao, H.S. Numerical Simulation and Experimental Investigation on Stilling Basin with Double Shallow-water Cushions. *J. Hydraul. Eng.* **2012**, 623–630. [CrossRef]
9. Hager, W.H.; Kawagoshi, N. Hydraulic jumps at rounded drop. *Proc. Inst. Civ. Eng. Part Res. Theory* **1990**, *89*, 443–470. [CrossRef]
10. Eroğlu, N.; Tokyay, N. Statistical approach to geometric properties of wave-type flow occurring at an abrupt drop. *J. Fac. Eng. Archit. Gazi Univ.* **2012**, *27*, 911–919.
11. Rice, C.E.; Kadavy, K.C. Riprap design upstream of straight drop spillways. *Trans. Asae* **1992**, *34*, 1715–1725. [CrossRef]
12. Ram, K.V.; Prasad, S.; Spatial, R. B-Jump at Sudden Channel Enlargements with Abrupt Drop. *J. Hydraul. Eng.* **1998**, *124*, 643–646. [CrossRef]

13. Sabzkoohi, A.M.; Kashefipour, S.M.; Bina, M. Investigation of effective parameters on stepped and straight drops energy dissipation using physical modeling. *J. Food Agric. Environ.* **2011**, *9*, 748–753. Retrieved from https://www.researchgate.net/publication/286803391.
14. Riazi, R.; Bejestan, M.S.; Kashkouli, H.; Khosrojerdi, A. Effect of roughness on characteristics of bed B-jump in stilling basin with abrupt drop. *Res. Crops* **2012**, *13*, 1137–1141. Retrieved from https://www.researchgate.net/publication/287059975.
15. Ilie, M. Fluid-structure interaction in turbulent flows; a CFD based aeroelastic algorithm using LES. *Appl. Math. Comput.* **2018**, *342*, 309–321. [CrossRef]
16. Wanik, A.; Schnell, U. Some remarks on the PISO and SIMPLE algorithms for steady turbulent flow problems. *Comput. Fluids* **1989**, *17*, 555–570. [CrossRef]
17. Wang, T.; Gu, C.G.; Yang, B. PISO algorithm for unsteady flow field. *J. Hydrodyn.* **2003**. [CrossRef]
18. Moeng, C.H. A Large-Eddy-Simulation Model for the Study of Planetary Boundary-Layer Turbulence. *J. Atmos. Sci.* **1984**, *41*, 2052–2062. [CrossRef]
19. Moin, P.; Kim, J. Numerical investigation of turbulent channel flow. *J. Fluid Mech.* **2006**, *118*, 1280–1284. [CrossRef]
20. Sarfaraz, M. Numerical Computation of Inception Point Location for Steeply Sloping Stepped Spillways. In Proceedings of the 9th International Congress on Civil Engineering, Isfahan University of Technology (IUT), Isfahan, Iran, 18–20 December 2012.
21. Smagorinsky, J.S.; Smagorinsky, J. General Circulation Experiments with the Primitive Equations. *Mon. Weather Rev.* **1963**, *91*, 99–164. [CrossRef]
22. Song, C.C.S. A weakly compressible flow model and rapid convergence methods. *J. Fluids Eng.* **1988**, *110*, 441–445. [CrossRef]
23. Canuto, V.M.; Cheng, Y. Determination of the Smagorinsky–Lilly constant CS. *Phys. Fluids* **1997**, *9*, 1368–1378. [CrossRef]
24. Issa, R.I. Solution of the Implicitly Discretised Fluid Flow Equations by Operator-Splitting. *J. Comput. Phys.* **1986**, *62*, 40–65. [CrossRef]
25. Hirt, C.W.; Nichols, B.D.; Romero, N.C. *SOLA: A Numerical Solution Algorithm for Transient Fluid Flows (LA-5852)*; Los Alamos Scientific Laboratory: Los Alamos, NM, USA, 1988. [CrossRef]
26. Nichols, B.D.; Hirt, C.W.; Hotchkiss, R.S. *SOLA-VOF: A Solution Algorithm for Transient Fluid Flow with Multiple Free Boundaries (LA-8355)*; Nasa Sti/recon Technical Report N.; Los Alamos Scientific Laboratory: Los Alamos, NM, USA, 1980.
27. Nichols, B.D.; Hirt, C.W.; Hotchkiss, R.S. A fractional volume of fluid method for free boundary dynamics. In *Seventh International Conference on Numerical Methods in Fluid Dynamics*; Springer: Berlin/Heidelberg, Germany, 1981.
28. Mossa, M.; Petrillo, A.; Chanson, H. Tail-water level effects on flow conditions at an abrupt drop. *J. Hydraul. Res.* **2003**, *41*, 39–51. [CrossRef]
29. Felder, S.; Chanson, H. Air-water flow patterns of hydraulic jumps on uniform beds macroroughness. *J. Hydraul. Eng.* **2017**, *144*. [CrossRef]
30. Pagliara, S.; Palermo, M. Hydraulic jumps on rough and smooth beds: Aggregate approach for horizontal and adverse-sloped beds. *J. Hydraul. Res.* **2015**, *53*, 243–252. [CrossRef]
31. Palermo, M.; Pagliara, S. Semi-theoretical approach for energy dissipation estimation at hydraulic jumps in rough sloped channels. *J. Hydraul. Res.* **2018**. [CrossRef]

Permissions

All chapters in this book were first published in MDPI; hereby published with permission under the Creative Commons Attribution License or equivalent. Every chapter published in this book has been scrutinized by our experts. Their significance has been extensively debated. The topics covered herein carry significant findings which will fuel the growth of the discipline. They may even be implemented as practical applications or may be referred to as a beginning point for another development.

The contributors of this book come from diverse backgrounds, making this book a truly international effort. This book will bring forth new frontiers with its revolutionizing research information and detailed analysis of the nascent developments around the world.

We would like to thank all the contributing authors for lending their expertise to make the book truly unique. They have played a crucial role in the development of this book. Without their invaluable contributions this book wouldn't have been possible. They have made vital efforts to compile up to date information on the varied aspects of this subject to make this book a valuable addition to the collection of many professionals and students.

This book was conceptualized with the vision of imparting up-to-date information and advanced data in this field. To ensure the same, a matchless editorial board was set up. Every individual on the board went through rigorous rounds of assessment to prove their worth. After which they invested a large part of their time researching and compiling the most relevant data for our readers.

The editorial board has been involved in producing this book since its inception. They have spent rigorous hours researching and exploring the diverse topics which have resulted in the successful publishing of this book. They have passed on their knowledge of decades through this book. To expedite this challenging task, the publisher supported the team at every step. A small team of assistant editors was also appointed to further simplify the editing procedure and attain best results for the readers.

Apart from the editorial board, the designing team has also invested a significant amount of their time in understanding the subject and creating the most relevant covers. They scrutinized every image to scout for the most suitable representation of the subject and create an appropriate cover for the book.

The publishing team has been an ardent support to the editorial, designing and production team. Their endless efforts to recruit the best for this project, has resulted in the accomplishment of this book. They are a veteran in the field of academics and their pool of knowledge is as vast as their experience in printing. Their expertise and guidance has proved useful at every step. Their uncompromising quality standards have made this book an exceptional effort. Their encouragement from time to time has been an inspiration for everyone.

The publisher and the editorial board hope that this book will prove to be a valuable piece of knowledge for researchers, students, practitioners and scholars across the globe.

List of Contributors

David Santillán, Luis Cueto-Felgueroso, Alvaro Sordo-Ward and Luis Garrote
Department of Civil Engineering, Hydraulics, Energy and Environment, Universidad Politécnica de Madrid, 28040 Madrid, Spain

Ruichang Hu and Jianmin Zhang
State Key Laboratory of Hydraulics and Mountain River Engineering, Sichuan University, Chengdu 610065, China

Dongmei Yang, Zhen Ren, Zhiqun Guo and Zeyang Gao
College of Shipbuilding Engineering, Harbin Engineering University, Harbin 150001, China

Yaru Ren and Pengzhi Lin
State Key Laboratory of Hydraulics and Mountain River Engineering, Sichuan University, Chengdu 610065, China

Min Luo
Zienkiewicz Centre for Computational Engineering, College of Engineering, Swansea University, Swansea SA1 8EN, UK

Weilin Xu, Chunqi Chen and Wangru Wei
State Key Laboratory of Hydraulics and Mountain River Engineering, Sichuan University, Chengdu 610065, China

Shicheng Li
State Key Laboratory of Hydraulics and Mountain River Engineering, Sichuan University, Chengdu 610065, China

Jing Luo, Weilin Xu, Jun Deng, Yanwei Zhai and Qi Zhang
State Key Laboratory of Hydraulics and Mountain River Engineering, Sichuan University, Chengdu 610065, China

Ruonan Bai, Dejun Zhu and Danxun Li
State Key Laboratory of Hydroscience and Engineering, Tsinghua University, Beijing 100084, China

Huai Chen
State Key Laboratory of Hydrology, Water Resources and Hydraulic Engineering, Nanjing Hydraulic Research Institute, Nanjing 210029, China

Ludek Bures, Petra Sychova, Petr Maca and Radek Roub
Department of Water Resources and Environmental Modeling, Faculty of Environmental Sciences, Czech University of Life Sciences Prague, Kamycka 1176, 165 21 Prague 6, Suchdol, Czech Republic

Stepan Marval
Research Institute for Soil and water Conservation, Zabovreska 250, 156 27 Prague 5, Czech Republic

Roman Gabl
School of Engineering, Institute for Energy Systems, FloWave Ocean Energy Research Facility, The University of Edinburgh, Max Born Crescent, Edinburgh EH9 3BF, UK
Unit of Hydraulic Engineering, University of Innsbruck, Technikerstraße 13, 6020 Innsbruck, Austria

Jeffrey Steynor, David I. M. Forehand, Thomas Davey, Tom Bruce and David M. Ingram
School of Engineering, Institute for Energy Systems, FloWave Ocean Energy Research Facility, The University of Edinburgh, Max Born Crescent, Edinburgh EH9 3BF, UK

Xiangyang Zhou
College of Resource and Environmental Engineering, Guizhou University, Guiyang 550025, China
State Key Laboratory of Hydraulics and Mountain River Engineering, Sichuan University, Chengdu 610065, China

Zhipan Niu
State Key Laboratory of Hydraulics and Mountain River Engineering, Sichuan University, Chengdu 610065, China
Institute for Disaster Management and Reconstruction, Sichuan University, Chengdu 610065, China

Wenjuan Lei
College of Architecture & Environment, Sichuan University, Chengdu 610065, China

Kwang Seok Yoon
Korea Institute of Civil Engineering and Building Technology, Goyang 10223, Korea

Seung Oh Lee
School of Urban and Civil Engineering, Hongik University, 94 Wausan-ro, Mapo-gu, Seoul 04066, Korea

Seung Ho Hong
Department of Civil and Environmental Engineering, West Virginia University, 1306 Evansdale Drive, Morgantown, WV 26506, USA

Anping Shu, Guosheng Duan, Lu Tian, Mengyao Wang and Shu Wang
School of Environment, Key Laboratory of Water and Sediment Sciences of MOE, Beijing Normal University, Beijing 100875, China

Matteo Rubinato
Department of Civil and Structural Engineering, The University of Sheffield, Sir Frederick Mappin Building, Mappin Street, Sheffield S1 3JD, UK

Qiulin Li and Lianxia Li
State Key Laboratory of Hydraulics and Mountain River Development and Protection, College of Water Resource and Hydropower, Sichuan University, Chengdu 610065, Sichuan, China

Huasheng Liao
Department Civil and Environmental Engineering, Michigan State University, East Lansing, MI 48824, USA

Index

www.ingramcontent.com/pod-product-compliance
Lightning Source LLC
LaVergne TN
LVHW060518210926
848485LV00005B/1102

9781641168274